AROMATIC AND MEDICINAL PLANTS

Editor

Luisella Profeta

scitus
academics

Aromatic and Medicinal Plants

Edited by **Luisella Profeta**

This Book is Printed on Acid-Free Paper. ♾

ISBN: 9781681179353

Library of Congress Control Number: 2017949073

SCITUS Academics LLC,

616, Corporate Way, Suite 2, 4766,

Valley Cottage, NY 10989

www.scitusacademics.com

Printed in United States of America.

Preface

Plants are important source of a variety of products found to be useful in human beings. These are also a major component of forming biodiversity in certain ecological areas. Plants are not only a source of varied products but also these are actively participating role in ecological balances in multifold directions. Medicinal and Aromatic plants are important source as a primary health care among the human society. Demand for medicinal plants increases with the market, which is broad. Herbal treatment of various disorders is now a day's going popular because it's potential application and effect, less side effect. Each one Medicinal and Aromatic plants includes a specific type and concentration of chemical compounds and are differ as per variation in plant species and also concentration changes by the age of plants. The utility of the Medicinal and Aromatic plants is depending on their presence as well as by the active treatment potential to control specific disorders. In one of the studies by the World Health Organization, it is estimated that 80 per cent of the population of developing countries relies on traditional plant based medicines for their health requirement. Even in many of the modern medicines, the basic composition is derived from medicinal plants and these have become acceptable medicines for many reasons that include easy availability, least side effects, low prices, environmental friendliness and lasting curative property. India and China are the two major producing countries, having 40 per cent of the global biodiversity and availability of rare species. These are well known as the home of medicinal and aromatic crops that constitute a segment of the flora, and provide raw materials to the pharmaceutical, cosmetic, fragrance, flavor etc. industries.

This volume 'Aromatic and Medicinal Plants' covers interested research topics dealing with isolation, purification, and identification of active ingredients from wild and medicinal plants. It covers interesting research topics and the use of natural resources for medical treatments in some severe diseases. Day by day there is an increase in ethnobotanical, taxonomic, or phytochemical studies, providing an encouraging picture of research to support its use in traditional medicine. This discovery will lead to an increase in the global pharmaceutical market as well as open such new gate for medicinal plant research. This book will add significant information to medical researchers, practitioners and can be used for postgraduate students.

Table of Contents

CHAPTER 1

From Medicinal Plant Raw Material to Herbal Remedies

Sofija M. Djordjevic*

[1] *Institute of Medicinal Plant Research "Dr Josif Pancic", Belgrade, Republic of Serbia*

ABSTRACT

The use of medicinal plants is old as the existence of mankind. According to World Health Organization (WHO) data, about 80% of world population are using products based on medicinal herbs. Phytotherapy is based on the use of herbal drugs and medicinal products for the purpose of prevention and treatment. Rational phytotherapy is a modern concept of herbal medicines using, which are made of standardized herbal extracts. The quality of each final product is guaranteed by the use of raw materials of a standard quality, defined process of production, and validated equipment. Quality control of herbal drugs and herbal isolates (tinctures, extracts, and essential oils) is done according to the requirements of Pharmacopoeia and other relevant regulations. The scope of phytopreparation quality control depends on its pharmaceutical form. The formulation of a new phytopreparation is a process that has strictly defined phases: from analysis of literature and market, through defining recipes, validation of the production process, quality control of a final product to the preparation of technological and registration documents. The aim of this chapter is to present the process of herbal preparations production from selecting plant raw materials to herbal remedies (on the examples of making tea, tea mixture, drops, gels, and capsules).

Keywords: medicinal plants, medicinal plant raw material, quality control, formulation of phytopreparation, regulations

1. INTRODUCTION

Since ancient times, medical plants and simpler herbal remedies have been used in all parts of the world for the treatment and alleviation of various ailments. Although the use of medicinal plants is as old as mankind itself, their controlled application, the isolation and characterization of active substances, started only in the early nineteenth century. It is a known fact that the extractive plant isolates and isolated active substances played a major role in the development of modern pharmacotherapy. Many of the isolated compounds are still used today, or they have served as a model for the synthesis of a large number of drugs [1].

The use of plants as medicines has a long history in the treatment of various diseases. Plants especially those with ethnopharmacological uses have been the primary sources of medicine for early drug discovery.

Herbal remedies, from simple to complex forms, should be made of the raw materials required for quality, because only then they could be safe and effective for use. The Pharmacopoeia monographs, Monographs European Medicinal Evaluation Agency (EMEA), which encompasses monographs World Health Organization (WHO), European Scientific Cooperative on Phythotherapy (ESCOP), and Commission E (The German Commission E is a scientific advisory board of the "Bundesinstitut für Arzneimittel und Medizinprodukte" formed in 1978. The commission gives scientific expertise for the approval of substances and products previously used in traditional, folk, and herbal medicine) national regulations, precisely defined parameters of control quality.

The process of drafting a new herbal remedies is very complex and strictly defined phase. Each step in the process is important, from the initial idea, market analysis, selecting high-quality plant material and ancillary pharmaceutical raw materials, recipe formulation, production preparation, quality control of product, preparation of documentation, protection of intellectual property rights, to the introduction of herbal drug in regular production. The drafting process must be validated and secure documentation.

In modern pharmacotherapy, despite the widespread use of drugs obtained by chemical synthesis, the importance of herbal medicines in the treatment and prophylaxis is still large. According to the latest WHO researches, 11% of the 252 basic medicines are in fact herbal preparations [1].

2. DEVELOPMENT AND MANUFACTURING OF HERBAL PREPARATIONS

2.1. Use of Medicinal Plants Through History

The use of medicinal plants in the prevention and treatment of various diseases is known since ancient times. Documents of exquisite value show that herbs were extensively used by human population throughout the history. Since ancient times, people have sought safety and relief for their health problems in medicines from nature. Prehistoric men have dared to use particular medicinal plants, based on careful observation of the behavior of animals who have been using them [2, 3]. Over time, the use of herbal medicines and other natural products has developed on the basis of both positive and negative experiences. The collected rich experiences have gradually developed into folk medicine, such as traditional European medicine, traditional Chinese medicine, Indian Ayurveda, Japanese Kampo, or traditional Arabic and Islamic medicine. They consist, not only of herbal remedies but also of other types of drugs, for example, from minerals or animals, or physical procedures [3].

Material evidence on the use of medicinal plants in the distant past is kept by many ethnographic and archaeological sources. The oldest of these sources are clay tables, discovered in Mesopotamia (2600 BC), which in addition to the description showed also therapeutic application and galenical form in which the plants were to be used. In those ancient times, medical plants mentioned were castor oil, grapes, coffee, oils of cedar and cypress, licorice, myrrh, and poppy juice [4, 5]. The ancient Egyptian papyrus, Ebers Papyrus (1550 BC), represented some kind of first Pharmacopoeia. Egyptians were known for their skill of embalming, distilling scented water, and making perfume of aromatic plants, and for those they were using many medicinal plants that are still in use today (aloe, peppermint, plantain, poppy seeds, and coriander) [6]. The first written records about the use of medical herbs in Chinese traditional medicine date from the third millennium BC. Emperor Shen Nung made a collection of wild medicinal plants. He is credited with the discovery of tea and many of which are used nowadays: cinnamon, ephedra, rhubarb, camphor, and great yellow gentian [6, 7]. The Indian holy books provide many examples of the treatments using medical plants, widespread in that country. A large number of aromatic herbs and spices that are still in use nowadays throughout the world, such as pepper, cloves, nutmeg, originate from India. According to data from the Bible and the holy Jewish book, the Talmud, during various rituals accompanying a treatment, aromatic plants were utilized such as myrtle and incense [7].

With comprehensive development of science in ancient Greece, the pharmacy also receives a special place. The most famous doctor of ancient Greece, which is considered to be the "father of medicine," is Hippocrates (460–377 BC). He was the first to systematize overall medical and pharmaceutical experience and publish them in the capital work Corpus Hyppocraticum. The most ancient botanist Theophrastus (371–286 BC) together with his students founded the first botanical garden in Athens. He described more than 500 most important medicinal plants. Among others, he referred to cinnamon, iris rhizome, false hellebore, mint, pomegranate, cardamom, fragrant hellebore, monkshood, and so on. In the description of the plant, its toxic action was also stated [4, 7]. The founder of the European pharmacognosy, a Roman doctor of Greek origin, Dioscorides, who lived in the first century BC., described medicinal plants which were used in the ancient world, in his capital work De Materia Medica. Dioscorides' most appreciated domestic plants were as follows: willow, camomile, garlic, onion, marsh mallow, ivy, nettle, sage, common centaury, coriander, parsley, sea onion, and false hellebore. The strong influence of Hippocrates and Dioscorides was notable in the school of Alexandria, where some of the major breakthroughs in medicine were made. Unfortunately, a great fire has destroyed the vast library with approximately two million books, and at the same time all the knowledge of medicinal plants of that era [4, 7]. By the Roman conquest of Greece, Romans took over all the medical and pharmaceutical knowledge and certainly the most important mind in this area was the Roman statesman and military leader Pliny, the Elder (23–79). He is the writer of the capital work Historia naturalis. The most famous Roman doctor and a pharmacist is well-known Galenius-Galen, who lived from 131 to 201 and is considered to be the father of galenic pharmacy. In his writings on the development of complex preparations or galenic preparations, he described 304 drugs of plant origin [4].

The Arabs preserved a large amount of the Greco-Roman knowledge during the Dark and Middle ages (i.e., fifth to twelfth centuries), and complemented it with their own medicinal expertise, and with herbs from Chinese and Indian traditional medicines [8]. The treatments during the Middle Ages were conducted in the restricted environment of monasteries. Skills of cultivation and collection of herbal medicines, as well as making simple herbal remedies, were reserved for doctors-monks. They used the different herbs: mint, sage, tansy, anise, fenugreek, savory, and so on [7].

At the time of Charles V, the famous medical school of Salerno was founded and started its rise by introducing and applying experiences of Arab medicine and pharmacy. Benedictine monks played an important role in the preservation of the Greco-Roman tradition. Their legacy was large botanical gardens where mainly medicinal plants were grown [4]. The Arab world has promoted many sciences including medicine and pharmacy. Certainly, the most famous Arabic doctor was Abu Ali Ibn Sina (Avicenna) and his famous book, The Canon of Medical Science, has been translated into Latin and other languages, and has been used in Europe for many years.

In medieval Europe, the level of medical knowledge was quite low. Arab medicine, starting from the twelfth century, began to penetrate into Europe, through Spain and Sicily. The Arabic books were translated into Latin medicine, and in this sense also the Arabic translations of ancient Greek and Roman books. Paracelsus (during the late Middle Ages) argued that the salubrity of plant originates from chemical compounds that are represented in it [4]. In the eighteenth century, Swedish botanist Carl von Linné (1707–1778) created the Latin nomenclature for each plant (the name of the genus and species), and a botanical system for determining the species, which due to its transparency and convenience is used even nowadays. Scientific pharmacy began only after the French Revolution, and with it the development of the science of medicinal plants. In this area, the most distinguished pharmacists became Lavoisier in France, Scheele in Sweden, Priestley in England, and so on. [4]. The turning point in the approach and the use of herbal medicines is considered to be the beginning of the nineteenth century, when a German pharmacist Sertürner managed to isolate the alkaloid of morphine in its pure form, from poppy (1806). In the period from 1817 to 1820, French scientists isolated a whole series of alkaloids: caffeine, emetine, quinine, cinchonine, and strychnine. Improvements of instrumental analytical methods have allowed further detection of other groups and complexes of active substances, such as heterosides, saponosides, tannins, vitamins, and so on [7].

In the twentieth century, a large number of synthetic drugs were created and it represented the beginning of commercial production of a large number of allopathic medicines, which significantly led to neglect the use of herbs in pharmacotherapy.

2.2. Plants are Valuable Sources of Drug Discovery

As already mentioned, herbal medicines have been an extremely important source for the discovery of many drugs. Morphine, which was the first purely natural product to be isolated, was introduced in pharmacotherapy in 1826 (Merck). The first semisynthetic pure substance of aspirin, salicylic acid-based, was isolated from the bark of *Salix alba* willow and was produced in 1899 (Bayer). This was followed by the isolation of active compounds from old herbal drugs, such as digitoxin, codeine, pilocarpine, quinine, and many others, some of which are still in use today. Many herbal remedies, emerged after extensive scientific tests of "old and well-known" medicinal plants, were introduced in the therapy. Silymarin, extracted from the seeds of *Silibum marianum*, is used as a hepatoprotective, Paclitaxel from the bark of *Taxus brevifolia* in the treatment of lung, ovarian, and breast cancer, and Artemisinin from *Artemisia annua* herb to combat multiple-resistant malaria [1].

In recent years, many herbal medicines have found their way into the official medicine. Some of them are Dronabinol and Cannabidiol isolated from *Cannabis sativa*, Tiotropium derivative of atropine from *Atropa belladonna* for combating obstructive and chronic bronchitis, Galantamine, alkaloid from *Galanthus nivalis* which is used to relieve symptoms of Alzheimer's disease, and Apomorphine, which is a semisynthetic compound based on morphine from *Papaver somniferum* and is intended for people suffering from Parkinson's disease [1]. We can certainly say that a large number of medicinal plants, which in the past, were used and represent an important raw material for the production of herbal medicines, or have served as a model for similar synthesis of new molecules.

Given that man is an integral part of nature, the human body is compatible with medicines coming from nature. Nature, much like a flawless, perfect complex of laboratories, has created a variety of sophisticated active compounds contained in herbal medicines, which have a huge range of remedial action. Perhaps, this fact will speed up serious research of old manuscripts related to herbal medicines and brings out the "old drugs" of pure historical curiosity [8].

2.3. Basic Terms Related to Herbal Medicines

Phytotherapy, as a complementary part of pharmacotherapy, has an important place in many areas of modern medicine. It represents a system of treatments based on the use of natural medicinal resources (drugs) and herbal remedies (herbal remedies) in the purposes of prevention and treatment.

Herbal drug is the whole or grained, dried part of a plant, algae, fungi, or lichen, which is used for its medicinal properties. In addition to the plant organs (above-ground part of the blooming plant as flower, leaf, root, bark, fruit, and seed), plant exudates can also be considered as a drug (resins, balsams, and rubber). Herbal medicines, herbal remedies, or herbal medicinal products (HMPs) contain as active ingredients exclusively herbal drugs or herbal drug preparations. Herbal drug preparations are obtained from drugs, with the procedures of distillation, extraction, filtration, and so on. This concept does not include powdered forms of drugs, essential oils, fatty oils, tinctures, and extracts [9, 10].

Rational phytotherapy is a modern concept of use of herbal medicines, which was designed in Germany at the end of the last century and soon widely accepted in other European countries. It was created from the need to improve phytotherapy, in order for herbal preparations to be more efficient, safer, and their use based on the results of clinical trials. Herbal medicines, which are used in rational phytotherapy, are prepared from standardized herbal extracts, the chemical nature of their active principles is known, they exhibit dose-dependent therapeutic effect, their adverse effects and contraindications are known, and their pharmaceutical quality is well defined and standardized [11, 12].

Herbal medicines are used preventively, in the treatment of milder forms of a disease, or as adjunctive therapy for the treatment of chronic diseases. Most commonly, they are applied with the dysfunction of the respiratory, digestive, urogenital tract, mild, and medium forms of anxiety and depression, as well as of different lesions of the skin and mucous membranes. Their healing effects accrue gradually, so that the maximum effect manifested 2–3 weeks after the application.

2.4. Regulations

According to the WHO, preparations based on medicinal herbs are used by 80% of the world population. Medical use of medicinal plants has a long tradition in Europe, while in some parts of the world (e.g., China and India), herbal remedies still represent a central link in the chain of health services [13].

Extractive isolates of herbal medicines and herbal preparations are extremely complex multicomponent mixtures, as opposed to synthetic drugs that are most commonly a single pure compound. In the production of herbal remedies, certain actions and procedures are needed to be undertaken (collecting medicinal plants from spontaneous flora and plantation cultivation, obtaining extractive isolates, and their characterization), which do not precede the production of synthetic drugs. Fortunately, the procedures of making herbal medicines are largely modernized and defined in all segments. There are a number of guidelines that prescribe standards in all aspects of making herbal medicines: The European Medicines Agency guidelines for the quality of herbal medicines, the WHO guidelines provide standards and guidelines for good agricultural practices, good laboratory practices, and so on. The development of new, sophisticated analytical, and technological methods and procedures within the development and characterization of extractive isolate has greatly improved the quality of the final plant products. On the other hand, the process of harmonization of the quality system for the production and herbal drugs control is present in many countries. But globally speaking, more effort is yet to be made in order to revive the prescribed guidelines and regulations in practice [3]. The main goal of the Committee for herbal products (Herbal Medicinal Product Committee—HPMC) is to prepare a detailed list of monographs and processed herbal substances and preparations, which are in medical use for long enough time that their use is safe under normal conditions. The monograph contains the professional opinion of the Committee on a particular plant products based on scientific data or traditional use within the European Union (EU). For each plant, the substances are stated indications, speed, usage, and other relevant data concerning its safe use or composition that contains it. List and versions of monographs are available for public consultation [14]. In Europe, companies can apply for three different types of market authorization of an herbal medicinal products (HMPs):

- Full implementation. Manufacturer of a herbal drug must provide documentation proving its efficiency and safety, and studies are identical to those submitted for the registration of a synthetic drug.
- Well-established use. Manufacturer of a herbal drug may be permitted to register, on the basis of the submitted detailed scientific literature, stating that the herbal medicinal preparation is in use for medical purposes not less than 10 years in Europe and has recognized efficiency and an acceptable level of safety.
- Traditional use. Efficiency and safety of a herbal drug can be accepted on the basis of long experience. Herbal remedies can be registered, if the documents prove their use in mitigating certain ailments, not less than 30 years, with at least 15 years in Europe.

The registration procedure for herbal medicines, at all levels of the European Union (EU), is done according to European Directive 2004/24/EC, which introduces simplified, but strictly defined procedures and affects the harmonization of existing national legislative regulations. Regarding the registration in the non-EU countries, despite the efforts made within the framework of national legislation and harmonization in larger systems, a limited number of herbal medicines have been registered. Therefore, the identification of problems and discrepancies and the systematic plan for overcoming them represent a major challenge for the presence of these herbal drugs on the market of EU countries [15].

In Republic of Serbia, legislation on plant products is harmonized with recommendations of The European Directive 2004/24/EC. Law on medicines and medical devices (Official Gazette of the Republic of Serbia No. 30/2010), Regulation on health safety of dietary products (Official Gazette of RS No. 45/2010), Guidelines of Good Manufacturing Practice, Annex 7- Manufacture of herbal medicines, are all in effect. According to the Law on medicines and medical devices, Herbal medicine, is each drug whose active ingredients are exclusively one or more substances of vegetable origin or one or more herbal preparations, or one or more substances of vegetable origin in combination with one or more herbal preparations. Traditional herbal medicine may be based on scientific principles and is the result of tradition or other traditional therapeutic approaches. The active components of a herbal medicine/traditional herbal medicine are herbal drugs and herbal preparations and their combinations, and this is widely accepted in all European and national documents. In the context of food supplements (dietary supplements), a new the Regulation defines the notion of herbal dietary supplements. These are supplements that contain medicinal plants, their parts or preparations and their quantity in a daily dose of the product should not be less than 15% and greater than 65% compared to a known therapeutic dose of these plant materials or preparations.

2.5. Parameters of Quality and Quality Control

The quality of each final product is ensured by the standard quality of raw materials, the application of validated production processes and procedures on validated equipment. It is similar with herbal remedies, which are made of high-quality herbal raw materials, extractive preparations (extracts and tinctures) and isolates (essential oil and fatty oil). The latest European Pharmacopoeia Ph Eur 8 comprises 270 Monographs on herbal drugs and herbal drug preparations [16]. Monographs define parameters of quality control.

2.5.1. Quality control of herbal drugs

The basis of high-quality herbal remedies is the plant material of a standard quality. Many factors affect the quality of plant material. Regardless of whether the medicinal plants are grown or collected from the wild, biogenetic factors are certainly important (species, variety, chemotype, and sorta). The following are the conditions in which a plant grows as air, climate, land, then agro-technical measures that are applied during the large-scale production (proper sowing, irrigation, fertilization, control against weed and pests), and then collection from the wild or harvest of the plantation, transport, proper storage, drying, and grinding. It is very important to educate people how to deal with the collection of herbal raw materials from spontaneous flora, as well as those who grow the plants, whether they are doing so in the conventional conditions or organic conditions of medicinal plants production. Quality control of herbal raw materials is strictly defined and traceable. First, the identification of plant raw materials is approached. Responsible and expert persons in laboratories for the pharmaceutical control or in other relevant institutions, conduct identification, and categorization under a certain number.

Table 1 gives a list of parameters of quality control of herbal drugs by Ph Eur 8.0, based on whose defined border values the quality of herbal drugs can be determined. For all the aforementioned parameters, Pharmacopoeia prescribes the procedure. Any organization deals with herbs, forming his specs-attests that rely on the requirements of the applicable European Pharmacopoeia, standards, national regulations, and internal regulations. Definition includes biological source of drug, the Latin name of the genus and species, and the minimum amount of essential oil in the case of aromatic drugs or the minimum quantities of active substances to which the drug is declared. Characteristics define appearance, odor, taste, and solubility. Identification in addition to macroscopic analysis (organoleptic inspection), which includes appearance, color, odor, and microscopic analysis, is carried out for certain herbal drugs, followed by chemical analysis (specific chemical reactions and thin-layer chromatography (TLC)). Predicted tests include water, loss on drying, total ash, foreign matter, swelling index, microbiological purity, bitterness value, starch, and broken drug. Assay includes essential oils, tannins, declared active substance (gas chromatography (GC), GC/mass spectrometry (GC/MS), liquid chromatography (LC), and UV-VIS spectroscopy). For the evaluation of herbal raw materials quality, in addition to the determination of microbiological safety, complete analysis of the health safety is often performed, according to the current regulations (which includes not only the results of physical, physicochemical, and chemical tests but also organoleptic findings, preservatives, sweeteners, mycotoxins, metals and metalloids, pesticide residue, microbiological tests, and radioactivity).

2.5.2. Quality and quality control of extracts

The extracts are one of the most widely used herbal preparations. The extracts can be liquid, semi-solid, or solid consistency. They are most commonly made of dried and grained plant material. For the extract production various processes may be applied: maceration, percolation, extraction of the continuous, and so on. Nowadays, the procedure for extracting plant material by super critical fluids is increasingly in use [17–20]. Tinctures are extractive products which are usually obtained by the method of maceration.

Table 1.Parameters of quality control of herbal drugs and herbal preparations (Ph Eur 8).

Parameters of quality control	
Herbal drugs	1. Definition: the name of the herbal drugs and content of active substances 2. Characters: appearance, taste, odor, solubility 3. Identification: macroscopic and microscopic examination, TLC 4. Tests: water, loss on drying, total ash, foreign matter, insoluble matter, extractable matter, swelling index, microbiological purity, bitterness value, broken drug 5. Assay: essential oils, tannins, declared active substances (GC, LC, UV-VIS)
Herbal drug preparations	**Dry extract** (*Extractum siccum*) 1. Definition: standardized dry extract prepared from..., content of active substances 2. Production: method of extraction and solvents 3. Characters: appearance 4. Identification: TLC 5. Tests: loss on drying, total ash, microbiological purity 6. Assay: declared active substances (LC) **Liquid extract** (*Extractum fluidum*) 1. Definition: liquid extract produced from ..., and content of active substances 2. Production: method of extraction and solvents 3. Characters: appearance, taste, odor 4. Identification: TLC 5. Tests: ethanol, methanol and 2-propanol, loss on drying, microbiological purity 6. Assay: declared active substances (LC, UV-VIS) **Tincture** (*Tinctura*) 1. Definition: tincture produced from..., and content of active substances 2. Production: method of extraction and solvents 3. Characters: appearance, taste, odor 4. Identification: TLC 5. Tests: ethanol, methanol and 2-propanol, dry residue, microbiological purity 6. Assay: declared active substances (LC, UV-VIS) **Essential oil** (*Aetheroleum*) 1. Definition: essential oil obtained by ..., and content of dominant components 2. Production: method of extraction and solvents 3. Characters: appearance, odor, solubility 4. Identification: TLC and chromatographic profile (GC and GC-MS) 5. Tests: relative density, refractive index, optical rotation, chromatographic profile

The most often used extragents are ethanol, water, mixtures of water, and ethanol. The choice of solvent depends on the nature of ingredients that need to pass into the extract. In the product declaration, the ratio of components of the solvent mixture used for extraction must always be given. The quality of obtained extract depends on the plant material, the solvent, the drug/solvent ratio, and the extraction process technology. The process of extracts standardization is common. After the quantitative analysis, the extract is adjusted in order to contain a particular amount of an active compound or the group of compounds (or a marker compound) using an inert material or another extract [21]. Parameters of quality control of extracts are shown in **Table 1**. Qualitative and quantitative analyses of the essential oils are carried out by methods GC and GC-MS according to regulation Ph Eur or modified method [22].

2.5.3. Quality and quality control of phytopreparations

When it comes to *mono-components teas,* their quality matches the one defined for the individual drugs (**Table 2**). Quality control refers to the verification of the identity, the declared weight of packaging, and microbiological safety. In the case of *multi-component (mixtures) teas,* their control refers to the verification of the identity of each herbal drug on the recipe, checking the declared weight relationship of the components and microbiological purity.

Table 2. Parameters of quality control of phytopreparation.

Phytopreparations	Parameters of quality control
Mono-component tea and tea mixture	1. Identification 2. Appearance 3. Verification of components, declared mass ratio of components 4. Verification of package weight 5. microbiological purity
Liquid herbal preparations (liquid extracts, tinctures, and mixtures of extracts or tinctures), **herbal drops, solutions, syrups**	1. Identification 2. Appearance 3. Loss on drying 4. Content of ethanol 5. Relative density 6. Refractive index 7. Verification of package weight 8. Qualitative and quantitative analysis 9. microbiological safety and/or complete health safety control
Semi-solid forms (herbal gels, cream, and unguent)	1. Identification 2. Appearance 3. Verification of package weight 4. pH value 5. microbiological safety and/or complete health safety control
Solid-dosage forms (capsules, tablets, etc.)	1. Identification 2. Appearance 3. Declared mass of single-dose preparations 4. Disintegration 5. Qualitative and quantitative analysis of the declared active components 6. microbiological safety and/or complete health safety control

When a herbal remedy represents a *mixture of plant extracts or tinctures,* it is usually difficult to perform the quantitative analysis of active ingredients for each individual extract. For these preparations, the content analysis of active substances for which the preparation is declared is conducted (LC, UV-VIS spectroscopy, GC and GC-MS spectrometry, infrared (IR), nuclear magnetic resonance (NMR), etc.). Also, in principal, a manufacturer appends the typical fingerprint of chromatogram, spectrum, or some other physical parameter, of the preparation ingredient, which may be used by a control laboratory for identification. This approach can also be used for the quantitative analysis of the preparation.

Semi-solid forms (herbal gel, cream, and ointment) are checked for authentication, filling, pH, and microbiological purity. For *solid-dosage galenic forms* (capsules, tablets, etc.), a control is performed for each individual dose. Control for these forms of herbal remedies includes authentication, appearance, the declared weight of the package, or of the each individual dose of the preparation. Qualitative and quantitative analyses of the declared active components, or a marker compound, of the preparation, are obligatory. For this analysis, similar (or appropriately modified) methods to the analysis of herbal drugs or extractive preparations are applied (if not according to Pharmacopoeia, these methods must be validated by the CPMP/ICH/281/95)s.

2.5.4. Monitoring of phytopreparation stability

Medicinal herbs and products based on medicinal herbs are very sensitive to external influences—the presence of elevated temperatures, moisture, and direct light. They are prone to reactions of oxidation, degradation, hydrolysis, and evaporation, so during the preparation of herbal remedies special attention must be placed to a number of factors that can affect the quality of the final product. In order to determine the stability of a product, to define storage conditions, and durability, stability tests are carried out, which include the investigation of the environmental factors effect on the change in the final product quality. Stability tests are performed at different stages of development and production. During the period of investigation, the first stress test is carried out, in order to select the most optimal, compatible excipients, and the best formulation. All raw materials used, excipients, active ingredients, and extraction products, are subject to stability testing and durability defining, for appropriate packaging and under certain environmental conditions. Stability is defined as the period during which the product remains within the set quality limits of the prescribed specification. In accordance with the requirements of the EMEA, various specific stability tests are placed for different herbal [23].

2.6. Planning and Development of New Herbal Preparation

The formulation of a new herbal products is a process that has strictly defined phases and includes the work of several sectors of the company that is developing products based on medicinal herbs. The dynamics of the process is divided into stages, and they are completed by performing a multitude of necessary activities. Of course, the scope of new product development depends on the complexity of the galenic form of the new herbal preparation. All steps are defined in the documents of quality standards and other regulations of the company (**Figure 1**).

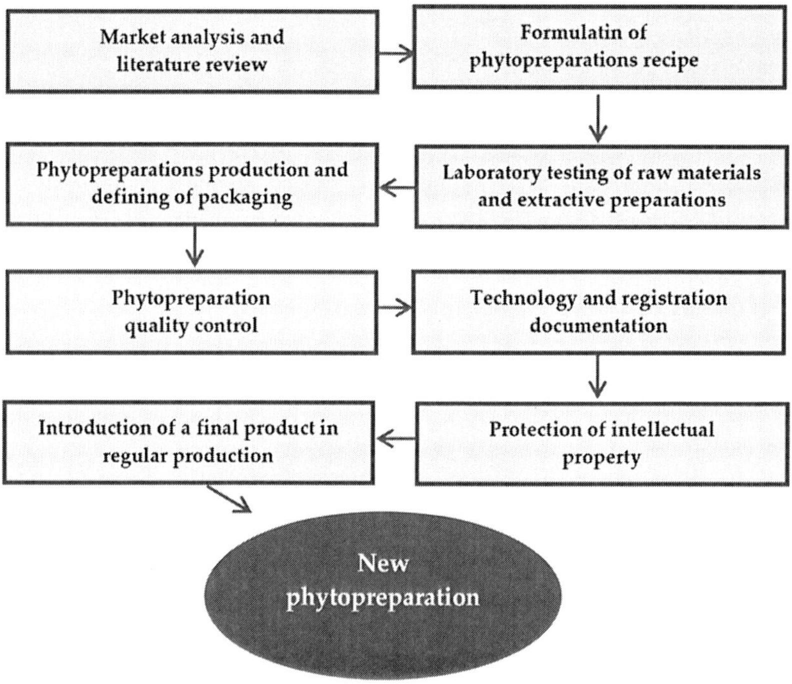

Figure 1. Schematic representation of the planning and development of new herbal preparation.

Furthermore, the procedures of developing a new herbal preparation will be explained on the example of the development of herbal remedies in the Institute for Medicinal Plant Research "Dr Josif Pancic," Belgrade, Republic of Serbia (the Institute).

Stages in the development of new herbal products have their logical sequence. Administration, management, the sales department, and the demand of patients expressed through visits to our herbal pharmacies may suggest the need for a new preparation, to the department of pharmaceutical research and development. Management gives the order to the department of pharmaceutical research and development, to create a team that will be allocated to the project and appoint a project manager.

Project manager, who is usually a doctor of pharmaceutical sciences in the field of pharmacognosy, forms a team, writes, and presents a plan for a new product development. The plan, according to the system of quality management, must contain general information: project name, the subject of research, description of development activities, the aim of the research, the necessary equipment, an indication of the place of realization of activities, project duration, and start time of the applicability of the project results. The second segment is represented through dynamic activities (through phases) and the engagement of researchers (throughout activities phases). The third segment is the planning of all material costs for the development of a new herbal remedy. When management of the Institute approves the presented plan for developing a new product, project manager starts with its execution.

The beginning phase represents preformulation studies. Market analysis and the review of relevant literature are very complex. It covers the activities of the Institution's herbal pharmacies that collect patients' requirements, commercial department, which collects data on herbal remedies from the observed therapeutic groups in the domestic and foreign markets. Analysis of relevant literature includes search for relevant directives and monographs, scientific, technical papers, patents, and so on.

When all relevant data are collected, an expert research team approaches the formulation of a recipe. The recipe for herbal remedies and the content of active substances, which will be declared in the preparation, depends on the plan for a herbal product registration. In addition to the selection of galenical form, auxiliary materials are also selected at this stage. Of course, the recipe is subject to small corrections during the production process.

Further on, laboratories are included in the process of preparations formulation. Analysis of active substances in the selected plant raw materials and quality control prescribed by Pharmacopoeia or other relevant document are performed. Then, the quality control of semi-final product is conducted. In the case of capsules, the analysis of the active substance in dry extract is performed, and also the analysis of other parameters that define its quality, and that is defined by Pharmacopoeia monographs or summarized in an internal specification or a certificate.

Afterwards, the preparation formulation is conducted (formulation of teas or herbal drops, herbal creams, capsules, etc.). In addition to the main plant raw materials, secondary raw materials are selected, which will synergistically facilitate the functioning of the dominant plant drugs, and auxiliary pharmaceutical raw materials are selected. Auxiliary raw materials are used as the basis for semi-solid galenic forms (creams, ointments, and gels), and capsule fillers (for the preparation of granules—mass for the capsule filling).

When the herbal remedy is designed, all laboratory examinations are carried out, as prescribed in specifications. The determination of the average capsule weight is performed, also the determination of active ingredients per capsule, stability testing for active substances per capsule, the capsule dissolution testing, and testing of complete health safety.

From the Intellectual Property Office, the search "recharge" is required and then the protection of the preparation name. The proposal for the primary and secondary packaging graphic layout is carried out. Technological documentation gets completed (processes specifications, recipes, and norms). After the test production of capsules, the production of capsules is introduced into regular production. The validation of technological processes, the validation of equipment, and the validation of laboratory methods are performed. Registration documentation, report writing, and the presentation of results are being prepared.

When the preparation is produced and packed in its primary and secondary packaging, quality testing of the final product is conducted (tea blends and herbal drops, ointments, and capsules). Quality testing is conducted according to the attest or the specification of the final product. It is necessary to examine the complete health safety of the preparation. The stability of the active substance is also monitored, to determine and define the expiration date.

Along with a new preparation production, the writing of technical documentation that represents the preparation file is carried out. The specifications of raw materials, semi-final products, and auxiliary materials are written. In order to define the dose for preparations, it is necessary to determine the range of content of the active substance. This is achieved by validation of the process, on the validated equipment.

Finally, the registration documents are prepared. The scope of the documentation is correlated with the desired herbal remedies group for registration (herbal medicine, traditional herbal medicine, or dietary supplement). If it is planned to register a new product as a herbal medicine or a traditional herbal medicine, then the instructions of the EMEA monographs need to be followed. If it is planned to register a new product as a dietary supplement, then the dose needs to be below 65% of the therapeutic dose.

Development of different galenic form of herbal remedies is presented in the following examples.

2.6.1. Mono-Component Tea

Herb yarrow *Achillea millefolium* L. is highly regarded medicinal and aromatic plant and has a long traditional use. According to WHO monograph, yarrow exhibits antibacterial, anticonvulsant, anti-inflammatory, antioxidant, antipyretic, antispasmodic, and antiviral activity. According to the Commission E monograph and the EMEA, yarrow is traditionally used only for temporary loss of appetite, mild spasmodic complaints of the digestive organs, bloating, flatulence, and externally as a bath with problems in the lower abdomen in women, and with superficial wounds [24].

The process of making mono-component tea from yarrow takes place in the following stages:

- Purchase of high-quality herbal material (overground top part of the blooming herb), which meets the quality parameters as prescribed by Pharmacopoeia.
- Grinding the herb up to a prescribed degree of fragmentation. Grinded herbal drug as a semi-final product is controlled on the microbiological safety, which is performed in accordance with the relevant regulations for tea and Pharmacopoeia requirements.
- Following the positive results obtained from the pharmaceutical and microbiological laboratories, chopped herbal drug is placed in packaging.
- Final product, mono-component yarrow tea, is sent for the control to pharmaceutical and microbiological laboratories. Pharmaceutical laboratory confirms the authentication and filling volume.
- When a controlled product receives the confirmation that corresponds to the standard quality, it is dispatched to the warehouse of final products and further distributed to pharmacies and other places.

The user manual is adapted to the prescribed use as a traditional herbal medicinal: as a means of relieving complaints of the digestive system, improve appetite, eliminate gases, regulating the secretion of bile, pains and cramps in the stomach, with the amenorrhea.

2.6.2. Tea For Weight Loss In Filter Bags (1.5 G)

For this indication, herbal drugs that have diuretic and laxative effect have been selected on one hand, and on the other hand, we have drugs that aid digestion and herbal drugs rich in polyphenols as potent antioxidants and vitamin C. The following herbal drugs are included in equal parts (ana partes) in the mixture composition: *Betulae folium, Frangulae cortex, Foeniculi fructus, Thea folium*, and *Cynosbati fructus*. The mixture of herbs whose active ingredients regulate digestion and excretion of urine, stimulate metabolism, and facilitate the breakdown of fat, and thus contribute to the cleansing and detoxification of body. Its is recommended as a supplement in weight loss diets and for helping to reduce and maintain a desired weight.

The process of making tea for weight loss in filter bags is as follows:

- Purchase of high-quality individual herbal material and quality control of parameters as prescribed by Pharmacopoeia.
- Grinding the herb up to a prescribed degree of fragmentation. Grinded herbal drugs are controlled on the microbiological safety and Pharmacopoeia requirements (appearance, moisture content, content of essential oils, impurities, and degree of fragmentation).
- Following the positive results obtained from the pharmaceutical and microbiological laboratories, homogeneous mixing of grinded herbal drugs is carried out, and then in the machine for tea bags, bags are filled with the contents. Afterwards, they are placed in special filter bags and packaging.
- Final product, herbal mixture in filter bags, is sent for the control to pharmaceutical and microbiological laboratories. Pharmaceutical laboratory confirms the authentication and filling volume of tea bags and the number of tea bags in a packaging.
- When a controlled product receives the confirmation that corresponds to the standard quality, it is dispatched to the warehouse of final products and further distributed to pharmacies and other places.

2.6.3. Herbal Drops For Weight Loss

Herbal drops, meant to regulate body weight, represent a combination of tinctures and herbal extracts, whose active ingredients stimulate the metabolism, have a beneficial impact on digestion, and eliminate the excess fluids from the body. The composition of herbal drops includes *Betulae tinctura, Frangulae tinctura, Foeniculi tinctura and Cynosbati extractum fluidum*.

The process of making herbal drops at the Institute, takes place as follows:

- Purchase of high-quality individual herbal material and quality control of parameters as prescribed by Pharmacopoeia.
- Grinding the herb up to a prescribed degree of fragmentation. Grinded herbal drugs are controlled on the microbiological safety and pharmacopoeia requirements.

- Following the positive results obtained from the laboratories, homogeneous mixing tinctures and extract, filtered, and then filled into glass bottles.
- Final product is sent for the control to pharmaceutical and microbiological laboratories.
- When a controlled product receives the confirmation that corresponds to the standard quality, it is dispatched to the warehouse of final products and further distributed to pharmacies and other places.

2.6.4. Herbal cream with extract of comfrey: comfrey gel

The root of comfrey *Symphyti radix* represents a very important medicinal herb raw material. According to the Commission E monograph, it is used for blunt injuries. According to EMEA, the traditional use of herbal preparations, it is used in semi-solid-dosage forms for cutaneous use. Traditional herbal medicinal remedy is also used for the symptomatic treatment of minor sprains and bruises. Gel with 10% propylene glycol extract of comfrey root (*Symphyti Extaractumfluidum* (1: 7)) containing mucus, tannins, saponosides, and allantoin improves epithelialization, drainage, and tissue regeneration. It has a beneficial effect with swellings, hematoma, fractures, sport injuries, and posttraumatic conditions. It should not be applied to open and infected wounds.

The production process of Comfrey gel is performed in the following stages:

- Purchase of high-quality herbal material *Symphyti radix* and quality control of parameters, as prescribed by Pharmacopeia selection of high-quality pharmaceutical raw materials according to the manufacturer attest.
- Grinding of comfrey root and producing the liquid extract with propylene glycol with a method of percolation. The resulting extract is controlled to authentication, relative density, and microbiological safety.
- Liquid extract of comfrey, after the processing procedure is incorporated into the semi-solid base, and then filled into tubes. The final product, herbal gel from the roots of comfrey, is sent to the control to laboratories.
- When a controlled product receives confirmation that corresponds to the standard quality, it is dispatched to the warehouse of final products and further distributed.

2.6.5. Herbal capsules odoval s®, herbal sedative

Herbal product is used for the defense of organism against the effects of daily stress.

Valerian root (*Valerianae radix*) has a long traditional use. It is approved by Commission E to help combat restless states and difficulties falling asleep caused by nervousness. According to EMEA monograph, it has a well-established use as a herbal medicinal remedy for the relief of mild nervous tension and sleep disorders.

Melissa leaf (*Melissae folium*) is a favorite vegetable drug that is used as a mild sedative, carminative, antispasmodic, and aromatic. The EMEA monograph only credits *Melissae folium* with a traditional use: it is considered to be a traditional herbal medicinal remedy for relief of mild symptoms of mental stress and to aid sleep and for symptomatic treatment of mild gastrointestinal. Capsules of this herbal sedative are designed to have the *Valerianae extractum siccum* as a dominant component, with the synergistic effects of *Melissae extractum siccum*.

Odoval S® is a herbal remedy meant for maintaining mental balance. It contains extracts of valerian root and lemon balm leaf, medicinal plants traditionally used for their calming properties. The active ingredients of these extracts have a favorable effect on alleviating anxiety and irritability, facilitate sleep, and establishment of natural sleep patterns. Odoval S® can be used as a valuable help in alleviating various symptoms caused by chronic stress (mental tension and anxiety during the day, irritability, and feelings of worry).

In the Institute , the production process of capsules is carried out in the following stages:

1. Purchase of high-quality herbal raw material of *Valerianae radix* and *Melissae folium* and quality control on parameters, as prescribed by Pharmacopeia.
2. Production of dry extracts: *Valerianae extractum siccum* and *Melissae extractum siccum* in the circular extractor, according to the specifications of the process.
3. The resulting dry extract is controlled as a semi-final product to authentication, moisture content, and microbiological safety.
4. Controlled dry extracts, along with additional pharmaceutical materials, are used to produce granulate (mass for encapsulation). Granulate, as a semi-final product, also goes to the control to pharmaceutical and microbiological laboratories. Pharmacopoeial method is used to determine the content of valeric acid [2].

5. Upon obtaining the results of quality control, encapsulation, that is, automatic filling of gelatin capsules with prepared granulate, is conducted. Capsules are forwarded in the glass jars, and then into the boxes.
6. The final phytopreparation is sent to the control to laboratories. The following parameters are determined: properties, number of capsules in a package, the average weight of the capsule content, uniformity of mass capsule, identification of valeric acid LC and TLC for lemon balm, disintegration of capsules, and microbiological safety. The product is sent to the analysis of the complete health safety.
7. When results confirm the quality according to certificates and specifications, the final product is dispatched to the warehouse of final products and afterwards distributed to pharmacies and other places.

3. CONCLUSIONS AND PERSPECTIVES

Natural products discovered so far have played a vital role in improving the human health and have been the drugs of choice despite facing a tough competition from compounds obtained by chemical procedures, due to their safety and efficacy. The most striking feature of natural products in connection to their long-lasting importance in drug discovery is their structural diversity that is still largely untapped [1].

Comprehensive development of science and technology, able to produce high-quality herbal medicines, is greatly improved in recent decades. The acceptance of herbal medicine as a natural and gentle alternative to synthetic drugs is very high in public in developed countries and, from a global perspective, unit sales of herbal medicines is constantly growing. However, we still face many problems in these areas [2].

A comprehensive approach to these problems, the state of the field of medicinal plants and herbal remedies, can be repaired. A better education of people is involved in the collection and cultivation of medicinal plants on the necessity of obtaining plant raw material of high quality. In particular, it should encourage the concept of organic production herbal products. Producers should be required to produce only quality-assured medicines.

Improved harmonization of regulatory classification of herbal preparations in the world would inevitably lead to greater transparency and consistency of the market.

Special attention should be paid to improving knowledge about the benefits of rational phytotherapy, particularly evidence-based phytotherapy, health workers, especially doctors.

The aim of all efforts would be to improve the overall awareness of the possibilities of choice in prevention and treatment and can judge the effectiveness of the use of medicinal herbs and herbal preparations.

Financial support of this work by the Serbian Ministry of Education, Science and Technological Development (Project III 46001) is gratefully acknowledged.

REFERENCES

1. Veeresham C. Natural products derived from plants as a source of drugs. Journal of Advanced Pharmaceutical Technology & Research. 2012;3(4):200–201.
2. Đorđević S, Dickov A, Pavkov S, Tadić V, Arsić I, Žugić A. Manufacturing process of high quality phytopreparation on example of herbal sedative. Medicinski Pregled. 2013;66(3–4):170–176.
3. Fürst R, Zündorf I. Evidence-based phytotherapy in Europe: where do we stand? Planta Medica. 2015;81(12–13):962–967.
4. Tasić S, Šavikin K, Menković N. Guide to Medicinal Herbs. 2nd ed. Belgrade: Alexandria; 2009. 6 p. ISBN: 978-86-6119-034-6
5. Kišgeci J. Medicinal Plants. Belgrade: Partenon; 2002. 7 p. ISBN: 978-86-7157-188-6
6. Cragg GM, Newman DJ. Natural products: A continuing source of novel drug leads. Biochimica et Biophysica Acta. 2013;1830(6):3670–3695.
7. Petrovska BB. Historical review of medicinal plants' usage. Pharmacognosy Reviews. 2012;6(11):1–5.
8. Atanasov AG, Waltenberger B, Pferschy-Wenzig EM, Linder T, Wawrosch C, Uhrin P, Temml V, Wang L, Schwaiger S, Heiss EH, Rollinger JM, Schuster D, Breuss JM, Bochkov V, Mihovilovic MD, Kopp B, Bauer R, Dirsch VM, Stuppner H. Discovery and resupply of pharmacologically active plant-derived natural products: A review. Biotechnology Advances. 2015;33(8):1582–1614.
9. Blumenthal M, Hall T, Goldberg A, Kunz T, Dinda K. The ABC Clinical Guide to Herbs. Austin, TX: American Botanical Council; 2003. 19 p. ISBN: 3-13-132391-4
10. Wichtl M. Herbal Drugs and Phytopharmaceuticals. Stuttgart: Medpharm; 2004. 17 p. ISBN 9780849319617.
11. Schulz V, Hänsel R, Tyler V. Rational Phytotherapy. Berlin: Springer-Verlag; 2001. 326 p.
12. Tadić V, Đorđević S, Arsić I. Herbal drugs in the treatment of nervous and cardiovascular disorders. HealthMED. 2011;5(5):1206–1210. ISSN: 1840-2291
13. Maridass M, John de Britto A. Origins of plant derived medicines. Ethnobotanical Leaflets. 2008;12:373–387. http://opensiuc.lib.siu.edu/ebl/vol2008/iss1/44
14. Cupara S, Janković S. Significance of herbal drugs legislation for their rational use. Vojnosanitetski pregled. Vojnosanit Pregled. 2009;66(1):53–56. UDC: 615.311:581.19
15. Qu L, Zou W, Zhou Z, Zhang T, Greef J, Wang M. Non-European traditional herbal medicines in Europe: A community herbal monograph perspective. Journal of Ethnopharmacology. 2014;156:107–114.

16. European Pharmacopoeia. 8th ed. vol. 1. Ph Eur 8.1. Council of Europe (EDQM). Strasbourg. 2013. 1133 p. ISBN: 978-92-871-7525-0
17. Žižović I, Stamenić M, Ivanović J, Orlović A, Ristić M, Đorđević S, Petrović DS, Skala D. Supercritical carbon dioxide extraction of sesquiterpenes from valerian root. Journal of Supercritical Fluids. 2007;43(2):249–258.
18. Žižović I, Ivanović J, Mišić D, Stamenić M, Đorđević S, Kukić-Marković J, Petrović DS. SFE as a superior technique for isolation of extracts with strong antibacterial activities from lichen *Usnea barbata* L. Journal of Supercritical Fluids. 2012;72:7–14.
19. Tadić V, Bojović D, Arsić I, Đorđević S, Aksentijević K, Stamenić M, Janković S. Chemical and antimicrobial evaluation of supercritical and conventional *Sideritis scardica* Griseb., Lamiaceae extracts. Molecules. 2012;17(3):2683–2703.
20. Ćujić N, Šavikin K, Janković T, Pljevljakušić D, Zdunić G, Ibrić S. Optimization of polyphenols extraction from dried chokeberry using maceration as traditional technique. Food Chemistry. 2016;194:135–142.
21. Kovačević N. Quality and quality control of herbal drugs, extracts and phytomedicines. Lekovite Sirovine. 2001;20:57–68. UDC: 615.322.07
22. Đorđević S, Petrović S, Ristić M, Đoković D. Composition of *Carlina acanthifolia* root essential oil. Chemistry of Natural Compounds. 2005;41(4):410–412.
23. Huynh-Ba K (ed.). Handbook of Stability Testing in Pharmaceutical Development: Regulations, Methodologies, and Best Practices. Delaware: Pharmalytik Newark; 2009. 139 p.
24. Arsić I, Tadić V, Đorđević S, Žugić A, Vujić Z, Petrović DS. Optimization of extraction of antioxidant components from Yarrow herb. Hemijska Industrija. 2014;68(4):511–517.

CHAPTER 2

Thymus Plants: A Review—Micropropagation, Molecular and Antifungal Activity

Fernanda Leal[1, 2*], Meriem Taghouti[3], Fernando Nunes[4], Amélia Silva[3], Ana Cláudia Coelho[5] and Manuela Matos[1, 2]

[1] *Genetics and Biotechnology Department, University of Trás-os-Montes and Alto Douro (UTAD), Vila Real, Portugal*

[2] *BioISI—Biosystems & Integrative Sciences Institute, Faculty of Sciences University of Lisboa, Lisboa, Portugal*

[3] *Centre for Research and Technology of Agro-Environmental and Biological Sciences (CITAB, UTAD), Vila Real, Portugal*

[4] *CQ-VR Chemistry Research Center, UTAD, Vila Real, Portugal*

[5] *Animal and Veterinary Research Centre (CECAV), UTAD, Vila Real, Portugal*

ABSTRACT

Medicinal and aromatic plants are important sources for plant secondary metabolites. The genetic manipulation of plants associated with in vitro plant regeneration systems facilitates efforts to engineer secondary product metabolic pathways. The fungal infections have been increasing in recent years due to several factors, namely, the increased incidence of high-risk patients, particularly immunocompromised hosts. Aromatic plants have been empirically used as antimicrobial agents, but the mechanisms of action are still unknown. Thyme has a great interest due to the possibility of its use in different applications, in medicine, in the cosmetic industry, or as food additives. Several studies have shown that thyme oils possess antimicrobial activity. Increasingly, plant breeding has taken advantage of molecular biology developments in order to genotype the species of interest to accelerate their selection. These approaches consist in choosing desired genotypes based on molecular markers or the knowledge of the genes involved in a particular trait. The in vitro culture techniques can be used to multiply plants selected after molecular and antifungal studies. The course of the investigation and the current state in relation to micropropagation, molecular studies, and antifungal action of the *Thymus* genus plants will be presented.

Keywords: *Thymus*, micropropagation, molecular, antifungal

1. INTRODUCTION

The genus *Thymus* L. belongs to the Lamiaceae family and consists of over 400 species of herbaceous annuals and perennial plants that are extensively used, for medicinal and non-medicinal purposes. These plants are widely distributed throughout the Old World [1, 2] and have been used for many centuries in traditional medicine due to their antiseptic, carminative, antiviral, and antioxidant properties [3]. *Thymus* species are also interesting as a source of pentacyclic triterpenoids with several properties, as anti-inflammatory, hepatoprotective, antimicrobial, anti-HIV-1 activity, antiulcer, gastroprotective, hypoglycemic, antihyperlipidemic activity, and specific cytotoxicity against a variety of tumor cell lines [4–6]. By all this it becomes increasingly important to know the bioactive components of the *Thymus*.

Furthermore, interests focusing mainly on few selected chemotypes for the cosmetic and food industries, among others, lead to the loss of other species in nature, such as *Thymus cariensis* Hub.-Mor. & Jalas, *Thymus cilicicus* Boiss. & Balansa, *Thymus sipyleus* Boiss., *Thymus pulvinatus* Čelak., and *Thymus cherlerioides* Vis. [7]. These species should be preserved to make available the access to a wide range of genetic diversity. On the other hand, as the plant has a low propagation rate in nature, a suitable method to obtain a high number of plants is needed [8].

2. *THYMUS* MAIN ACTIVE COMPOUNDS

Among the *Lamiaceae* family, the genus *Thymus* is one of the most studied genera, due to the use of these plants as a remedy in folk medicine and as a condiment, mainly in the Mediterranean zone [1]. High bioactivity (e.g., antioxidant, anti-inflammatory, antiproliferative, and antimicrobial effects) is linked to the considerable content in phytochemicals of different *Thymus* species. Volatile compounds are extracted in essential oils (EOs), while nonvolatile compounds are found in alcoholic and aqueous extracts obtained by maceration, decoction, or infusion [9].

Noticeable interest was given to the study of EO in different *Thymus* species. High intraspecific chemical polymorphism was reported, especially in *Thymus vulgaris* L., which was shown to have a polymorphic variation in monoterpene production [10, 11]. Phenolic monoterpenes, such as carvacrol and thymol (**Figure 1**, compounds 1 and 2), seem to explain the important biological activity of the majority of *Thymus* spp. essential oils characterized by phenolic chemotypes. However, the presence of other terpenes in the EO results in the enhancement or reduction of EO bioactivity due to synergistic or antagonistic effects [12]. In the *Lamiaceae* family, EOs are produced in the glandular trichomes and stored in their cavities [13]. Several enzymes are involved in the biosynthetic pathways of terpenes, such as the terpene synthases enzyme class and the cytochrome P450s that are involved in the core terpene molecule functionalization [14].

Carvacrol and thymol have *p*-cymene as common precursor (**Figure 1**, compound 5). Carvacrol is frequent in some Lamiaceae aromatic plant genera like *Thymus*, *Origanum*, and *Satureja* [15], while thymol is mainly frequent in *Thymus* plants. Phenolic chemotypes are characteristic of *T. vulgaris* [10], *Thymus hyemalis* Lange [16], *Thymus capitatus* L. [17], *Thymus zygis* L. [18], and some populations of *Thymus pulegioides* L. [19]. Thymol's antibacterial effect is enhanced by the presence of carvacrol in the essential oil [12]. Both thymol and carvacrol have important antifungal effect; they act by causing damage in the cell membrane and interacting with ergosterol [18, 20–22]. This mechanism was also proved with other monoterpenes, namely, linalool (**Figure 1**, compound 3), which acts also by interfering with biofilm formation and stability [23].

Linalool and geraniol (**Figure 1**, compounds 3 and 4, respectively) are monoterpene alcohols, which characterize several chemotypes of *Thymus* species such as *Thymus pulegioides*, *T. x. citriodorus* (Pers.) Schreb., *Thymus algeriensis* Boiss. & Reut., and some populations of *T. vulgaris* [11, 19, 24, 25]. Eucalyptol, or 1,8-cineole (**Figure 1**, compounds 6), a monoterpene cyclic ether monoterpene that is mainly present in *Eucalyptus* genus, is the major component of *Thymus mastichina* (L.) L. EO [26] and was also reported in populations of *T. algeriensis*, *Thymus hirtus* Raf., and *Thymus glabrescens* [27, 28].

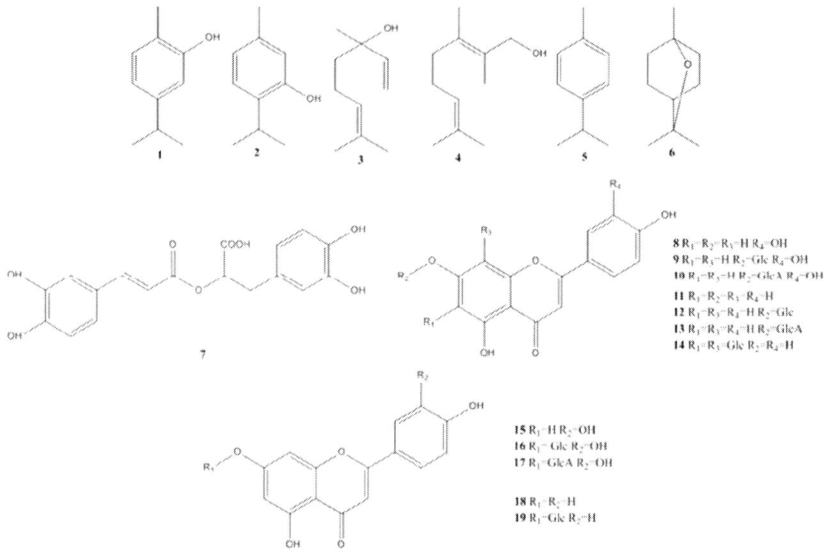

FIGURE 1. Chemical structure of main active compounds in *Thymus* spp. (1) carvacrol, (2) thymol, (3) linalool, (4) geraniol, (5) *p*-cymene, (6) eucalyptol, (7) rosmarinic acid, (8) luteolin, (9) luteolin-*O*-glucoside, (10) luteolin-*O*-glucuronide, (11) apigenin, (12) apigenin-7-*O*-glucoside, (13) apigenin -7- *O*-glucuronide, (14) apigenin-6,8-di-*C*-glucoside, (15) eriodictyol, (16) eriodictyol-*O*-glucoside, (17) eriodictyol-*O*-glucuronide, (18) naringenin, (19) naringenin-7-*O*-glucoside. Glc, glucoside unit; GlcA, glucuronide unit.

Although more interest has been given to study thyme essential oils, nonvolatile extracts contain highly active secondary metabolites which are mainly phenolic compounds. Aromatic amino acids L-phenylalanine and L-tyrosine, produced by the shikimic acid pathway, are the precursors of the biosynthesis of polyphenols, namely, phenolic acids and flavonoids [29, 30]. Rosmarinic acid and luteolin (**Figure 1**, compounds 7 and 8, respectively) are the main frequent phenolic compounds found in thyme plants and which are related to their extracts' biological activity. Rosmarinic acid, caffeic acid esterified with 3,4-dihydroxyphenyllactic acid, is the most abundant phenolic acid in several *Thymus* species: *T. vulgaris*,

T. pulegioides, *T. zygis*, *T. mastichina*, *T. capitatus*, *Thymus longicaulis* C. Presl, and *T. x. citriodorus* [9, 31–36].

From the class of flavones, luteolin and apigenin (**Figure 1**, compounds 8 and 11, respectively) are the most important in *Thymus*. Luteolin is frequently present also in fruits and other *Lamiaceae* genera like *Mentha*. Its glycosylated derivatives such as luteolin-*O*-glucoside (**Figure 1**, compound 9) are also frequent in thyme species. Apigenin and its derivatives (**Figure 1**) are reported in some species such as *T. x. citriodorus*, *T. vulgaris*, and *Thymus herba-barona* Loisel. [37]. Eriodictyol and naringenin (**Figure 1**, compounds 15 and 18) are flavanones which were described as well as their glucoside and glucuronide derivatives (**Figure 1**, compounds 16 and 19 and 17, respectively) in some *Thymus* species such as *T. x. citriodorus* and *T. vulgaris* [34].

3. MICROPROPAGATION OF *THYMUS*

Advances in biotechnological approaches provide a set of techniques that contribute to solving problems of extinction or genetic erosion in particular of plants. Alternatives for fast multiplication, like "in vitro micropropagation" that enables propagation of plants under controlled environmental conditions, can help in multiplying selected plants after molecular and antifungal studies or subjected to excessive demand by the people.

Furthermore, it was also possible to develop techniques that allow the maintenance of germplasm for a long time, like "cryopreservation" that make available long-term storage of *Thymus* germplasm at ultra-low temperatures [8].

Optimizations for micropropagation process involve the use of different growth regulators like cytokinins, auxins, or gibberellic acid for the induction of multiple shoots. The first works in *Thymus* micropropagation are from Furmanowa and Olszowska dated from 1980 [38] to 1992 [39]. Since that date, a lot of works for the micropropagation of different species of *Thymus* (**Table 1**), as well as the cryopreservation of the same plants, were developed (**Figure 2**).

Cryopreservation procedures, PVS2 vitrification, encapsulation—vitrification and droplet—and freezing methods showed to be effective to induce cryotolerance and long-term conservation of thyme shoot tips obtained from in vitro propagated plantlets [49]: Different *Thymus* species have already been studied, *Thymus moroderi* [50], *T. vulgaris* and *T. longicaulis* [2], *T. cariensis* and *T. vulgaris* [41], and *Thymus lotocephalus* [51].

Figure 2. *Thymus caespititius* multiple shoots produced. (A) Outgrowth of the axillary buds maintained on MS medium with 0.4 mg L^{-1} BA and 0.1 mg L^{-1} IBA. (B) Detail of spontaneous root formation in shoot cultures (bars = 1 cm) [43].

Table 1. Approaches in *Thymus* plant micropropagation.

Species	Achievements	Growth regulators or others	References
Thymus mastichina	Propagation of thyme from mature field-grown plants	Nodal segments—MS + 0.1 mg L^{-1} BAP Roots—MS + 1 mg L^{-1} NAA	[40]
Thymus vulgaris and *Thymus longicaulis*	In vitro propagation protocol	Shoots, semisolid MS + 1 mg L^{-1} KN and 0.3 mg L^{-1} GA3; roots, MS + 0.05 mg L^{-1} 2,4-D	[41]
Thymus lotocephalus	Propagation protocol using seeds as explants	MS BAP induce high % of hyperhidric shots	[42]
Thymus caespititius	Shoots with high proliferation capacity	MS + 0.4 mg L^{-1} BA + 0.1 mg L^{-1} IBA	[43]
Thymus persicus	System for regeneration via direct organogenesis	Shoot, 8.9 μM BAP + 2.7 μM NAA; roots, 1/2 MS + 2.5 μM IBA	[44]
Thymus hyemalis	Regeneration of plants through somatic embryogenesis	MS + 4.44 μM BAP, 0.54 μM NAA, and 4.65 μM KIN	[45]
Thymus daenensis	Colloidal silver nanoparticles reduce hyperhydricity	2.5 mg L^{-1}AgNPs (silver nanoparticles)	[46]
Thymus moroderi	Propagation disinfection processDouble phase culture system	growth regulators did not improve the morphogenic response	[47]
Thymus persicus	Callus induction and micropropagation	Callus, MS + 2.0 mg L^{-1} NAA and 0.5 mg L^{-1} KN Shoot, MS + 2.0 mg L^{-1} BAP + 1.0 mg L^{-1} NAA	[48]

4. GENETIC ANALYSIS

Researchers, from different areas, use the genetic analysis of plants as the basis of their work in order to identify and to characterize plant materials in nature, to detect genetic diversity or the genetic homogeneity, and to select plants with desired compounds.

The pool of genetic variation in plants, namely, the medicinal and the aromatic ones, serves as the base for plant breeding as well as for selection. Molecular markers are very useful in breeding program allowing germplasm screening independent to the developmental stage of the plants and/or environmental factors [52].

Applications of DNA methods, with different purposes, in *Thymus* genus are few when compared with other economically important plant species. A few reports are available about genetic characterization of species of *Thymus* genus (**Table 2**), using different molecular markers, being the techniques of random amplified polymorphic DNA markers (RAPD) and inter-simple sequence repeat (ISSR) the most commonly used. Amplified fragment length polymorphisms (AFLPs) and microsatellites are also utilized for genetic analysis and genetic relationships in *Thymus* species [56, 63].

Beyond the genetic characterization of different species, the knowledge of genetic diversity within species is necessary for any improvement of cultivars and biodiversity maintenance and restoration [64]. Yousefi et al. [54] studied ecotypes grown in different parts of Iran using ISSRs and verified that the accessions were relatively grouped according to their location and conclude that ISSRs provided a powerful and reliable molecular tool for detecting genetic variation and relationships. A similar study was done by Rahimmalek et al. [55] with the purpose to assess the genetic diversity of *Thymus daenensis* Celak accessions toward the conservation of the endangered aromatic species. Solyman and Alkowni [59] studied genetic diversities of five Palestinian *Thymus* species using RAPD markers and concluded that it could be useful tools for identifying *Thymus* species in any putative breeding programs that will be carried in the country. A study, with AFLPs, in *Thymus*, section *Serpyllum* [63], demonstrated that this markers could be suitable for complex genetic relationships analysis, including frequent interspecies hybridization events. Recently Karaca et al. [56] present the first report of genomic microsatellite markers for the genus *Thymus*, used in 48 individuals representing 9 species and subspecies of *Thymus*.

Table 2. Employment of molecular markers in genetic characterization of *Thymus* genus plants.

Species	Achievements	Markers	References
Thymus caespititius	Identification of terpene synthase genes in *Lamiaceae*	TPS gene	[53]
Thymus daenensis *Thymus kotschyanus* *Thymus vulgaris*	Assessment of genetic diversity and relationships	ISSRs	[54]
Thymus daenensis subsp. *daenensis*	Assessment of genetic diversity and geographic differentiation	ISSRs	[55]
Thymus cilicicus *Thymus revolutus* *Thymus cherlerioides* *Thymus leucotrichus* *Thymus zygioides* *Thymus sipyleus* *Thymus longicaulis*	Development of 23 microsatellite primer pairs for *Thymus* genus and assessment of genetic diversity and relationships of 48 samples representing nine species and subspecies of the genus *Thymus*	Microsatellites	[56]
Thymus persicus	Confirmation of genetic homogeneity in in vitro regenerated plants	RAPDs	[48]
Thymus kotschyanus	Assessment of genetic diversity of wild populations	RAPDs	[57]
Thymus daenensis *Thymus fallax* *Thymus fedtschenkoi* *Thymus migricus* *Thymus vulgaris*	Assessment of genetic diversity and chemical polymorphism of *Thymus* species	RAPDs	[58]
Thymus syriacus *Thymus fruticosus* *Thymus incanus* *Thymus majorana* *Thymus capitatus*	Assessment of thyme genetic diversity in Palestine	RAPDs	[59]
Thymus vulgaris	Correlation between the chemical and genetic relationships among commercial *Thymus* cultivars	RAPDs	[60]
Thymus glabrescens *Thymus pannonicus* *Thymus praecox* *Thymus pulegioides*	Essential oil composition and molecular analysis	RAPDs	[61]
Thymus caramanicus	Assessment of genetic and chemical variability	ISRRs	[62]
Thymus pulegioides *Thymus glabrescens* *Thymus marschallianus* *Thymus pannonicus* *Thymus balcanus* *Thymus moesiacus* *Thymus praecox*	Assessment of genetic diversity and relationships among species of the genus *Thymus* L. (section *Serpyllum*)	AFLPs	[56]

The overexploitation of wild plants for commercial purposes (and consequent decreased of populations) associated to the increasing demand for secondary metabolites has intensified the application of biotechnological methods to propagate and reproduce high-yielding plants under controlled growing conditions and/or to obtain homogenous and stable genotypes. Other application of molecular markers in this genus is the analysis of the reliability of the in vitro propagation regarding the genetic homogeneity, most of times associated to the phytochemical productivity of the produced plantlets, as the experiment reported by Bakhtiar et al. [48] using RAPDs. In this work RAPD profiles confirmed the homogeneity and high-performance liquid chromatography (HPLC) confirmed the phytochemical productivity of the in vitro regenerated plants.

Mendes et al. [53], using in vitro genotypes, characterized *Thymus caespititius* terpene synthase 2 (*Tctps2*) gene and identified other terpene synthase genes responsible for the chemical polymorphism observed in *T. caespititius* essential oils.

Another genetic approach is the analysis of the chemical and the genetic relationships among species as the study described by [58] that also determinate the correlation between these two sets of data, the essential-oil composition and genetic variability of six populations of *Thymus*. RAPD markers allowed a perfect distinction between the six species, based on their distinctive genetic background: however, they did not show identical clustering with the volatile-oil profiles [58]. Contrary Echeverrigaray et al. [60], also based in RAPD profiles, observed that the cultivars (*T. vulgaris* L.) could be divided into two clusters, which coincided with results obtained by oil GS-MS analysis. Chemical and genetic differences of four *Thymus* species were studied by Pluhár et al. [61] in order to determine whether molecular characters (RAPDs) and essential oil components could be used as taxonomic markers and obtained a partial correlation between molecular and chemical assessments. In *Thymus caramanicus* Jalas, Hadian et al. [62] assessed the genetic (using ISSRs) and chemical variability and observed a relationship between genetic and chemical variability and geographic distribution.

5. FUNGAL INFECTIONS

Fungal infections are a serious problem of public health concern and have been increasing in recent years due to several factors given increased international travel, immigration, changing climate conditions, and the increased incidence of high-risk patients [65]. Invasive mycoses are especially problematic for immunocompromised individuals and patients in intensive care units, and some conditions can predispose as organ transplantation, the use of drugs in treatments as corticosteroids and antineoplastics, and complex surgical procedure acts [66, 67]. Other cases may be found in patients suffering from diabetes mellitus, patients with human immunodeficiency virus infection, patients with neoplasias after receiving chemotherapy, patients with transplantation surgeries, or those with prolonged antibiotherapy [68].

Oral and vulvovaginal candidiases caused by *Candida albicans* are the most common fungal diseases [69]. Invasive fungal diseases can be less frequent but much more severe. Other mycotic diseases or complications associated included asthma, bronchopulmonary and invasive aspergillosis, pneumocystosis, meningeal cryptococcosis, mucormycoses, or invasive candidiasis [70]. Deaths related with fungal infections are mostly associated with *Cryptococcus*, *Candida*, *Aspergillus*, and *Pneumocystis* [71].

Aromatic plants have been empirically used as antimicrobial compounds, but the mechanisms of action are still under study [72]. Inhibitory action of aromatic plants possibly includes cytoplasm granulation, cytoplasmic membrane lesion, and inhibition and/or inactivation of extracellular and intercellular enzymes [72, 73] and might be due to different compounds, including phenolics, terpenoids, and alkaloids. These compounds together or independently use different levels of antifungal effect ending with mycelium germination inhibition [73]. Also, it is described that plant lytic enzymes act in the fungal cell wall causing breakage of β-1,6 [72]glycan, β-1,3 glycan, and chitin polymers [74]. The antimicrobial activity of the aqueous extracts could be due to the anionic components such as chlorides, thiocyanate, nitrate, and sulfates besides other water-soluble constituents which are naturally occurring in the plant material [75].

6. ANTIFUNGAL ACTIVITY

Only limited numbers of new antifungal drugs were developed in recent years, and there are only small numbers of drugs available for their treatment [68]. Toxicity and drug resistance have become an increasing problem. The resistance to antifungal drugs, the high costs associated with treatment, and the fungistatic activity of most of the antifungal drugs are problems making their treatment difficult and expensive [67]. So, alternatives for treating invasive fungal infections are necessary [67].

The spread of multidrug-resistant strains of fungi is a medical problem worldwide, and the reduced number of drugs available led to a search for therapeutic substitutes, namely, among aromatic and medicinal plants and compounds isolated from them used for their antifungal [76, 77].

Numerous molecules obtained from the natural environment are investigated and described in bibliography with antimycotic activity. Several extracts are investigated for antifungal activities like crude extracts or isolated constituents as essential oils, saponins, terpenoids, alkaloids, phenolic compounds, peptides, and proteins [78, 79].

The in vitro evaluation methods of antifungal activity can be divided into diffusion methods and dilution methods. The diffusion methods yielding to inhibition diameters are used mostly for the qualitative screening. Examples are the agar overlay technique [80]. Dilution methods offer more quantitative results regarding the action of essential oils. Serial dilutions are used for detecting the minimal inhibitory concentration (MIC in mg ml^{-1}) of an essential oil in a liquid medium its antimicrobial properties and rank it among the most potent essential oils in this respect [81].

A wide range of aromatic and medicinal plants with therapeutic properties have been explored and used for the extraction of essential oils all over the world due to their antimicrobial capacity against fungal pathogens [79]. Several studies have shown that thyme oils possess antimicrobial activity [82–84].

Thymus spp., most of them possessing a large quantity of phenolic monoterpenes, revealed activity against fungi [83].

Medicinal plants have the capacity to inhibit the growth of a wide range of opportunistic or pathogenic microorganisms due to the presence of essential oils [85]. Essential oils are natural, volatile liquid, complex compounds characterized by rarely colored and a strong odor, soluble in lipid and organic solvents [85]. About 60% of essential oils show antifungal activity [84].

The essential oils and their components have been used broadly against molds [85]. The essential oil extracts from many plants have shown their considerable antifungal activity against the wide range of fungal pathogens [73].

Thyme essential oils are apparently among the greatest inhibitors of fungal microorganisms because of the presence of the phenolic compounds such as thymol as main components which might disrupt the fungal cell membrane [85]. Another component that appears to show antimicrobial activity is terpene hydrocarbons (γ-terpinene) [86]. *p*-Cymene is a compound of essential oil that does not show antibacterial efficacy when used alone which suggest a synergistic effect of the compounds [86, 87].

Thyme essential oils may in the future represent a new source of natural antiseptics with applications in industry of pharmaceutics and food [86]. The essential oils have the ability to penetrate and disrupt the fungal cell wall and cytoplasmic membranes, permeabilizing them and finally causing damage to mitochondrial membranes [88].

Variability in essential oil compounds might be linked to differences in concentration and amount recovered based on several factors, including species of plant used, method of extraction, solvents, and extraction time, which in turn may differ in their antifungal potency [68, 89]. Differences can also be attributed to raw materials used (dried or fresh), types of soils used for cultivation, the harvesting time in the year, or differences in oil extraction techniques [90].

These oils have been used in folk medicine in different communities for patients suffering from mycotic infections [91].

Differences in essential oil compounds might be related to variability in the dried or fresh materials used, to the harvesting time in the year, to types of soils used for cultivation, or to differences in oil extraction techniques. *T. vulgaris* has medicinal properties and is widely used in traditional medicine for its expectorant, antispasmodic, antibronchiolitic, antitussive, anthelmintic, carminative, and diuretic properties [92].

Many studies have shown that thyme (*T. vulgaris*) has antifungal activities and have suggested their integration into pharmaceutical preparations in the treatment of candidiasis [72, 86, 93]. Previous studies showed fungistatic activity of the essential oils of *T. vulgaris* as carvacrol, *p*-cymene, and thymol [94]. The antifungal mechanism of action by which thymol or carvacrol acts is actually not well understood, although the antifungal activity is probably because thymol is lipophilic and together with carvacrol can act in the fatty acyl chains of membrane lipid bilayers, and alters the fluidity and permeability of cell membranes [95]. Other mechanisms have been theorized as damaged of membrane and cell wall with disruption associated with morphological deformation, deterioration, collapse, and of the conidia and/or hyphae [96]. Thymol and carvacrol have a stronger antifungal capacity, indicating more susceptibility of *Aspergillus* spp. than that of *Penicillium* spp. [97]. Thyme essential oil showed its capabilities of inhibiting fungal development causing leakage of the cytoplasm of *Aspergillus flavus* and was responsible for degenerative alterations in hyphae alterations, which appeared degraded or with the complete absence of conidia [92]. A previous report showed capabilities of inhibiting aflatoxin production [92].

Thymus essential oil causes irreversible damage to cell wall, cell membrane, and cellular organelles which affects *Aspergillus niger* growth and morphology [98].

The fungal activity of *Thymus* plants has been summarized in **Table 3**.

Table 3. Effect of *Thymus* plants on the fungal microorganisms.

Thymus plant	Inhibited microorganisms	References
Thymus vulgaris	*Candida albicans*	[72, 86, 94, 99, 100]
Thymus vulgaris	*Aspergillus flavus*	[92]
Thymus vulgaris	*Aspergillus ochraceus*	[96]
Thymus eriocalyx *Thymus x-porlock*	*Aspergillus niger*	[98]
Thymus vulgaris	*Penicillium chrysogenum*	[96]
Thymus serpyllum	Dermatophytes	[101]
Thymus schimperi	*Penicillium chrysogenum*	[66]
Thymus schimperi	*Verticillium* sp.	[66]
Thymus schimperi	*Aspergillus tubingensis*	[66]
Thymus schimperi	*Aspergillus minutus*	[66]
Thymus schimperi	*Beauveria bassiana*	[66]
Thymus schimperi	*Microsporum gypseum*	[66]

7. CONCLUSIONS

Thymus plants have been playing, in recent years, an increasingly important role in the intense study being targeted to run through the different areas of biotechnology. So here we intend to present the latest developments involving this plant in such diverse areas as active compound analysis and their effect on antifungal activity, genetic analysis, micropropagation, and finally cryopreservation techniques, one of the most recent methodologies to plant preservation.

The utilization of new plant breeding technologies will be the future in aromatic and medicinal plant manipulation and production, in order to obtain a multitude of valuable characteristics like increased nutrient and metabolite production and resistance to different stresses. Next-generation sequencing (NGS) technology and the associated bioinformatics tools will allow general profiling of RNA expression in plant species with limited molecular genetics studies as the majority of aromatic and medicinal plants.

REFERNCES

1. Morales, R. The history, botany and taxonomy of the genus Thymus. In: Stahl-Biskup, E. & Sáez, F. (eds). Thyme: The Genus Thymus. London: CRC Press Taylor & Francis Group; 2002. pp. 1–43. ISBN 0203216857
2. Ozudogru, E. A., Kaya, E. & Kirdok, E. Development of protocols for short-, medium- and long-term conservation of thyme. Acta Horticulturae. 2011; 918: 43–50.
3. Stahl-Biskup, E. Thyme as a herbal drug pharmacopoeias and other product characteristics. In: Stahl-Biskup E. & Sáez F. (eds). Thyme: The Genus Thymus. London: CRC Press Taylor & Francis Group; 2002. pp. 293–316. ISBN 0203216857
4. Huang, M. T., Ho, C. T., Wang, Z. Y., Ferraro, T., Lou, Y. R., Stauber, K., et al. Inhibition of skin tumorigenesis by rosemary and its constituents carnosol and ursolic acid. Cancer Research. 1994; 54(3): 701–708.
5. Liu, J. Pharmacology of oleanolic acid and ursolic acid. Journal of Ethnopharmacology. 1995; 49: 57–68.
6. Liu, J.Oleanolic acid and ursolic acid: research perspectives. Journal of Ethnopharmacology. 2005; 100: 92–94.
7. Rey, C. & Sáez, F. Field culture, in vitro culture and selection of Thymus. In: Stahl-Biskup E. & Sáez F. (eds) Thyme, the Genus Thymus. London: Taylor & Francis; 2002. pp. 177–196.
8. Shabnum, S. & Wagay, M. G. Micropropagation of different species of Thymus. Journal of Research & Development. 2011; 11: 71–81. ISSN 0972-5407
9. Martins, N., Barros, L., Santos-Buelga, C., Silva, S., Henriques, M. & Ferreira, I. C. F. R. Decoction, infusion and hydroalcoholic extract of cultivated thyme: antioxidant and antibacterial activities, and phenolic characterisation. Food Chemistry. 2015; 167: 131–137.
10. Thompson, J. D., Chalchat, J. C., Michet, A., Linhart, Y. B. & Ehlers, B. Qualitative and quantitative variation in monoterpene co-occurrence and composition in the essential oil of *Thymus vulgaris* chemotypes. Journal of Chemical Ecology. 2003; 29: 859–880.
11. Kaloustian, J., Abou, L., Mikail, C., Amiot, M. J. & Portugal, H. Southern French thyme oils: chromatographic study of chemotypes. Journal of the Science of Food and Agriculture. 2005; 85: 2437–2444.
12. Cosentino, S., Tuberoso, C. I. G., Pisano, B., Satta, M., Mascia, V., Arzedi, E. & Palmas, F. In-vitro antimicrobial activity and chemical composition of Sardinian Thymus essential oils. Letters in Applied Microbiology. 1999; 29: 130–135.
13. Turner, G., Gershenzon, J., Nielson, E. E., Froehlich, J. E. & Croteau, R. Limonene synthase, the enzyme responsible for monoterpene biosynthesis in peppermint, is localized to leucoplasts of oil gland secretory cells. Plant Physiology. 1999; 120: 879–886.
14. Degenhardt, J., Kollner, T. G. & Gershenzon, J. Monoterpene and sesquiterpene synthases and the origin of terpene skeletal diversity in plants. Phytochemistry. 2009; 70: 1621–1637.
15. De Vincenzi, M., Stammati, A., De Vincenzi, A. & Silano, M. Constituents of aromatic plants: carvacrol. Fitoterapia. 2004; 75: 801–804.
16. Rota, M. C., Herrera, A., Martinez, R. M., Sotomayor, J. A. & Jordan, M. J. Antimicrobial activity and chemical composition of *Thymus vulgaris*, *Thymus zygis* and *Thymus hyemalis* essential oils. Food Control. 2008; 19: 681–687.
17. Bounatirou, S., Smiti, S., Miguel, M. G., Faleiro, L., Rejeb, M. N., Neffati, M., Costa, M. M., Figueiredoe, A. C., Barroso, J. G. & Pedro, L. G. Chemical composition, antioxidant and antibacterial activities of the essential oils isolated from Tunisian *Thymus capitatus* Hoff. et Link. Food Chemistry. 2007; 105: 146–155.
18. Pina-Vaz, C., Rodrigues, A. G., Pinto, E., Costa-De-Oliveira, S., Tavares, C., Satgueiro, L., Cavaleiro, C., Goncalves, M. J. & Martinez-De-Oliveira, J. Antifungal activity

of *Thymus* oils and their major compounds. Journal of the European Academy of Dermatology and Venereology. 2004; 18: 73–78.

19. Loziene, K., Vaiciuniene, J. & Venskutonis, P. R. Chemical composition of the essential oil of different varieties of thyme (*Thymus pulegioides*) growing wild in Lithuania. Biochemical Systematics and Ecology. 2003; 31: 249–259.

20. Lima, I. O., Pereira, F. D., De Oliveira, W. A., Lima, E. D., Menezes, E. A., Cunha, F. A. & Diniz, M. D. F. M. Antifungal activity and mode of action of carvacrol against *Candida albicans* strains. Journal of Essential Oil Research. 2013; 25: 138–142.

21. Chavan, P. S. & Tupe, S. G. Antifungal activity and mechanism of action of carvacrol and thymol against vineyard and wine spoilage yeasts. Food Control. 2014; 46: 115–120.

22. Nobrega, R. O., Teixeira, A. P., Oliveira, W. A., Lima, E. O. & Lima, I. O. Investigation of the antifungal activity of carvacrol against strains of *Cryptococcus neoformans*. Pharmaceutical biology. 2016; 54: 2591–2596.

23. Hsu, C. C., Lai, W. L., Chuang, K. C., Lee, M. H. & Tsai, Y. C. The inhibitory activity of linalool against the filamentous growth and biofilm formation in *Candida albicans*. Medical Mycology. 2013; 51: 473–482.

24. Houmani, Z., Azzoudj, S., Naxakis, G. & Skoula, M. The essential oil composition of Algerian Zaâtar: *Origanum* spp. and *Thymus* spp. Journal of Herbs Spices & Medicinal Plants. 2002; 9: 275–280.

25. Omidbaigi, R., Sefidkon, F. & Hejazi, M. Essential oil composition of *Thymus x citriodorus* L. cultivated in Iran. Flavour and Fragrance Journal. 2005; 20: 237–238.

26. Salgueiro, L. R., Vila, R., Tomas, X., Canigueral, S., Dacunha, A. P. & Adzet, T. Composition and variability of the essential oils of *Thymus* species from Section *Mastichina* from Portugal. Biochemical Systematics and Ecology. 1997; 25: 659–672.

27. Elhadj Ali, I. B., Zaouali, Y., Bejaoui, A. & Boussaid, M. Variation of the chemical composition of essential oils in Tunisian populations of *Thymus algeriensis* Boiss. et Reut. (*Lamiaceae*) and implication for conservation. Chemistry and Biodiversity. 2010; 7: 1276–1289.

28. Fatma, G., Mouna, B., Mondher, M. & Ahmed, L. In-vitro assessment of antioxidant and antimicrobial activities of methanol extracts and essential oil of *Thymus hirtus* sp. *algeriensis*. Lipids in Health and Disease. 2014; 13: 114.

29. Koes, R. E., Quattrocchio, F. & Mol, J. N. M. The flavonoid biosynthetic-pathway in plants – function and evolution. BioEssays. 1994; 16: 123–132.

30. Riceevans, C. A., Miller, N. J. & Paganga, G. Structure-antioxidant activity relationships of flavonoids and phenolic acids. Free Radical Biology and Medicine. 1996; 20: 933–956.

31. Loziene, K., Venskutonis, P. R., Sipailiene, A. & Labokas, J. Radical scavenging and antibacterial properties of the extracts from different *Thymus pulegioides* L. chemotypes. Food Chemistry. 2007; 103: 546–559.

32. Jordan, M. J., Martinez, R. M., Martinez, C., Monino, I. & Sotomayor, J. A. Polyphenolic extract and essential oil quality of *Thymus zygis ssp gracilis* shrubs cultivated under different watering levels. Industrial Crops and Products. 2009; 29: 145–153.

33. Jabri-Karoui, I., Bettaieb, I., Msaada, K., Hammami, M. & Marzouk, B. Research on the phenolic compounds and antioxidant activities of Tunisian *Thymus capitatus*. Journal of Functional Foods. 2012; 4: 661–669.

34. Pereira, O. R., Peres, A. M., Silva, A. M. S., Domingues, M. R. M. & Cardoso, S. M. Simultaneous characterization and quantification of phenolic compounds in *Thymus x citriodorus* using a validated HPLC-UV and ESI-MS combined method. Food Research International. 2013; 54: 1773–1780.

35. Delgado, T., Marinero, P., Asensio-S-Manzanera, M. C., Asensio, C., Herrero, B., Pereira, J. A. & Ramalhosa, E. Antioxidant activity of twenty wild Spanish *Thymus mastichina* L. populations and its relation with their chemical composition. LWT--Food Science and Technology. 2014; 57: 412–418.

36. Galasso, S., Pacifico, S., Kretschmer, N., Pan, S. P., Marcianci, S., Piccolella, S., Monaco, P. & Bauer, R. Influence of seasonal variation on *Thymus longicaulis* C. Presl chemical composition and its antioxidant and anti-inflammatory properties. Phytochemistry. 2014; 107: 80–90.
37. Pereira, O. R. & Cardoso, S. M. Overview on *Mentha* and *Thymus* polyphenols. Current Analytical Chemistry. 2013; 9: 382–396.
38. Furmanowa, M. & Olszowska, O. *Thymus vulgaris* L. propagation through tissue culture. Acta Poloniae Pharmaceutica. 1980; 37: 243–247
39. Furmanowa, M. & Olszowska, O. Micropropagation of Thyme (*Thymus vulgaris* L.). In: Bajaj, Y. P. S. (ed) Biotechnology in Agriculture and Forestry. Vol. 19. Berlin: Springer; 1992. pp. 230–243.
40. Mendes, M. L. & Romano, A. In vitro cloning of *Thymus mastichina* L. field-grown plants. Acta Horticulturae. 1999; 502: 303–306.
41. Ozudogru, E. A., Kaya, E., Kirdok, E. & Issever-Ozturk, S. In vitro propagation from young and mature explants of thyme (*Thymus vulgaris* and *T. longicaulis*) resulting in genetically stable shoots. In Vitro Cellular & Developmental Biology – Plant. 2011; 47(2): 309–320.
42. Coelho, N., Gonçalves, S., González-Benito, M. E. & Romano, A. Establishment of an in vitro propagation protocol for *Thymus lotocephalus*, a rare aromatic species of the Algarve (Portugal). Plant Growth Regulation. 2012; 66: 69–74.
43. Mendes, M. D., Figueiredo, A. C., Oliveira, M. M. & Trindade, H. Essential oil production in shoot cultures versus field-grown plants of *Thymus caespititius*. Plant Cell, Tissue and Organ Culture (PCTOC). 2013; 113: 341–351.
44. Bakhtiar, Z., Mirjalili, M. H., Sonboli, A., Sonboli, A., Farimani, M. M. & Ayyar, M.. In vitro propagation, genetic and phytochemical assessment of Thymus persicus — a medicinally important source of pentacyclic triterpenoids. Biologia. 2014; 69(5): 594–603.
45. Nordine, A., Tlemcani, C. R. & Meskaoui, A. Regeneration of plants through somatic embryogenesis in *Thymus hyemalis* Lange, a potential medicinal and aromatic plant. In Vitro Cellular & Developmental Biology – Plant. 2014; 50: 19–25.
46. Bernard, F., Navab Moghadam, N. & Mirzajani, F. The effect of colloidal silver nanoparticles on the level of lignification and hyperhydricity syndrome in *Thymus daenensis* vitro shoots: a possible involvement of bonded polyamines. In Vitro Cellular & Developmental Biology – Plant. 2015; 51(5): 546–553
47. Marco-Medina, A. & Casasi, J. L. In vitro multiplication and essential oil composition of *Thymus moroderi* Pau ex Martinez, an endemic Spanish plant. Plant Cell, Tissue and Organ Culture (PCTOC). 2015; 120(1): 99–108.
48. Bakhtiar, Z., Mirjalili, M. H. & Sonboli, A. In vitro callus induction and micropropagation of Thymus persicus (*Lamiaceae*), an endangered medicinal plant. Crop Breeding and Applied Biotechnology. 2016; 16: 48–54.
49. Kaya, E. Development of efficient micropropagation and cryopreservation protocols for *Thymus* germplasm. In Vitro Biology Meeting held in collaboration with the 12th IAPB Congress, 2010, P283
50. Marco-Medina, A., Casasi, J. L., Swennen, R., Panis, B. Cryopreservation of *Thymus moroderi* by droplet vitrification. Cryoletters. 2010; 31(1): 14–23. ISSN 0143-2044
51. Coelho, N., González-Benito, M. E., Martín, C. & Romano, A.. Cryopreservation of *Thymus lotocephalus* shoot tips and assessment of genetic stability. Cryoletters. 2014; 35(2): 119–128.
52. Kumar, P., Gupta, V. K., Misra, A. K., Modi, D. R. & Pandey, B. K.. Potential of molecular markers in plant biotechnology. Plant Omics Journal. 2009; 2(4): 141–162. ISSN 1836-3644
53. Mendes, M. D., Barroso, J. G., Oliveira, M. M. & Trindade, H. Identification and characterization of a second isogene encoding γ-terpinene synthase in *Thymus caespititius*. Journal of Plant Physiology. 2014; 171: 1017–1027.

54. Yousefi, V., Najaphy, A., Zebarjadi, A. & Safari, H. Molecular characterization of *Thymus* species using ISSR markers. The Journal of Animal & Plant Sciences. 2015; 25(4): 1087–1094. ISSN 1018-7081
55. Rahimmalek, M., Bahreininejad, B., Khorrami, M. & Tabatabaei, B. E. Genetic variability and geographic differentiation in *Thymus daenensis* subsp. *daenensis*, an endangered medicinal plant, as revealed by inter simple sequence repeat (ISSR). Markers. Biochemical Genetics. 2009; 47(11–12): 831–842.
56. Karaca, M., Ince, A. G., Aydin, A., Elmasulu, S. Y. & Turgut, K. Microsatellites for genetic and taxonomic research on thyme (*Thymus* L.). Turkish Journal of Biology. 2015; 39: 147–159.
57. Khoshsokhan, F., Babalar, M., Fatahi, M. & Poormeidani, A.. Assessment of genetic diversity of some wild populations of *Thymus Kotschyanus* using RAPD molecular markers. Cercetări Agronomice în Moldova. 2014; 3(159): 71–81.
58. Rustaiee, A. R., Yavari, A., Nazeri, V., Shokrpour, M., Sefidkon, F. & Rasouli, M. Genetic diversity and chemical polymorphism of some *Thymus* species. Chemistry & Biodiversity. 2013; 10: 1088–1098.
59. Solyman, E. & Alkowni, R. RAPD for assessment of thymes genetic diversity in Palestine. Palestine Technical University Research Journal. 2014; 2(2): 01–08.
60. Echeverrigaray, S., Agostini, G., Atti-Serfini, L., Paroul, N., Pauletti, G. F. & Atti dos Santos, A. C. Correlation between the chemical and genetic relationships among commercial thyme cultivars. Journal of Agricultural and Food Chemistry. 2001; 49: 4220–4223.
61. Pluhár, Z., Kocsis, M., Kuczmog, A., Csete, S., Simkó, H., Sárosi, S., Molnár, P. & Horváth, G.. Essential oil composition and preliminary molecular study of four Hungarian *Thymus* species. Acta Biologica Hungarica. 2012; 63(1): 81–96.
62. Hadian, J., Bigdeloo, M., Nazeri, V. & Khadivi-Khub, A.. Assessment of genetic and chemical variability in *Thymus caramanicus*. Molecular Biology Reports. 2014; 41(5): 3201–3210.
63. Sostaric, I., Liber, Z., Grdisa, M., Marin, P. D., Stevanovi, Z. D. & Satovic, Z.. Genetic diversity and relationships among species of the genus *Thymus* L. (section *Serpyllum*). Flora. 2012; 207: 654–661.
64. Karp, A., Edwards, K. J., Bruford, M., Funk, S., Vosman, B., Morgante, M., Seberg, O., Kremer, A., Boursot, P., Arctander, P., Tautz, D. & Hewitt, G. M. Molecular technologies for biodiversity evaluation: opportunities and challenges. Nature Biotechnology. 1997; 15(7): 625–628.
65. Kollipara, R., Peranteau, A. J., Nawas, Z. Y., Tong, Y., Woc-Colburn, L., Yan, A. C, Lupi, O. & Tyring, S. K. Emerging infectious diseases with cutaneous manifestations: Fungal, helminthic, protozoan and ectoparasitic infections. Journal of the American Academy of Dermatology. 2016; 75(1): 19–30.
66. Pagiotti, R., Angelini, P., Rubini, A., Tirillini, B., Granetti, B. & Venanzoni, R.. Identification and characterisation of human pathogenic filamentous fungi and susceptibility to *Thymus schimperi* essential oil. Mycoses. 2011; 54(5): e364–e376.
67. Voltan, A. R., Quindós, G., Alarcón, K. P. M., Fusco-Almeida, A. M., Mendes-Giannini, M. J. S. & Chorilli, M. Fungal diseases: could nanostructured drug delivery systems be a novel paradigm for therapy? International Journal of Nanomedicine. 2016; 11: 3715–3730.
68. Labib, G. S. & Aldawsari, H. Innovation of natural essential oil-loaded Orabase for local treatment of oral candidiasis. Drug Design, Development and Therapy. 2015; 9: 3349–3359.
69. Dorgan, E., Denning, D. W. & McMullan, R. Burden of fungal disease – Ireland. Journal of Medical Microbiology. 2015; 64(pt 4): 423–426.
70. Chander, J., Stchigel, A. M., Alastruey-Izquierdo, A., et al. Fungal necrotizing fasciitis, an emerging infectious disease caused by *Apophysomyces* (Mucorales). Revista Iberoamericana de Micología. 2015; 32(2): 93–98.

71. Brown, G. D., Denning, D. W., Gow, N. A., Levitz, S. M., Netea, M. G. & White, T. C. Hidden killers: human fungal infections. Science Translational Medicine. 2012; 4: 165rv13
72. Pinto, E., Pina-Vaz, C., Salgueiro, L., Gonçalves, M. J., Costa-de-Oliveira, S., Cavaleiro, C., Palmeira, A., Rodrigues, A. & Martinez-de-Oliveira, J. Antifungal activity of the essential oil of *Thymus pulegioides* on Candida, *Aspergillus* and dermatophyte species. Journal of Medical Microbiology. 2006; 55: 1367–1373.
73. Cowan, M. M.. Plant products as antimicrobial agents. Clinical Microbiology Reviews. 1999; 12(4): 564–582.
74. Brull, S. & Coote, P.. Preservative agents in foods: mode of action and microbial resistance mechanisms. International Journal of Food Microbiology. 1999; 50(1–2): 1–17. ISSN 0168–1605
75. Darout, I., Cristy, A., Skaug, N. & Egeberg, P. Identification and quantification of some potential antimicrobial anionic components in miswak extract. Indian Journal of Pharmacology. 2000; 32: 11–14.
76. Sharma, R., Sharma, C. L. & Kapoorl, B. Antibacterial resistance: current problems and possible solutions. Indian Journal of Medical Sciences. 2005; 59(3): 120–129. ISSN 0019–5359
77. Moghaddam, K. M., Arfan, M., Rafique, J., Rezaee, S., Fesharaki, P. J., Gohari, A. R. & Shahverdi, A. R. The antifungal activity of Sarcococca saligna ethanol extract and its combination effect with fluconazole against different resistant *Aspergillus* species. Applied Biochemistry and Biotechnology. 2010; 162(1): 127–133. ISSN 1559–0291
78. Edeoga, H. O., Okwu, D. E. & Mbaebie, B. O. Phytochemical constituents of some Nigerian medicinal plants. African Journal of Biotechnology. 2005; 4(7): 685–688. ISSN 1684–5315
79. Panghal, M., Kaushal, V. & Yadav, J. P. In vitro antimicrobial activity of ten medicinal plants against clinical isolates of oral cancer cases. Annals of Clinical Microbiology and Antimicrobials. 2011; 10: 21
80. Janssen, A. M., Chin, N. L., Scheffer, J. J. & Baerheim Svendsen, A. Screening for antimicrobial activity of some essential oils by the agar overlay technique. Pharmaceutisch Weekblad. Scientific Edition. 1986; 8(6): 289–292.
81. Rios, J. L., Recio, M. C. & Villar, A. Screening methods for natural products with antimicrobial activity: a review of the literature. Journal of Ethnopharmacology. 1988; 23: 127–149.
82. Bruneton, J. Partie 3. Terpeenes et steroids. In Pharmacognosie, Phytochimie, Plantes médicinales. 4a Édition. Paris: E´ ditions Tec&Doc & Éditions Médicales Internationals; 2009, pp. 547–934.
83. Pina-Vaz, C., Sansonetty, F., Rodrigues, A., Costa-de-Oliveira, S., Martinez-de-Oliveira, J. & Fonseca, A. F. Cytometric approach for rapid evaluation of susceptibility of Candida strains to antifungals. Clinical Microbiology and Infection. 2001; 7: 609–618.
84. Stahl-Biskup, E. & Sáez, F. Thyme – The Genus Thymus. London: Taylor & Francis; 2002.
85. Mohd, S., Birhanu, D. & Tanweer, A.. Antimicrobial activity of essential oils extracted from medicinal plants against the pathogenic microorganisms: a review. Issues in Biological Sciences and Pharmaceutical Research. 2014; 2(1): 001–007.
86. Dorman, H. J. D. & Deans, S. G. Antimicrobial agents from plants: antibacterial activity of plant volatile oils. Journal of Applied Microbiology. 2000; 88(2): 308–316.
87. Arnal-Schnebelen, B., Hadji-Minaglou, F., Peroteau, J. F., Ribeyre, F. & de Billerbeck, V. G. Essential oils in infectious gynaecological disease: a statistical study of 658 cases. International Journal of Aromatherapy. 2004; 14(4): 192–197.
88. Villanueva Bermejo, D., Angelov, I., Vicente, G., Stateva, R. P., Rodriguez García-Risco, M., Reglero, G., Ibañez, E. & Fornari, T. Extraction of thymol from different varieties of thyme plants using green solvents. Journal of the Science of Food and Agriculture. 2015; 95(14): 2901–2907.

89. Wohlmuth, H., Smith, M. K., Brooks, L. O., Myers, S. P. & Leach, D. N. Essential oil composition of diploid and tetraploid clones of ginger (*Zingiber officinale* Roscoe) grown in Australia. Journal of Agricultural and Food Chemistry. 2006; 54: 1414–1419.
90. Kalemba, D. & Kunicka, A. Antibacterial and antifungal properties of essential oils. Current Medicinal Chemistry. 2003; 10(10): 813–829.
91. Kohiyama, C. Y., Yamamoto Ribeiro, M. M., Mossini, S. A., Bando, E., Bomfim Nda, S., Nerilo, S. B., Rocha, G. H., Grespan, R., Mikcha, J. M. & Machinski, M. Jr. Antifungal properties and inhibitory effects upon aflatoxin production of *Thymus vulgaris* L. by *Aspergillus flavus* link. Food Chemistry. 2015; 15(173): 1006–1010.
92. Manou, I., Bouillard, L., Devleeschouwer, M. J. & Barel, A. O.. Evaluation of the preservative properties of *Thymus vulgaris* essential oil in topically applied formulations under a challenge test. Journal of Applied Microbiology. 1998; 84(3): 368–376.
93. Tampieri, M. P., Galuppi, R., Macchioni, F., Carelle, M. S., Falcioni, L., Cioni, P. L. & Morelli, I.. The inhibition of *Candida albicans* by selected essential oils and their major components. Mycopathologia. 2005; 159: 339–345.
94. Ahmad, A., Khan, A., Akhtar, F., Yousuf, S., Xess, I., Khan, L. A. & Manzoor, N.. Fungicidal activity of thymol and carvacrol by disrupting ergosterol biosynthesis and membrane integrity against Candida. European Journal of Clinical Microbiology and Infectious Diseases. 2011; 30: 41–50.
95. Sharifzadeh, A., Javan, A. J., Shokri, H., Abbaszadeh, S. & Keykhosravy, K. Evaluation of antioxidant and antifungal properties of the traditional plants against foodborne fungal pathogens. Journal of Medical Mycology. 2016; 26(1): e11–e17.
96. Markovic, T., Chatzopoulou, P., Siljegovic, J., Nikolic, M., Glamoclija, J. & Ciric, A. Chemical analysis and antimicrobial activities of the essential oils of *Satureja thymbra* L. and *Thymbra spicata* L. and their main components. Archives of Biological Sciences. 2011; 63: 457–464.
97. Rasooli, I., Rezaei, M. B. & Allameh, A. Growth inhibition and morphological alterations of *Aspergillus niger* by essential oils from *Thymus eriocalyx* and Thymus x-porlock. Food Control. 2006; 17: 359–364.
98. Mandras, N., Nostro, A., Roana, J., Scalas, D., Banche, G., Ghisetti, V., Del Re, S., Fucale, G., Cuffini, A. M. & Tullio, V. Liquid and vapour-phase antifungal activities of essential oils against *Candida albicans* and non-albicans Candida. BMC Complementary and Alternative Medicine. 2016; 16(1): 330.
99. Leal, F., Matos, M., Coelho, A. C. & Pinto-Carnide, O. *In vitro* multiplication of aromatic and medicinal plants and fungicide activity. In: Dhanasekaran D., Thajuddin, N. & Panneerselvam, A. (eds). Fungicides for Plant and Animal Diseases. Croacia: Intech Open Access Publisher; 2011. pp. 119–138. ISBN 978-953-307-804-5
100. Nardoni, S., Giovanelli, S., Pistelli, L., Mugnaini, L., Profili, G., Pisseri, F. & Mancianti, F. In vitro activity of twenty commercially available, plant-derived essential oils against selected dermatophyte species. Natural Product Communications. 2015; 10(8): 1473–1478.
101. Lima, O. I., Oliveira, R. A. G., Lima, E. O., Porto Farias, N. M. & Souza, E. L. Antifungal activity from essential oils on Candida species. Revista Brasileira de Farmacognosia. 2006; 16: 197–201.

CHAPTER 3

Tahitian Vanilla (*Vanilla ×tahitensis*): A Vanilla Species with Unique Features

Christel Brunschwig[1, 2*], François-Xavier Collard[2], Sandra Lepers-Andrzejewski[2] and Phila Raharivelomanana[1]

[1] Integrated Study of Secondary Metabolites (EIMS), Research Unit UMR 241 Insular Ocean Ecosystems (EIO), University of French Polynesia, Faaa, Tahiti, French Polynesia

[2] Research and Development Department, Institute "Vanille de Tahiti", Raiatea, French Polynesia

ABSTRACT

This chapter reviews the main findings on Tahitian vanilla (*Vanilla ×tahitensis*) over the last 10 years. It brings new insights into the hybrid origin of *V. ×tahitensis* and its diversification in French Polynesia. It details then the different analytical methods used to characterize the flavour properties and the aroma impact compounds of Tahitian vanilla, with a special emphasis on how they can be used to differentiate Tahitian vanilla from other vanillas. Finally, the effect of the curing process on the chemical composition and the sensory properties is discussed. These results highlight the need to include some of the key volatile compounds into a more adapted quality control, in order to describe the characteristic sensory properties of Tahitian vanilla but also those from other origins.

Keywords: *V. ×tahitensis*, Tahitian vanilla, genetic biodiversity, aroma chemistry, high performance liquid chromatography, gas chromatography, sensory properties, quality control, curing process

1. INTRODUCTION

Though Tahitian vanilla production represents less than 1% of the world vanilla production [1], it possesses a unique aura, due to an original anise flavour, highly prized especially in gastronomy and perfumery. In order to better develop and organize the vanilla sector and to promote Tahitian vanilla specificities abroad, a dedicated institute "Vanille de Tahiti" was created in 2003 in French Polynesia [2]. One of its main objectives is to provide support to local actors and help controlling pests and diseases threatening vanilla vines, with the sponsored supply of insect-proof shade houses for instance. A research department was also established in order to characterize Tahitian vanilla in comparison to other vanilla species, through the study of the flavour composition and genetic diversity. The institute has been part of a larger project, which focused on characterizing, protecting and promoting the sustainable use and valorisation of vanilla biodiversity (VaBiome 2012–2015). This chapter aims at capturing the most significant results that have been acquired over the last 10 years at the institute "Vanille de Tahiti" through different collaborations inside and outside French Polynesia, with a particular focus on genetic diversity and aroma chemistry.

2. ORIGINS AND DIVERSIFICATION OF *VANILLA ×TAHITENSIS*

The introduction of vanilla in French Polynesia is estimated to have occurred during the mid-nineteenth century. The elucidation of the origin of Tahitian vanilla is difficult due to different reports of at least three introduction events from various origins (Philippines, France and West Indies) [3–5]. On top of that, the origin of *Vanilla* introduced in the Philippines was not solved and believed to be in Central America [5, 6]. Furthermore, the number of different vanilla morphotypes or species that were introduced is unknown, as authors described the introduction of *Vanilla planifolia* or *Vanilla pompona*, and even used different names

(*Vanilla aromatic* and *Vanilla sativa*) [3, 4] for species known today as synonymous of *V. planifolia*. So it remains unclear, whether the Tahitian vanilla specificities appeared before its introduction in a still indeterminate cultivation area, or after through cultivation in French Polynesia.

Even if the origin of Tahitian vanilla is still confused, it is obvious that Tahitian vanilla, as we know it today, is different from other vanillas and in particular *V. planifolia,* the most cultivated species in the world. Since 1915, about 60 years after the first well-documented introduction events, various new vanilla morphotypes were described in French Polynesia [7]. Four morphotypes were early distinguished as "Tahiti", "Haapape", "Tiarei" and "Potiti" [3, 5, 7, 8]. Based on morphological differences between vanillas found in French Polynesia and other cultivated vanillas, Moore went a step further and claimed that Tahitian vanilla was a new species and therefore called it *Vanilla tahitensis* [9]. Indeed, Tahitian vanilla differs morphologically by having thinner stems, narrower leaves and shorter pods when compared to *V. planifolia* [5, 9] (**Figure 1**).

Figure 1. *Vanilla ×tahitensis* from French Polynesia: (a) vine in an insect-proof shade house, (b) flower, (c) mature pods and (d) cured vanilla pods.

In 1951, Porteres suggested that Tahitian vanilla could be a hybrid offspring between *V. planifolia* and a complex *V. pompona – V. odorata* [10]. More recent assessment of the Tahitian vanilla genetic diversity gave new insight into its hybrid origin. A genetic analysis based on universal markers widely dispersed through the genome [amplified fragment length polymorphism (AFLP)] was carried out to compare the presence/absence of a set of AFLP markers. It showed that 31% of *V. ×tahitensis* AFLP markers were shared only with *V. planifolia*, while 6% of them were shared only with *V. odorata,* and other markers were shared with more than two other *Vanill*a species [11, 12] (**Figure 2**). Another study based on the analysis of a part of the genome preferentially inherited from the maternal parent (non-coding region of chloroplast DNA trnH-psbA) confirmed that *V. planifolia* was the maternal parent of Tahitian vanilla [6]. Since then, Tahitian vanilla is called *Vanilla ×tahitensis*; the "×" referring to its hybrid nature [13].

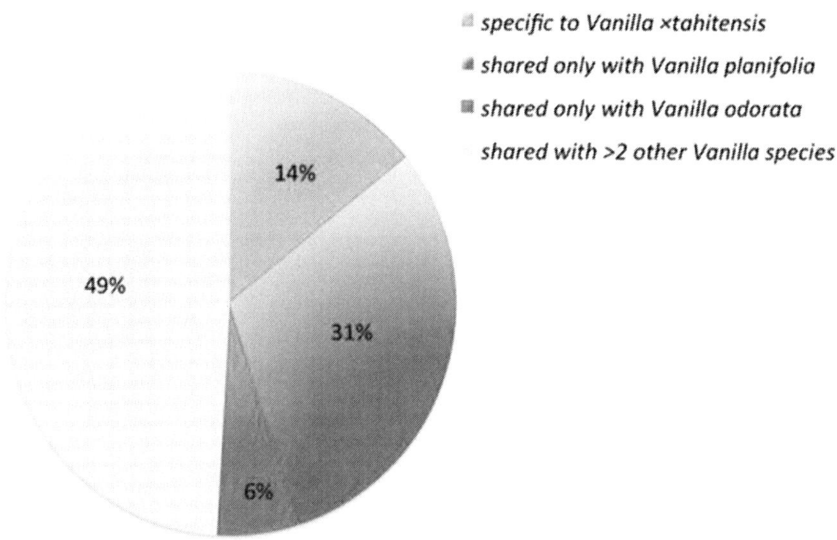

Figure 2.. Specificity of the AFLP markers amplified from Tahitian vanilla accessions (*Vanilla ×tahitensis*) compared to other *Vanilla* species.

Coming back to the analysis of genetic diversity based on AFLP patterns, all Tahitian morphotypes were (i) distinct from the other *Vanilla* species studied (*V. planifolia*, *V. pompona* and *V. odorata*), (ii) genetically close to each other (mean dissimilarity index D = 0.0596) [11] and strongly grouped (**Figure 3**). Amongst Tahitian vanilla cultivars, the genetic diversity was higher than expected. While only 16 vanilla morphotypes were distinguished, more than 30 AFLP patterns were found. "Parahurahu" was the morphotype most genetically different from "Tahiti". It also differentiated by its morphological traits such as leaf size, pod shape and its aromatic composition. The genetic diversity amongst Tahitian vanilla was also explained by ploidy level differences. Indeed, the two mainly cultivated Tahitian vanilla morphotypes "Tahiti" and "Haapape" shared the same AFLP pattern, however "Tahiti" appeared to be diploid while "Haapape" was tetraploid [14]. Since vanilla is mainly propagated by stem cuttings, such variability was unexpected. The comparison of molecular markers patterns led to suggest that the recent genetic diversification in French Polynesia originated from self-pollination rather than from mutation events.

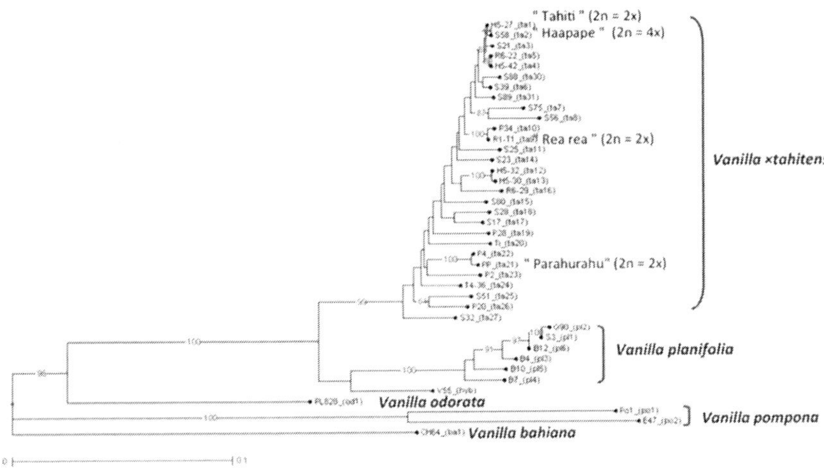

Figure 3. Phylogenetic tree of *Vanilla* species based on AFLP markers. Obtained from the Sokal and Michener genetic dissimilarity index and neighbour joining clustering; bootstrap values are expressed in percentage

Biodiversity of Tahitian vanilla is preserved in the Genetic Resources Centre "Vanille de Tahiti" created on the island of Raiatea. About 140 accessions of *V. ×tahitensis* collected on different Polynesian islands are grown under insect-proof shade houses. These accessions, which are part of a vanilla-breeding programme, can be diffused on demand to scientists from any research institute for genetic or new flavour sourcing projects. Accessions are continuously assessed for various agronomical traits, resistance to diseases and aroma composition of their pods for further improvement.

3. TAHITIAN VANILLA FLAVOUR AND DIFFERENTIATION FROM OTHER VANILLAS

V. ×tahitensis is mainly cultivated in French Polynesia but is also found, together with *V. planifolia*, in New Guinea (Papua New Guinea and Indonesia). The origin of the introduction of *V. ×tahitensis* is unknown and the cultivar(s) grown have not been identified. Moreover the pods are sometimes not differentiated according to their species and commercialized as a mixture of *V. planifolia* and *V. ×tahitensis* [15]. It results in products with sensory properties completely different, when compared to those of Tahitian vanilla (see Section 3.2).

In French Polynesia, the quality of Tahitian vanilla is guaranteed by the training of different actors of the sector and several controls, after harvest and before exportation, are realized by sworn experts, as specified in the designation of origin "Vanille de Tahiti" recently released [16, 17]. Until 2014, there were no official analytical criteria to authenticate Tahitian vanilla in the international trade context, which would enable to differentiate it with vanilla from Papua New Guinea.

In order to determine what specifically differentiates Tahitian vanilla from other vanillas (*V. planifolia* the most cultivated in the world, *V. pompona* and vanilla from Papua New Guinea), in terms of flavour properties, the content of some chemical compounds of interest was assessed using different analytical techniques. Due to ease of implementation, the analysis by high performance liquid chromatography (HPLC) of vanilla extracts is still a method of reference. However, other analytical methods developed to assess the olfactory (gas chromatography-olfactometry) and gustative properties of vanilla, appear more appropriate to characterize the sensory properties of the pods. In order to correlate sensory properties and chemical composition, gas chromatography/mass spectrometry (GC/MS) analysis was also considered because of its ability to quantify a much larger number of compounds, when compared to HPLC.

3.1. Authentication Based On Hplc Profile

To authenticate natural vanilla (i.e. *V. planifolia*), quality control is usually based on the analysis by HPLC of some characteristic compounds (vanillyl and p-hydroxybenzyl derivatives) (**Figure 4**) extracted with ethanol as recommended by the International Organization for Standardization and the AOAC International [18, 19]. The ratios between those compounds are quite stable for the different origins of *V. planifolia* and throughout years of production, and are therefore recommended as quality markers by the Direction Générale de la Concurrence, de la Consommation et de la Répression des Fraudes (DGCCRF) guidelines [20]. However, these ratios are not appropriate to authenticate Tahitian vanilla, as data fall without ranges. It is worth noting that typical chromatograms obtained from Tahitian vanilla analysis are composed not only of p-hydroxybenzyl or vanillyl derivatives but also and predominantly of anisyl derivatives, which are actually not taken into account for those criteria (for compound structures see **Figure 4**).

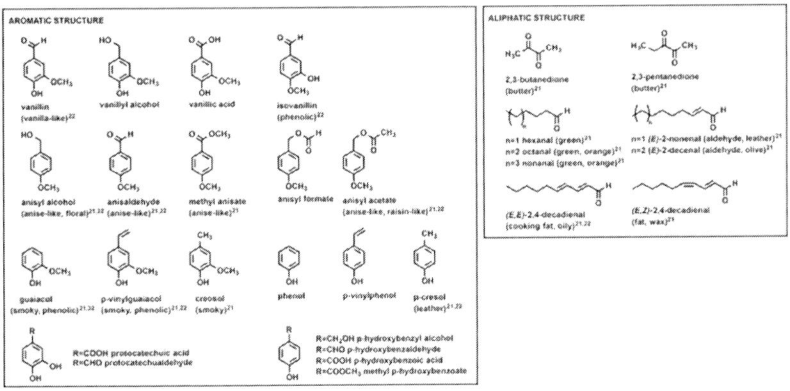

Figure 4. Structure of key compounds in the aroma chemistry of Tahitian vanilla (*V. ×tahitensis*). Odour descriptors are specified into brackets when compounds have been detected by GC-olfactometry in Refs. [21, 22].

This is exemplified by a study carried out between 2005 and 2007 at the Institute "Vanille de Tahiti". More than 300 vanilla samples, collected from vanilla curers based on the islands of Tahaa and Raiatea, where most of the vanilla was produced in French Polynesia at that time, were analysed by HPLC, together with 22 samples of *V. planifolia* and 9 samples from Papua New Guinea. The results showed that contrary to *V. planifolia,* which was composed almost only of vanillin (80% of the total quantified), Tahitian vanilla was characterized by a more subtle distribution: vanillin (25%), anisyl alcohol (30%), anisic acid (15%) as well as p-hydroxybenzyl compounds (20%) and protocatechuyl derivatives (5%) for a total content of 47,000 ppm (**Figure 5**), which exceeded the values typically reported for *V. planifolia* (40,000 ppm) [23–25].

The HPLC composition of Tahitian vanilla was homogeneous between the three years of production, for compounds whose concentration was higher than 1000 ppm (relative standard deviation <15%). As a result of this, low variability shown by principal component analysis using the HPLC composition and vanilla samples from various origins, it was possible to differentiate a set of Tahitian vanilla samples from *V. planifolia* of different origins and also from vanilla of Papua New Guinea, independently of the species (**Figure 6**).

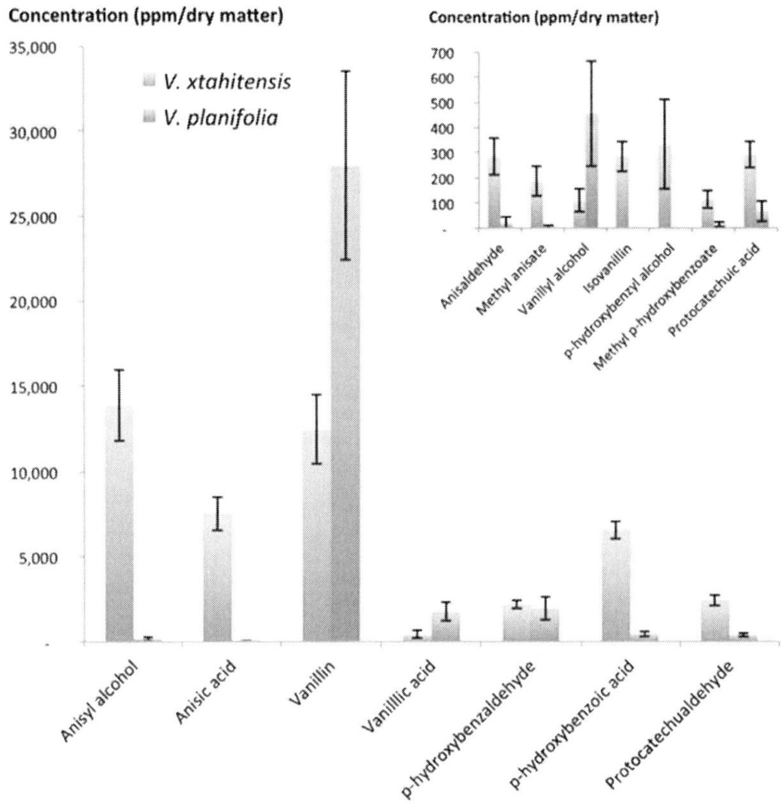

Figure 5. HPLC composition of Tahitian vanilla (*V. ×tahitensis* from French Polynesia; *n* = 300 samples over 3 years of production 2005–2007) and *V. planifolia* (20 samples from various origins: Madagascar, la Réunion, Mexico, Costa Rica).

The HPLC compositions of "Tahiti" and "Haapape" cultivars, which are the two main cultivars produced in French Polynesia, were found to be very close. The slight variations consisted in more vanillin and less p-hydroxybenzyl compounds in "Tahiti" pods. Consequently to this study, recommendations based on the contents of characteristic HPLC aroma compounds and their ratios have been made to assess Tahitian vanilla quality in French Polynesia. They have been integrated into official decrees published recently [16, 17]. These decrees define the Tahitian vanilla's designation of origin ("Vanille de Tahiti"), its production area, the cultivars to be produced ("Tahiti" and "Haapape") with their genetic characteristics, the production and the curing process methods, as well as the pods quality based on moisture content (45–50%), aroma content and characteristic ratios using the concentrations of anisyl, vanillyl, p-hydroxybenzyl and protocatechuyl compounds determined by HPLC. These criteria can also be used to authenticate Tahitian vanilla abroad.

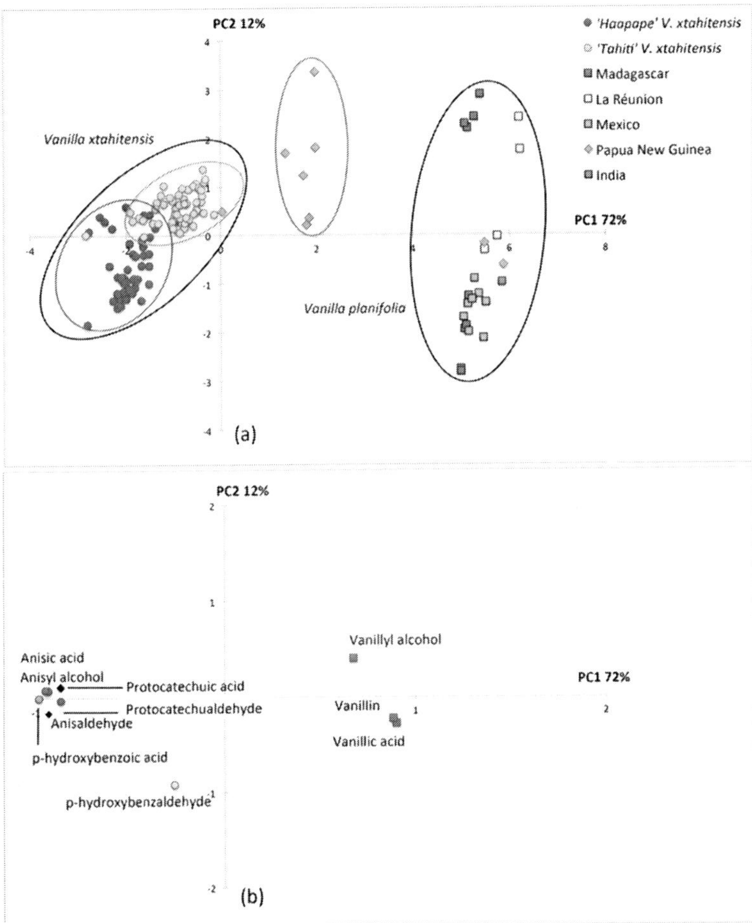

Figure 6. Projection plots on Principal Components of (a) observations scores (dataset of vanilla samples from different origins and (b) variables coefficients (compound concentrations analysed by HPLC). PC1 and PC2 explained, respectively, 72% and 12% of the total variance in the data.

3.2. Towards a Better Authentication Based on The Relevant Flavour Compounds

Though HPLC is a reliable analytical technique to analyse some characteristic compounds of the vanilla pods, it appears as more convenient to use complementary analytical and sensory techniques to quantify more compounds, in particular those likely to contribute/to affect the vanilla flavour. The perception of a flavour is due to the detection of a complex mixture of compounds having various physico-chemical properties. Non-volatile (or less volatile) compounds, like essential fatty acids, have the ability to fix some non-polar aroma constituents. Volatile compounds, which have various odour-detection thresholds, can sometimes have a preponderant impact on the aroma, even when present in low amounts. Tahitian vanilla aroma chemistry has been therefore investigated using additional analytical techniques, such as gas chromatography-mass spectrometry (GC-MS), sensory analysis and a hyphenated technique, gas chromatography coupled with olfactometry (GC-O) designed to detect more volatile compounds.

3.2.1. Gas Chomatography-Olfactometry and Volatile Compounds

GC-olfactometry enables to decompose the individual odours of an aroma extract and attribute these odours to specific compounds, which are called aroma impact compounds. GC-olfactometry was applied to the analysis of different vanilla extracts: *V. planifolia* from Mexico [26], *V. planifolia* from Madagascar and Uganda [27], *V. ×tahitensis* from French Polynesia [21, 22] and *V. pompona* from Mexico [28]. The unique sensory fingerprint of Tahitian vanilla was highlighted by the detection of about 60 notes using the odour-specific magnitude estimation (OSME) method for both cultivars "Tahiti" and "Haapape" [21]. Amongst them, 38 were attributed to a specific compound. Anisaldehyde and guaiacol were found to be the main aroma impact compounds of Tahitian vanilla. These two compounds were also found to be primary impact aroma compounds of *V. pompona*, although the overall aroma characteristics of the two species were quite different. Main aroma impact compounds of *V. planifolia* were vanillin derivatives and phenolic compounds such as guaiacol, creosol, p-cresol or phenol. It is noteworthy that despite a relatively low content in the pods, anisyl compounds such as anisaldehyde and anisylalcohol were also identified as odour-impact compounds in *V. planifolia* [26, 27]. The GC-O profile of *V. ×tahitensis* was more balanced compared to other species; enabling the co-expression of multiple notes such as phenolic-vanilla like notes, anise-spicy and floral notes, while *V. planifolia* flavour was mainly characterized by phenolic-vanilla notes and *V. pompona* by floral notes (**Figure 7**).

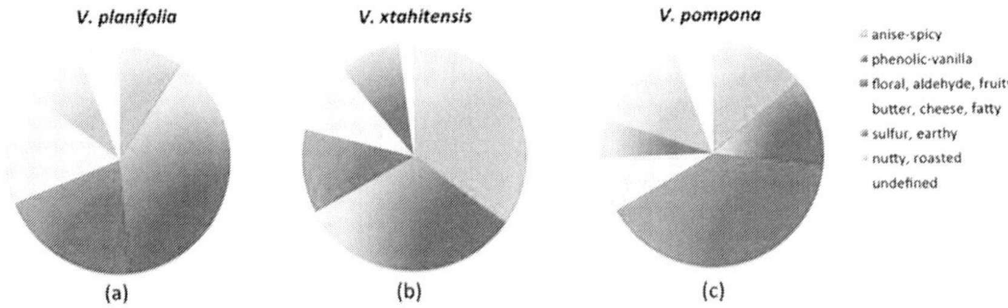

Figure 7. Distribution of odour vanilla notes based on the analysis by GC olfactometry of the impact aroma compounds of (a) *V. planifolia*, (b) *V. xtahitensis*, and (c) *V. pompona* (adapted with permission from Ref. [**21**, **27**, **28**]).

3.2.2. Sensory Properties and Volatile Compounds

In addition to the compounds analysed by HPLC, various volatile compounds such as phenolic compounds, aliphatic aldehydes, ketones and esters contributing to the flavour of Tahitian vanilla were identified and quantified by GC-MS [29]. Still, this method highlighted that compared to *V. planifolia*, *V. ×tahitensis* had a higher content of anisyl compounds, which represented 70% of the total volatile content against 7% for *V. planifolia*; amongst them anisyl alcohol, anisaldehyde, methyl anisate and anisyl acetate were the major ones (for compound structures see **Figure 4**). The other characteristics of *V. ×tahitensis* were a much lower vanillin content than *V. planifolia* (5–10% against 30%, respectively), as well as lower contents of phenolic compounds and aliphatic aldehydes (less than 10% against more than 40% and 0.5–1% against 2%, respectively). The volatile composition of Papua New Guinea vanilla showed some similarities to that of Tahitian vanilla especially regarding anisyl compounds, but some key compounds could differentiate them such as methyl esters, anisyl formate, p-cresol or p-vinylguaiacol [29].

In relation to their volatile composition, it was also possible to clearly differentiate the sensory properties of Tahitian vanilla from *V. planifolia* and also from vanilla from Papua New Guinea. The method used, called quantitative descriptive analysis, consisted in a panel of judges who tasted aroma extracts. Compared to the other vanillas, Tahitian vanilla displayed more intense anise, caramel and vanilla notes despite a relatively low vanillin content. *V. planifolia* was characterized by more intense phenolic, woody and smoky notes (**Figure 8**) [29]. Similarly, another sensory study depicted Tahitian vanilla aroma as less resinous, less dried fruit-like and more floral compared to *V. planifolia* from Madagascar [22]. The sensory profile of vanilla from Papua New Guinea was well differentiated from that of Tahitian vanilla with strong fruity, spicy and brown rum notes [29], the latter being probably related to the curing method as those were found to become stronger throughout the curing process (see Section 4).

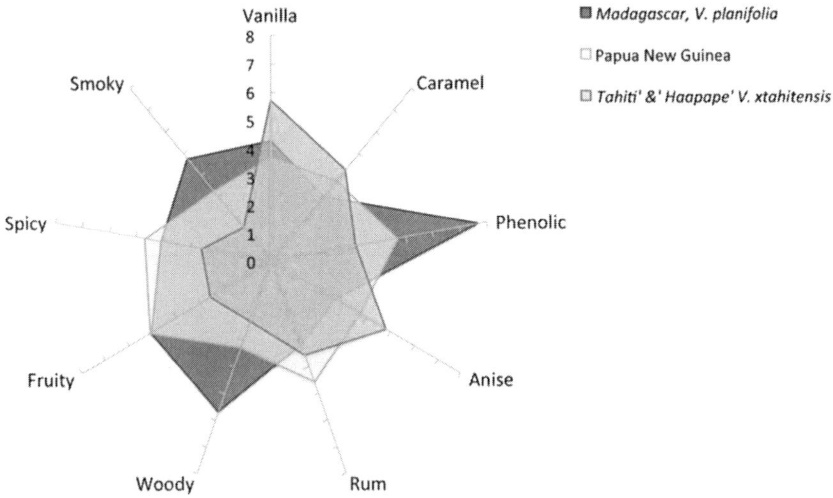

Figure 8. Sensory profiles of Tahitian vanilla (*V. ×tahitensis,* mix of "Tahiti" and "Haapape" cultivars) compared to vanilla from other origins (adapted with permission from Ref. [**29**]).

As we tried to correlate the sensory properties of vanilla from different origins to their volatile composition, it remained difficult to link single volatile compounds to specific notes, as there are many interactions between volatile and non-volatile compounds. However, we were able to show that the strongest correlations were between phenolic, woody, smoky notes and a pool of phenolic compounds such as guaiacol and creosol. Such correlation between guaiacol and woody notes has also been highlighted for *V. planifolia* [25] and led to the development of more appropriate analysis methods to characterize vanilla flavour also based on the detection of negative compounds such as guaiacol [30]. Anisyl compounds, especially methyl anisate were well correlated with anise notes. p-Vinylguaiacol was another compound of interest, as it was in higher amounts in vanilla from Papua New Guinea compared to Tahitian vanilla and differentiated the two origins [29]. These results stressed the need to integrate volatile compounds belonging to the phenolic series (guaiacol, creosol, p-cresol and p-vinylguaiacol) and to the anisyl series (anisyl alcohol, anisaldehyde and methyl anisate but also anisyl formate and anisyl acetate) in a more appropriate quality control, be it to authenticate *V. ×tahitensis* or more generally to assess vanilla quality independently of its origin.

4. THE EFFECT OF CURING PROCESS ON VANILLA FLAVOUR

Since we started our journey studying Tahitian vanilla characteristics, there was a question that needed to be answered: how much of the aroma is inherent to the vanilla species and what is the influence of the curing process on the aroma development? To get an answer, we first have to go back to the curing process as it is performed in French Polynesia. First, Tahitian vanilla pods are harvested when fully ripe, then they are cured following three main steps [31]:

- Shade browning: vanilla pods have 80% moisture and are exposed on the shade until being entirely brown.
- Sun drying and sweating: for several weeks, pods are alternatively exposed for a short period of time (2–4 h) in the sun every day, then wrapped into a cotton fabric and leave to sweat overnight in closed wooden cases. As pods become increasingly flexible and glossy due to water loss, they are massaged to ensure seeds are spread lengthwise.
- Air drying and refining: finally pods are left in the shade so that the moisture and aroma contents stabilize in order to obtain homogeneous batches (around 50% moisture).

Unlike other vanilla species, *V. ×tahitensis* pods do not split when fully ripe. It means they can be kept on the vine until full maturity when aroma content is at its optimum. Late harvest is also likely to limit the risk of mould formation during storage. Another consequence is that there is no need for any initial heat treatment in hot water or in an oven, as it is performed commonly in the Indian Ocean region, in Central America or in New Guinea. During curing, many biotransformations take place due to the presence of intrinsic enzymes located in the pods and/or the action of colonizing microorganisms on the surface as evidenced by the latest findings on *V. planifolia* [32]. Even if the different biotransformations involved are still unclear, they will definitely contribute to the development of the final vanilla flavour that one can experience when smelling cured vanilla pods.

4.1. Effect of Curing on Aroma Composition Using Hplc

The evolution of the aroma composition of Tahitian vanilla pods was monitored by HPLC analysis of "Tahiti" and "Haapape" samples collected at various steps of the curing process. It was observed that through drying, the aroma content of the pods decreased by around 30%, from around 70,000 ppm at 80% moisture content to 50,000 ppm for a final moisture content of 50–55% (**Figure 9**). This loss could be attributed to co-evaporation of the compounds with water while sun drying and/or loss in liquid form during pod sweating, as evidenced by the oily substance observed at the surface of the fabric wrap.

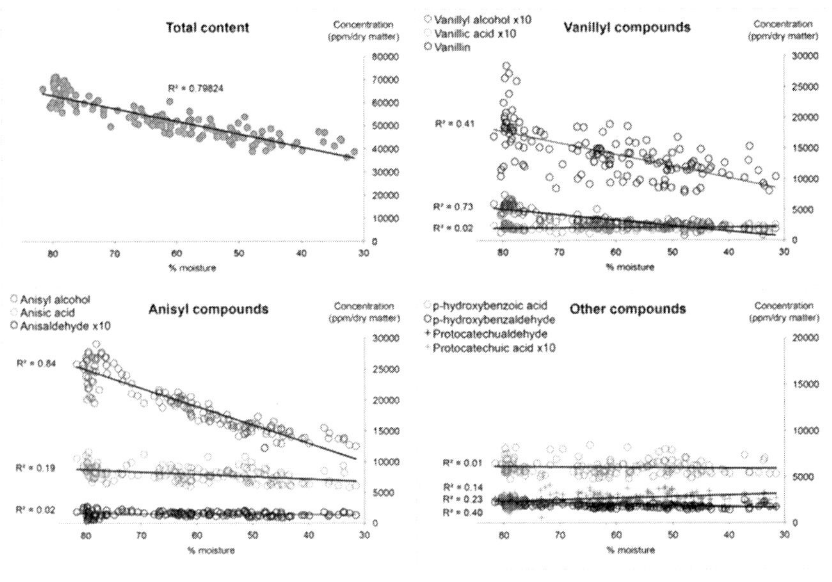

Figure 9. Variations of Tahitian vanilla composition obtained by HPLC during the curing process (*V. ×tahitensis*, "Tahiti" and "Haapape" cultivars).

Within a same compound series, acids concentrations (in red) remained stable during the curing process; while alcohols concentrations (in blue) tended to decrease linearly over time. The fate of aldehydes (in black) varied across the series, as anisaldehyde concentration remained stable, while vanillin concentration decreased linearly. These different trends were related to many interdependent factors such as the oxidation state of the compounds (alcohols can be oxidized into aldehydes then into acids within a series), their physico-chemical properties (volatility and lipophilicity), and the fact that the compounds can be potential substrates or metabolites of intrinsic enzymes or colonizing microorganisms. The relatively high variability of the compounds concentration, especially vanillin, observed at the beginning of the curing process progressively decreased, resulting in more homogeneous vanilla batches at the end of the curing process.

4.2. Effect of Curing on Sensory Properties and Volatile Compounds

In order to get a better understanding of how the pod aroma was developed, we studied the evolution of sensory properties by quantitative descriptive analysis and of relative volatile compounds concentration by GC-MS at different steps during the curing process: (i) step S1—shade browning (80% moisture), (ii) step S2—sun drying (60–65%, pods are wrinkled); (iii) step S3—air drying (50–55% moisture); (iv) step S3*—enhanced drying: sun drying was extended to mimic curing methods of vanillas from other origins and obtain vanilla pods at 40% moisture.

Some of the sensory notes of "Tahiti" cultivar varied significantly during the curing process (**Figure 10**), while the variations for "Haapape" were not significant (data not shown). Regarding the "Tahiti" cultivar, vanilla, anise, rum and caramel notes remained as intense as they were initially or slightly decreased while woody, phenolic, smoky, spicy and fruity notes progressively (and significantly for most of them) built up during the curing process. It is noteworthy that even when *V. ×tahitensis* pods were taken to the enhanced drying step until reaching 40% moisture, they still had a sensory profile different from other vanillas. However, some sensory characteristics, especially phenolic, woody, fruity and spicy notes tended towards to that of *V. planifolia* or vanilla from Papua New Guinea, which is not required as the development of such notes is not seen as an improvement. Based on these findings, curing should be processed judiciously in order to enable the perfect balance of phenolic and anise-floral notes to develop. Drying up to 50% appeared to be the optimal conditions in order to retain as much as possible the initial organoleptic properties of Tahitian vanilla (vanilla and caramel) while limiting the woody, smoky and phenolic notes.

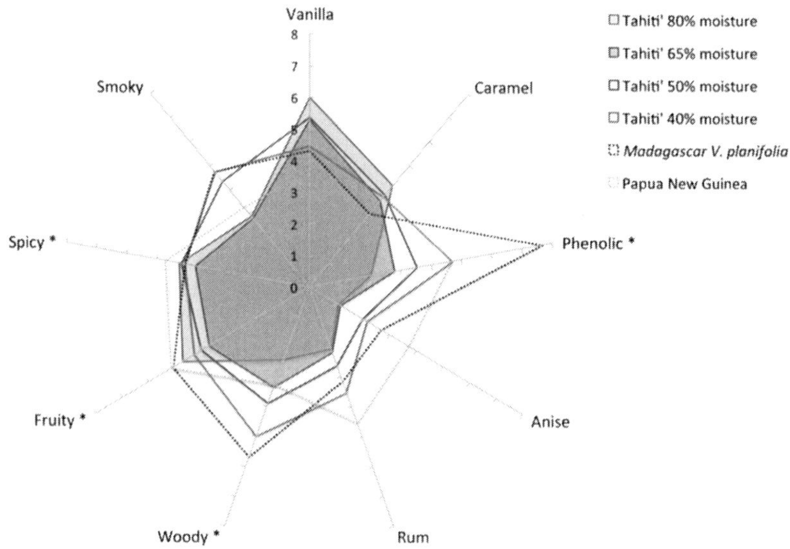

Figure 10. Sensory scores of Tahitian vanilla (*V. ×tahitensis,* "Tahiti" cultivar) at different moisture contents during the curing process compared to other vanillas. *indicate that sensory scores were significantly different between Tahitian vanilla samples during the curing process ($p < 0.05$).

The curing method induced variations in the volatile composition, which impacted greatly the final flavour. Despite a relatively low concentration in the Tahitian vanilla extracts, some aliphatic aldehydes and ketones played a pivotal role by GC-olfactometry [21] due to a low odour threshold [29]. They originate from the oxidation of essential fatty acids such as linoleic and oleic acids, the major fatty acids in Tahitian vanilla pods [33]. Indeed, the fatty acid content was found to decrease from 2.7% of dry matter to 1.6% during the curing process (cultivar "Tahiti"). The increase in the concentration of saturated aldehydes (hexanal, heptanal, octanal and nonanal) linked to fruity notes was seen as positive, contrary to the increase of monounsaturated (heptenal, octenal, nonenal and decenal) and diunsaturated aldehydes (2,4 decadienal) (**Figure 11a**), which displayed less pleasant notes like leather, olive, wax and cooking fat by GC-O [21]. Thus, drying the pods until 40% moisture is not recommended. Similar variations of aliphatic aldehydes were observed during the curing process of *V. planifolia* [34]. Aliphatic ketones, such as 2,3 butanedione and 2,3 pentanedione perceived as butter, were found to have an optimal content at 50% moisture, while 3-hydroxybutanone decreased, being metabolized by bacili bacteria into 2,3-butanediol as suggested for cocoa [35].

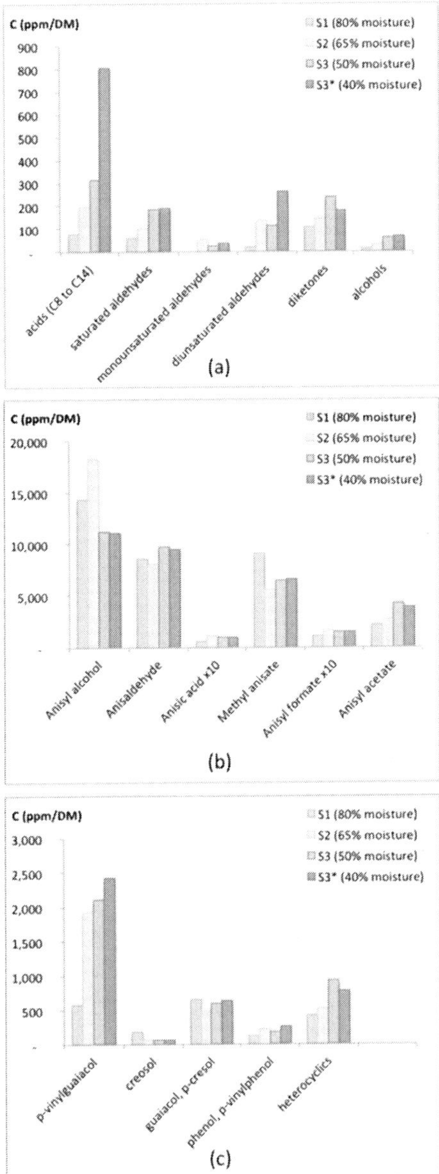

Figure 11. Evolution of key volatile compounds of Tahitian vanilla during the curing process (*V. ×tahitensis,* Tahiti cultivar) (a) aliphatic compounds, (b) anisyl derivatives, (c) phenolic compounds and heterocyclics. DM: dry matter.

The overall content of odour-active anisyl compounds was stable during the curing process of Tahitian vanilla, even though individual compounds contents varied (**Figure 11b**). When comparing the evolution between steps S1 and S3*, anisyl alcohol and methyl anisate contents were found to decrease, while anisaldehyde, anisyl acetate and anisyl formate contents slightly increased. The overall content of phenolic compounds (guaiacol, p-cresol, creosol, phenol, p-vinyl-phenol and p-vinyl guaiacol) increased during the curing process, in tune with the development of phenolic notes. Particularly, there was a dramatic increase of p-vinyl guaiacol by five-fold (**Figure 11c**). The increase of such phenolic compounds was also observed when Tahitian vanilla was stored for a long period of time (five years, data not shown). Overdrying vanilla pods to 40% moisture, as usually performed in other countries, was not beneficial to the Tahitian vanilla aroma, due to the concomitant increase of phenolic compounds concentration and the development of phenolic notes. The higher levels of p-vinylguaiacol and other phenolic compounds detected in vanilla from Papua New Guinea can find its roots in the way the curing process is performed.

Single compounds variations such as anisaldehyde, methyl anisate, guaiacol or p-vinyl guaiacol could not explain all the variations observed while monitoring sensory properties during the curing process, as aroma compounds also interact with less volatile components and between them. However, they were found to be overall good indicators of the development of the targeted sensory properties and it would be advisable to monitor their concentrations.

5. CONCLUSION

Tahitian vanilla has come a long way to be as it is known today. Various vanilla vines from different species have travelled around the world and been introduced in French Polynesia, to give the hybrid *V. ×tahitensis,* which is currently grown in French Polynesia. Polynesian vanilla growers have selected over time the best cultivars to be produced, "Tahiti" and "Haapape" and have refined the curing method according to the specificity of Tahitian vanilla. This study showed that HPLC analysis and vanillin content of the pods were not always appropriate to assess the flavour properties of the pods. Modern analytical techniques have highlighted the subtle aroma chemistry of Tahitian vanilla. The use of sensory techniques has enabled to show that the flavour of Tahitian vanilla differentiated from other vanilla by stronger anise, vanilla and caramel notes. The weaker phenolic and woody notes appeared to be linked to the curing process, which plays a very important role in the development of volatile compounds. Curing vanilla pods at 50% moisture seems to be the optimal content so that Tahitian vanilla aroma develops fully and keeps its distinctive characteristics compared to other origins.

New authentication criteria of Tahitian vanilla based on HPLC profiles and specific anisyl compounds have been published recently in French Polynesia and could be used to certify its origin. In order to maintain Tahitian vanilla originality, quality control should also be orientated towards odour-active compounds, which impact definitely the aroma. This will help protecting the specificities of this unique spice.

REFERENCES

1. FAOSTAT. Vanilla production in French Polynesia between 2000 and 2013 [Internet]. 2015. Available from: http://faostat3.fao.org/browse [Accessed: 2016-10-25]
2. Institute "Vanille de Tahiti". About us [Internet]. 2015. Available from: http://vanilledetahiti.com [Accessed: 2016-10-25]
3. Chevalier A. La systématique du genre Vanilla et l'amélioration des plants de vanille [Systematics of Vanilla genus and improvement of vanilla seedlings]. Revue de botanique appliquée et d'agriculture tropicale. 1946;26(281):107–111.
4. Cuzent G. Archipel de Tahiti: recherches sur les productions végétales [Tahiti archipelago: research on plant production], 2nd edition (1860 reed. 1983), Papeete: Eds Haere po no Tahiti, 1983. 208 p.
5. Henry C. Caractères généraux des différents vanilliers cultivés à Tahiti [Systemic characteristics of various vanilla plants cultivated in Tahiti]. L'agronomie tropicale. 1924;79:133–136.
6. Lubinsky P, Cameron KM, Molina MC, Wong M, Lepers-Andrzejewski S, Gomez-Pompa A, et al. Neotropical roots of a polynesian spice: the hybrid origin of Tahitian vanilla, Vanilla tahitensis (Orchidaceae). American Journal of Botany. 2008;95(8):1040–1047.
7. Costantin J, Bois D. Note sur trois types de vanilles commerciales de Tahiti [Note on three commercial vanilla types from Tahiti]. Compte rendu de l'académie des Sciences de Paris. 1915;61:196–202.
8. Bouriquet G, Hibon E. Quelques vanilliers cultivés dans les établissements français d'Océanie [Some vanilla plants cultivated in the french settlements in Oceania]. L'agronomie tropicale. 1950;1–2:54–61.
9. Moore JW. New and critical plants from Raiatea. Bernice P Bishop Museum Bulletin. 1933;102:25.
10. Portères R. Nouvelles espèces de vanilliers [New vanilla plant species]. Bulletin de la Société Botanique de France. 1951;98(4–6):92–95.
11. Lepers-Andrzejewski S, Causse S, Caromel B, Wong M, Dron M. Genetic linkage map and diversity analysis of Tahitian vanilla (Vanilla ×tahitensis, Orchidaceae). Crop Science. 2012;52:795–806.
12. Duval MF, Bory S, Andrzejewski S, Grisoni M, Besse P, Causse S, et al. Diversité génétique des vanilliers dans leurs zones de dispersion secondaire [Genetic diversity of vanilla plants observed in secondary dispersal areas]. Les Actes du Bureau des Ressources Génétiques. 2006;6:181–196.
13. Soto Arenas MA, Dressler RL. A revision of the Mexican and Central American species of Vanilla Plumier ex Miller with a characterisation of their ITS region of the nuclear ribosomal DNA. Lankesteriana. 2010;9(3):285–354.

14. Lepers-Andrzejewski S, Siljak-Yakovlev S, Brown S, Wong M, Dron M. Diversity and dynamics of plant genome size: an example of polysomaty from a cytogenetic study of Tahitian vanilla (Vanilla ×tahitensis, Orchidaceae). American Journal of Botany. 2011;98(6):1–12.
15. Ranadive AS. Vanilla cultivation. Vanilla First International Congress, East Brunswick. 2003, 25–32.
16. Conseil des Ministres de la Polynésie française. Arrêté n° 960 du 26 juin 2014 portant définition de l'appellation d'origine "Vanille de Tahiti" [Order n° 960 from the Executive Council, June 26th 2014, defining the designation of origin "Vanille de Tahiti"]. Journal Officiel de la Polynésie française. 2014:8210–8214.
17. Conseil des Ministres de la Polynésie française. Arrêté n°1111 CM du 10 août 2016 portant définition de l'appellation d'origine "Vanille de Tahiti" [Order n° 1111 from the Executive Council, August 10th 2016, defining the designation of origin "Vanille de Tahiti"]. Journal Officiel de la Polynésie française. 2016:9169–9175.
18. The International Organization for Standardization. ISO 5565–2 Vanilla (Vanilla fragrans (Salisbury) Ames)—Part 2: tests methods. 1999. Geneva : ISO. 11 p.
19. AOAC International. AOAC Official method 990:25—Vanillin, vanillic acid, p-hydroxybenzaldehyde, and p-hydroxybenzoic acid and ethyl vanillin in vanilla extract and artificial vanilla flavor. Liquid chromatographic method. 2013. AOAC Official Methods of Analysis AOAC: Rockville, MD.
20. Ministère de l'Economie, des Finances et de l'Industrie, République Francaise. Direction Générale de la Concurrence de la Consommation et de la Répression des Fraudes. Note d'information n°2003–61 sur la vanille: Définition des critères analytiques d'authenticité de la vanille, des extraits et des arômes naturels vanille [Information note n°2003–61 on vanilla: Definition of analytical criteria of authenticity of vanilla, vanilla extracts, and natural vanilla flavours]. 2003. 6 p.
21. Brunschwig C, Senger-Emonnot P, Aubanel ML, Pierrat A, George G, Rochard S, et al. Odor-active compounds of Tahitian vanilla flavor. Food Research International. 2012;46(1):148–157.
22. Takahashi M, Inai Y, Miyazawa N, Kurobayashi Y, Fujita A. Identification of the key odorants in Tahitian cured vanilla beans (Vanilla tahitensis) by GC-MS and an aroma extract dilution analysis. Bioscience, Biotechnology, and Biochemistry. 2013;77(3):601–605.
23. Fayet B, Tisse C, Guérère M, Estienne J. Nouveaux critères analytiques dans l'étude des gousses de vanille. Analusis [New analytical criteria for the study of vanilla pods]. 1987;15(5):217–226.
24. Derbesy M. Gousses de vanille Comores. Banque de données pour la récolte 1987 [Vanilla pods from Comoros. Database from harvest year 1987]. Parfums, cosmétiques, arômes. 1989;87:77–90.
25. Gassenmeier K, Riesen B, Magyar B. Commercial quality and analytical parameters of cured vanilla beans (Vanilla planifolia) from different origins from the 2006–2007 crop. Flavour and Fragrance Journal. 2008;23:194–201.
26. Pérez-Silva A, Odoux E, Brat P, Ribeyre F, Rodriguez-Jimenes G, Robles-Olvera V, et al. GC-MS and GC-olfactometry analysis of aroma compounds in a representative organic aroma extract from cured vanilla (Vanilla planifolia G. Jackson) beans. Food Chemistry. 2006;99:728–735.
27. Zhang S, Mueller C. Comparative analysis of volatiles in traditionally cured Bourbon and Ugandan vanilla bean (Vanilla planifolia) extracts. Journal of Agricultural and Food Chemistry. 2012;60(42):10433–10444.
28. Del Pilar Galeas M. Gas-chromatography-mass spectrometry and gas-chromatography-olfactometry analysis of aroma compounds of Vanilla pompona Schiede. Master thesis. Graduate School-New Brunswick, Rutgers, The State University of New Jersey. 2015.
29. Brunschwig C, Rochard S, Pierrat A, Rouger A, Senger-Emonnot P, George G, et al. Volatile composition and sensory properties of V. ×tahitensis bring new insights for

vanilla quality control. Journal of the Science of Food and Agriculture. 2016;96:848–858.

30. Takahashi M, Sakamaki S, Fujita A. Simultaneous analysis of guaiacol and vanillin in a vanilla extract by using high-performance liquid chromatography with electrochemical detection. Bioscience, Biotechnology, and Biochemistry. 2013;77(3):595–600.
31. Lepers-Andrzejewski S, Brunschwig C, Collard F-X, Dron M. Morphological, chemical, sensory and genetic specificities of Tahitian vanilla. In: Grisoni M, Odoux E, editors. Vanilla, medicinal and aromatic plants-industrial profiles. CRC Press, Taylor and Francis: Boca Raton, USA. 2011. pp. 205–228.
32. Chen Y, Gu F, Li J, He S, Xu F, Fang Y. Involvement of colonizing Bacillus isolates in glucovanillin hydrolysis during the curing of Vanilla planifolia Andrews. Applied and Environmental Microbiology. 2015;81(15):4947–4954.
33. Brunschwig C, Collard F-X, Bianchini J-P, Raharivelomanana P. Evaluation of chemical variability of cured vanilla beans (Vanilla tahitensis and Vanilla planifolia). Natural Product Communications. 2009;4(10):1393–1400.
34. Pérez-Silva A, Günata Z, Lepoutre J-P, Odoux E. New insight on the genesis and fate of odor-active compounds in vanilla beans (Vanilla planifolia G. Jackson) during traditional curing. Food Research International. 2011;44(9):2930–2937.
35. Rodriguez-Campos J, Escalona-Buendía HB, Orozco-Avila I, Lugo-Cervantes E, Jaramillo-Flores ME. Dynamics of volatile and non-volatile compounds in cocoa (Theobroma cacao L.) during fermentation and drying processes using principal components analysis. Food Research International. 2011;44(1):250–258.

CHAPTER 4

Essential Oil Chemotypes of Three Ocimum Species Found in Sierra Leone and Nigeria

Olugbade TA*

Department of Pharmaceutical Chemistry, Obafemi Awolowo University, Osun State, Nigeria

ABSTRACT

Chemical composition of the essential oil of *Ocimum basilicum L., O. gratissimum L. and O. americanum L.* varieties found in Sierra Leone and Nigeria were determined by GC-FID and GC-MS. The antimicrobial test was evaluated by agar diffusion. The Sierra Leone and Nigeria *O. americanum L.* varieties were identified as the linalool chemotype and similarly, varieties of O. gratissimum L. from both countries are thymol chemotype. The high thymol content is consistent with the relative high antimicrobial activity of the *O. gratissimum L* oils. The Sierra Leone *O. basilicum L.* sample was established as the methyl eugenol chemotype while the Nigerian collection is predominantly methyl chavicol in composition. The high methyl eugenol content of the Sierra Leone collection is consistent with the observation of attraction of fruit flies to the distillate. To the best of our knowledge, there has been no report on essential oils of Ocimum species in Sierra Leone, hitherto.

Keywords: Chemotypes; Ocimum species; O. basilicum L; Lamiaceae

INTRODUCTION

The family Lamiaceae of which the genus Ocimum belongs is composed of diverse and rich source of essential oil containing plants. Ocimum contains between fifty to one hundred and fifty species of herbs and shrubs from the tropical regions of Asia, Africa, and Central and South America [1]. O. gratissimum L. is referred to as holy basil. In Sierra Leone, it is known as "Tea bush" (Krio) and "orgbethor" (Themne) while it is known as "efirin nla" in the Yoruba speaking culture of Nigeria. O. basilicum L. is also referred to as sweet basil. It is known as "Patmenji" (Krio) and "Sorow" (Themne) of Sierra Leone and "efinrin wewe" by the Yorubas. O. americanum L. (synonym of O. canum Sims) [2,3] is known as American basil or hoary basil. In Sierra Leone both the available varieties of O. americanum L and O. basilicum L. are used in culinary as "Patmengy" indiscriminately. The different varieties of O. gratissimum L. [implying different chemotypes] have been used extensively in the traditional system of medicine in many countries as had been reviewed [4-10]. The main issues of concern with the use of herbal drugs remain safety, validation of claims and standardisation of product. There exist the problems of significant variation in the content of Ocimum plants across and within species, with implication of varied biological activities. In spite of the popular use of the genus Ocimum in food and confectionery as spices and application in the treatment of gastrointestinal infections and conjunctivitis in Sierra Leone, the varieties of the plant growing in Sierra Leone have not been, hitherto, investigated. The present study therefore determined the essential oil constituents of the Ocimum species collected in Freetown, Sierra Leone compared with varieties growing in Ile-Ife, Nigeria in order to establish the specific chemotypes in the specific regions.

EXPERIMENTAL MATERIALS

Plant Materials

Aerial parts of cultivated *O. basilicum* (OBS), *O. gratissimum* (OGS) and (OAS) at full flowering stage were collected from Krootown Road, Central Freetown, Sierra Leone while the Nigerian varieties *O. basilicum* (OBN), *O. gratissimum* (OGN) and *O. americanum*(OAN) were collected in Ile-Ife, Osun- State, Nigeria. The plants were identified at the Department of Botany, Faculty of Sciences, OAU, Ile-Ife, Nigeria where Voucher specimens were deposited.

Extraction of Essential Oils

Fresh or air-dried leaves of the plants were subjected to hydrodistillation using a Clevenger-type apparatus for 4 hours. The volatile oils were dried over anhydrous magnesium sulphate and stored in amber sealed vials at 4°C until analysis.

Analysis of The Essential Oils

The GC-MS analyses of the essential oils were carried out on Agilent 6890N coupled to a quadrupole MSD 5973. Column: DB-5MS (5% phenyl 95% dimethylarylene siloxane) 30 m × 0.25 mm × 1 mm film thickness (Agilent). Temperature program: from 50°C (5 min) to 200°C (10 min) at 2°C/min. Injection temperature: 25°C. Injection volume: 1.0 µL. Inlet pressure: 10.3 kPa PSI. Carrier gas: He, Linear velocity: 40 cm/sec. Injection mode: split (10:1). MS interface temp.: 230°C; MS mode: EI at 70 eV; mass range: 30-400. The oil sample was dissolved in dichloromethane before injection.

The relative compositions of the essential oils were determined on Agilent 7890B coupled to FID and an autoinjector (Agilent G45138). Column: 19091J-413 HP-5 (5% phenyl methyl siloxane) 30 m × 0.32 mm × 0.25 mm film thickness (Agilent). Temperature program: from 50°C (5 min) to 200°C (10 min) at 2°C/min. Injection temperature: 250°C. Injection volume: 1.0 µL. Inlet pressure: 66.7 kPa. Carrier gas:He, Linear velocity: 40 cm/sec. Injection mode: split (50:1). FID temp.: 230°C; H2 flow: 40 mL/min; air flow: 400 mL/min. The oil sample was dissolved in dichloromethane before injection.

Identification of The Compounds

Compounds were identified by their retention indices determined using C8-C40 alkane standards and compared with literature data [11].

The identities were confirmed by the mass spectral data observed (**TABLE 1**) supported with library (National Institute of Standards and Technology).

Antimicrobial Tests

Overnight broth cultures of the test organisms were diluted to give cell suspensions of 10/6 -10/7 CFU which were used to surfaceinoculate nutrient agar (Oxoid, London) (bacteria) and Sabouraud dextrose agar (Oxoid, London) (C. albicans). The antimicrobial activity tests of the oils were performed using paper disc method with Whatman No. 1 paper of size 6 mm which were each saturated with the test neat essential oil and subsequently placed on the surface of the inoculated plates [12]. The plates were kept at 4°C for 1 h to allow for diffusion before subsequent incubation at 37°C for 24 h for bacteria and 25°C for 48 h for C. albicans. The reference antibacterial agents were Streptomycin (1 mg/ml) or acriflavine (1%). The tests were carried out in duplicate and the average zones of inhibition were determined.

Table 1: Observed Mass spectral features of the prominent components of the essential oils of Ocimum plants from Sierra Leone and Nigeria.

NO.	m/z (abundance)	Compound
1	m/z 134 (35), M+. ; 119 (100); 91 (20)	p-Cymene
2	m/z 154 (100), M+ ; 139 (70); 43 (75)	1,8-Cineole (Eucalyptol)
3	m/z 136 (50), M+. ; 121 (48); 93 (100); 91 (60)	γ-Terpinene
4	m/z 152 (10), M+. ; 137 (2); 81 (100), 69 (46)	L-Fenchone
5.	m/z 154 (1), M+. ; 139 (3); 136 (20) ; 121(40); 93 (100); 71 (95)	Linalool
6	m/z 152 (50), M+. ; 108 (40); 95 (100)	Camphor
7	m/z 154 (40), M+. ; 136 (20); 111 (90); 93 (75); 71 (100)	Terpinen-4-ol
8	m/z 148 (100), M+.; 133 (20); 117 (30)	Methyl chavicol (Estragole)
9.	m/z 150(35), M+. ; 135 (100), 115 (12.5)	Thymol
10.	(m/z 164 (100), M+. ; 149 (30)	Eugenol
11	m/z 178 (100), M+. ; 163 (28), 147 (28)	Methyl eugenol
12	m/z 204 (12), M+. ; 189 (24); 161 (45), 133 (100)	β-Caryophylene (E)
13	m/z 204 (3), M+. ; 189 (5); 119 (100); 93 (82)	α-Bergamotene (trans)
14	m/z 204 (45), M+. ; 189 (45); 161 (49); 93 (47); 28 (100)	β-Selinene (β-Eudesmene)
15	m/z 204 (60); M+. ; 189 (100); 161 (37); 133 (45)	α-Selinene
16	m/z 220 (3), M+. ; 205 (10); 177 (23); 93 (90), 79 (100), 43 (70).	Caryophyllene oxide
17	m/z 222 (2), M+. ; 204 (36); 189 (10); 179 (10); 161 (70); 95 (100)	α-Muurolol
18.	m/z 204 (48); 189 (20); 161 (100); 119 (75)	α-Cadinol

RESULTS AND DISCUSSION

The yields of the essential oils from the six plants are 0.34% (OAS), 0.98% (OAN), 0.71% (OBS), 0.96% (OBN), 0.93% (OGS) and 0.84% (OGN). The results of the antimicrobial test of the oils are presented below. The compounds identified in the different oil samples are presented in **TABLE 2**.

The chemotyping of *Ocimum* plants had been defined in terms of the combination of all major components constituting more than 20% rather than a dominant single component of the essential oil of the plant [13]. In this respect, both the Sierra Leone and Nigeria *O. americanum L.* leaf materials were thus identified as the linalool chemotype, constituting 49.1% and 39.6% respectively, in the present study. This chemotype of *O. americanum L.* is consistent with those previously reported in Benin [14],

Rwanda [15], Cameroon [6] and Brazil [16]. The high content of linalool in some varieties of *O. basilicum* characterises the varieties as the best grade of commercial sweet basil for culinary use [17,18].

Table 2: The composition of essential oils of Ocimum species collected from Sierra Leone and Nigeria.

% Composition of Sierra Leone collections			% Composition of Nigeria collections			Compounds	Retention index
O. ame*	O. basil*	O. grat*	O. ame	O. basil	O. grat		
-	-	-	Trace	-	0.5	3-Hexen-1-ol	855.8
Trace	-	-	Trace	Trace	2.5	α-Thujene	924.1
Trace	-	Trace	0.2	Trace	0.7	α-Pinene	928.7
-	-	-	-	-	0.2	Camphene	940.9
-	-	-	-	-	0.5	Sabinene	967.1
Trace	-	-	Trace	0.3	1.9	β-Pinene	969.5
Trace	-	Trace	Trace	-	1.8	β-Myrcene	993.5
-	-	-	-	-	0.2	α-Phellandrene	1001.4
-	-	-	-	-	0.2	(+)-3-Carene	1006.7
Trace	-	Trace	0.2	0.5	1.9	α-Terpinene	1014.5
2.5	-	1.5	0.5	-	4.1	p-Cymene	1022.2
-	-	-	-	-	0.6	D-Limonene	1030.7
0.7	0.6	-	2.4	4.3	-	1,8-Cineole (Eucalyptol)	1033.3
-	-	-		-	0.5	β-Ocimene (Z)	1039.2
			1.5		0.2	β-Ocimene (E)	1049.1
1.2	-	0.8	0.5	-	10.8	Y-Terpinene	1058.3
0.6	-	0.8	-	-	-	Cis-Sabinene hydrate	1065.1
-	-	-	2.3	-	-	L-Fenchone	1084.6
-	-	-	-	-	0.2	Terpinolene	1085.7
-	-	-	-	-	1.2	P-Cymenene	1089.9
49.1	0.6	10.3	39.6	-	0.4	Linalool	1102.7
-	0.6	-	-	-	-	Fenchol (endo)	1120.1
10.4	-	5.7	0.2	-	-	Camphor	1144.7
6.3	0.8	3.2	7.5	0.1	1.3	Terpinen-4-ol	1175.1
0.7	-	-	0.5	-	-	α- Terpineol	1184.4
-	0.4	-	-	89.8	-	Estragole (Methyl chavicol)	1197.4
0.5	-	0.9	-	Trace	1.0	Methyl thymol	1235.0
-	-	-	0.8	-	-	Geraniol	1260.1
-	-	-	-	-	0.7	Unidentified	1270.8
-	-	-	0.2	-	-	Bornyl acetate	1283.3
5.6	-	60.5	-	-	42.2	Thymol	1294.4
-	-	1.7	-	-	0.6	Carvacrol	1300.5
1.0	-	-	-	-	-	Terpinen-4-ol acetate	1301.1

-	-	-	0.2	-	-	Unidentified	1329.7
3.4	-	-	18.0	-	0.5	Eugenol	1354.8
-	-	-	1.9	0.2	-	β-Elemene	1391.0
1.1	89.7	0.9	0.4	-	-	Methyl eugenol	1402.1
1.7	-	3.1	-	1.2	1.5	β-Caryophyllene (E)	1419.9
2.4	-	-	6.6	-	0.3	α-Bergamotene (trans)	1436.6
0.7	-	-	-	-	0.5	Humulene (α-Caryophyllene)	1453.4
-	-	-	0.5	-	-	(E) - β-Farnesene	1457.4
-	-	-	1.0	-	-	Y-Muurolene	1475.5
-	-	-	0.7	-	-	Germacrene D	1480.3
-	-	-	0.44	-	-	Bicyclogermacrene	1489.5
0.7	-	3.6	-	-	7.1	β-Selinene	1491.0
-	-	1.3	-	-	2.5	α -Selinene	1496.0
-	-	1.4	-	-	0.9	Unidentified	1499.5
-	-	-	0.3	-	-	α-Bulnesene	1507.8
1.3	0.9	-	0.9	-	1.7	Y-Cadinene	1509.6
-	-	-	-	-	1.2	α-selinene-7-epi	1515.4
-	-	-	0.2	-	0.2	δ-Cadinene	1519.1
-	-	-	0.4	-	-	β-Sesquiphellandrene	1521.4
-	0.9	-	0.3	-	-	Spathulenol	1578.5
0.6	0.9	3.8	-	-	3.2	Caryophyllene oxide	1583.1
-	-	-	-	-	0.6	Humulene epoxide II	1604.7
-	-	-	0.6	-	-	Unidentified	1609.4
1.2	0.7	-	-	-	-	Cubenol	1635.2
-	-	-	3.57	-	-	α-Muurolol	1641.1
7.7	3.3	-	2.3	-	-	α-Cadinol	1651.9
99.4 %	99.4 %	99.5%	95.7 %	96.4 %	94.3 %	TOTAL COMPOSITION	

The Sierra Leone *O. basilicum* examined in the present study was clearly identified as the methyl eugenol chemotype (89.7%). This chemotype had been reported in Togo [19] and Turkey [20]. Methyl eugenol has been known in literature as a powerful insect attractant, [7,21-23], known to attract fruit flies from a distance as far as 0.8 km [7]. In the present study, attracted dead fruit insects (*Drosophilia melanogaster*) were found in the waste distillate (aqueous portion) drained into a bowl and left overnight. This observation underscores the potential use of methyl eugenol, or even this chemotype of *O. basilicum* co-formulated with an insecticide, as a powerful insecticidal product. On the other hand, the essential oil of the Nigerian *O. basilicum*variety was the methyl chavicol (89.8%) chemotype.

Both the Sierra Leone and Nigerian *O. gratissimum L.* varieties examined in this study are identified as the thymol chemotype having 60.5% and 42.2% thymol content respectively. This chemotype is widespread in West Africa (Nigeria [24], Cameroon [25], Togo [26] and Sao Tome [27]) although the presence of a Thymol-Cymene- Terpinene chemotype has been reported from the sub-region (Benin [14]). While *O. gratissimum L.* is the commercial source of eugenol in India [18,28], this component was not detected in the present material. The second most abundant constituent of Sierra Leone variety of L. is linalool (10.3%) which is a minor component in the Nigerian variety [0.4%] while the second most abundant constituent of the Nigerian variety is γ-terpinene (10.8%) which is only a minor component of the Sierra Leone variety (0.8%). Caryophyllene oxide constitutes a prominent sesquiterpene in both varieties of *O. gratissimum L.* The predominant thymol content probably explains the superior antibacterial activities of both *O. gratissimum* collections in the antimicrobial test (**TABLE 3**).

Table 3: The zones of inhibition (mm) of essential oils from Ocimum species collected from Sierra Leone and Nigeria (average of duplicate) compared with reference standards. Note: R=completely resistant, nt=not tested; OAN=*O. americanum* (Nigeria); OBS=*O. basilicum* (Sierra Leone); OBN=*O. basilicum* (Nigeria); OGS=*O. gratissimum* (Sierra Leone) and OGN=*O. gratissimum*(Nigeria).

Microorganism	OAN	OBS	OBN	OGS	OGN	Streptomycin	Acriflavine
S. aureus (NCTC 6571)	13	17	8.5	44	48	23	Nt
B. subtilis (NCTC 8236)	16	Nt	R	Nt	42	26	Nt
E. coli (ATCC 25922)	11.5	Nt	8	24	27	15	Nt
Ps. aeruginosa (ATCC 10145)	R	Nt	R	Nt	10	R	Nt
C. albicans	27	Nt	10	Nt	44	Nt	8

CONCLUSION

Both Sierra Leone and Nigeria *O. americanum L.* varieties were identified as linalool chemotype. The Sierra Leone *O. basilicum L.* variety is the methyl eugenol chemotype, while the Nigerian variety is predominantly methyl chavicol. Both varieties of *O. gratissimum L* collected from Sierra Leone and Nigeria are of thymol chemotype.

ACKNOWLEDGEMENTS

MIK is grateful to Obafemi Awolowo University Central Science Laboratory - Carnegie Corporation of New York Project for Visiting Fellowship. TAO is grateful to COMAHS - Ministry of Health and Sanitation World Bank Assisted Project RCHP for visiting appointment. The support of Nigeria TetFund for TETF/NRF/OAU grant is also acknowledged. The assistance of

Mr. I. I. Ogunlowo of the Department of Pharmacognosy, OAU, for the collection of plants, is similarly acknowledged.

REFERENCES

1. Simon JE, Quinn J, Murray RG [1990] Basil: A source of essential oils: Advances in new crops. Janick J and Simon JE [eds.]. Timber Press, Portland, OR.
2. Tchoumbougnang F, Zollo PHA, Avlessi F, Alitonou GA, Sohounhloue DK, et al. [2006] Variability in the Chemical Compositions of the Essential Oils of Five Ocimum Species from Tropical African Area. J Essential Oil Res 18: 194-199.
3. Xaasan CC, Cabdulraxmaan AD, Passannanti S , Piozzi F, Schmid JP [1981] Constituents of the essential oil of Ocimum canum. J. Nat. Prod. 44: 752-753.
4. Prabhu KS, Lobo R, Shirwaikar AA, Shirwaikar A [2009] Ocimum gratissimum: A Review of its Chemical, Pharmacological and Ethnomedicinal Properties. The Open Complementary Med J 11-15.
5. Vieira RF, Grayer RJ, Paton A, Simon JE [2001] Genetic diversity of Ocimum gratissimum L. based on volatile oil constituents, flavonoids and RAPD Markers. Biochem Syst Ecol 29: 287-304.
6. Ngassouma MB, Ousmailaa H, Ngamob LT, Maponmetsemb PM, Jirovetzc L, et al. [2004] Aroma compounds of essential oils of two varieties of the spice plant Ocimum canum Sims from Northern Cameroon. J Food Composition Anal 17: 197-204.
7. Holm Y [1999] Bioactivity of Basil. University of Helsinki, Finland, Overseas Publishers Association.
8. Nakamura CV, Ueda-Nakamura T, Bando E, Melo AFN, Cortez DAG, et al. [1999] Antibacterial Activity of Ocimum gratissimum L. Essential Oil. Mem Inst Oswaldo Cruz, Rio de Janeiro, 94: 675-678.
9. Matasyoh LG, Matasyoh JC, Wachira FN, Kinyua MG, Muigai AWT, et al. [2007] Chemical composition and antimicrobial activity of the essential oil of Ocimum gratissimum L. growing in Eastern Kenya. Afr J Biotechnol 6: 760-765.
10. Nweze EI, Eze EE [2009] Justification for the use of Ocimum gratissimum L in herbal medicine and its interaction with disc antibiotics. BMC Complem Altern Med 9: 37.
11. Babushok VI, Linstrom PJ, Zenkevich IG [2011] Retention Indices for frequently reported compounds of Essential oils. J. Phys. Chem. Ref. Data 40: 1-47.
12. Serban ES, Ionescu M, Matinca D, Maier CS, Bojita MT [2011] Screening of the Antibacterial and Antifungal activity of eight Volatile essential oils. FARMACIA 59: 440-446.
13. Grayer RJ, Kite GC, Goldstone FJ, Bryan SE, Paton A, et al. [1996] Intraspecific taxonomy and essential oil chemotypes in sweet basil, Ocimum basilicum. Phytochemistry 43: 1033-1039.
14. Yayi E, Moudachirou M, Chalchat JC [2001] Chemotyping of three Ocimum species from Benin: O.basilicum, O. canum and O. gratissimum. J Essential Oil Res 13: 13-17.
15. Ntezurubanza L, Scheffer JJ, Looman A [1985] Composition of the essential oil of Ocimum canum grown in Rwanda. Pharm Weekbl Sci 17: 273-276.
16. Nascimento JC, Barbosa LCA, Paula VF, David JM, Fontana R, et al. [2011] Chemical composition and antimicrobial activity of essential oils of Ocimum canum Sims. and Ocimum selloi Benth. An Acad Bras Cienc 83: 789-799.
17. Sajjadi SE [2006] Analysis of the essential oils of two cultivated basil [Ocimum basilicum L.] of Iran. Daru 14: 128-130.
18. Jirovetz L, Buchbauer G, Shafi MP, Kaniampady MM [2003] Chemotaxonomical analysis of the essential oil aroma compounds of four different Ocimum species from southern India. Eur Food Res Technol 217: 120-124.

19. Ozcan M, Chalchat J [2002] Essential Oil Composition of Ocimum basilicum L. and Ocimum minimum L. in Turkey. Czech J Food Sci. 20: 223-228.
20. Koba K, Poutouli PW, Raynaud C, Chaumont J, Sanda K [2009] Chemical composition and antimicrobial properties of different basil essential oils chemotypes from Togo. Bangladesh J Pharmacol 4: 1-8.
21. Siderhurst MS, Jang EB [2006] Attraction of Female Oriental Fruit Fly, Bactrocera dorsalis, to Terminalia catappa Fruit Extracts in Wind Tunnel and Olfactometer Tests. Formosan Entomol 26: 45-55.
22. Vargas RI, Pinero JC, Mau RFL, Stark JD, Hertlein M, et al. [2009] Attraction and mortality of oriental fruit flies to SPLAT-MAT-methyl eugenol with spinosad. Entomol Experimentalis et Applicata 131: 286-293.
23. Gang D, Wang J, Dudareva N, Nam KH, Simon JE, et al. [2001]. An investigation of the storage and biosynthesis of phenylpropenes in sweet basil. Plant Physiol. 125: 539-555.
24. Ekundayo O [1986] Essential Oils VIII. Volatile Constituents of the Leaves of Ocimum viride. Planta Medica 52: 200-202.
25. Ngassoum MB, Essia-Ngang JJ, Tatsadjieu N, Jirovetz L, Buchbauer G, et al. [2003] Antimicrobial study of essential oils of Ocimum gratissimum leaves and Zanthoxylum xanthoxyloides fruits from Cameroon. Fitoterapia 74: 284-287.
26. Sanda K, Koba k, Nambo P, Gaset A [1998] Chemical investigation of Ocimum species growing in Togo. Flavour Fragrance J 13: 226-232.
27. Martins AP, Salgueiro LR, Vila R, Tomi F, Canigueral S, et al. [1999] Composition of the essential oils of Ocimum canum, O. gratissimum and O. minimum. Planta Med. 65: 187-189.
28. Verma RS, Kumar A, Mishra P, Kuppusamy B, Padalia R, et al. [2016] Essential oil composition of four Ocimum spp. from the Peninsular India. J Essential oil Res 28: 35-41.

CHAPTER 5

Toxicity and Antiviral Activities of Some Medicinal Plants Used by Traditional Medical Practitioners in Zimbabwe

Deniz Iklim Viol[1], Lameck Shoriwa Chagonda[1*], Sylvester Rodgers Moyo[2], Ali Hikmet Mericli[3]

[1]*School of Pharmacy, College of Health Sciences, University of Zimbabwe, Harare, Zimbabwe*

[2]*Faculty of Health and Applied Sciences, Namibia University of Science and Technology, Windhoek, Namibia*

[3]*Department of Pharmacognosy, School of Pharmacy, University of Istanbul, Istanbul, Turkey*

ABSTRACT

Genital herpes, usually caused by Herpes Simplex Virus type-2 (HSV-2), is the commonest sexually transmitted disease especially amongst rural women in Southern Africa including Zimbabwe. This predisposes them to HIV/AIDS infection, cancer and opportunistic infections (OIs). Current antiviral treatments are often cytotoxic and/or ineffective. This motivates active research to find alternative safer drugs or lead drugs from traditional medicinal sources. Twenty six (26) methanol extracts from commonly used and often endangered plant species (14) used by communities and traditional medical practitioners for treating illnesses and sexually transmitted diseases from 5-selected districts of Zimbabwe were investigated for toxicity by Brine shrimp lethality test (BSLT) and by 50% Cytopathic effect on VERO cultured cells. The extracts were also tested for antiviral activity against Herpes Simplex Virus-2 (HSV-2) by the End Point Titration Technique (EPTT) and Neutralisation Test (NT) on VERO cells. Results from the BSLTs ranged 66.66 - 4304 μg/ml; 50% Cytopathic effect from 19.53 - 312 μg/ml whilst the NT ID_{50} values ranged from 10.41 - 125 μg/ml. The antiviral EPTT reduction factor (RF) was 1 - 10^4 with 13 extracts showing RF ≥ 10^3. All the plant extracts had moderate to high toxicity (LC_{50}, 789 - 66 μg/ml) in the BSLT. Six extracts had LC_{50} values greater than 1000 μg/ml. All 26 extracts were cytotoxic with CC_{50} values < 500 ug/ml of which 19 were more toxic CC_{50} < 100 ug/ml. Nine extracts had in vitro therapeutic indexes ≥ 3.7. Cassia abbreviata, Dichrostachys cinerea and Hypoxis hemerocallidea had therapeutic indexes (TI) 7.5 - 15.0. The more active plant extracts were from roots and root tubers. The results confirm the rationale for the use of traditional medicinal plants by traditional medical practitioners for treating various diseases and could bring awareness for their better use and improve conservation. The results also provide an opportunity to develop more efficacious drugs by isolating lead compounds and determining their mode of action.

Keywords: Medicinal Plants, Toxicity, Antiviral Activity, Herpes Simplex Virus-2, Zimbabwe

1. INTRODUCTION

Third world countries are often endowed with rich flora and fauna which are put to good use in their traditional medical practices. Zimbabwe has over 5000 plant species of which 500 are established in Traditional Medicine (TM) since being legalised in 1981 through an Act of Parliament [1] . Traditional Medical Practitioners (TMPs) treat common illnesses and chronic diseases with claims to treat HIV/AIDS and opportunistic infections (OIs); this is not uncommon in Africa as a whole. Medicinal plants (MPs) still play a central role as sources of drugs for drug development into modern medicines and are still the mainstay for TMPs and for Complementary and Alternative Medicines (CAM) [2] - [4] . The combined effect of poverty and unemployment, civil unrest, land hunger and disease prevalence in developing countries has pushed more and more people towards the traditional medical practice. With over 55,000 registered TMPs in Zimbabwe alone, the protection of the 500 medicinal plants identified as playing a significant role in the traditional practice becomes a problem for conservationists [1] . The ever increasing demand for more land for agriculture and land reforms, human settlements, urban and industrial expansion has put many African governments on the back foot as they seek to address these challenges. Governments are therefore challenged to add value to MPs through research to achieve the multiple goals of conversation awareness and health improvement for the poor. Through technical initiatives, protocols, and/or encouragements, research into traditional medicine and medicinal plants is being pursued vigorously [5] . In Zimbabwe, a study was set up with the help of GEF/UNDP/MET (Ministry of Environment and Tourism) to carry out scientific laboratory studies on 50 extracts from 26 medicinal plants commonly used by communities to assess their conservation status in five selected districts from two regions of the country and to raise awareness of their importance and promote their sustainable use. Subsequently, we reported the antioxidant properties of extracts from these plant species [6] [7] as well as their antimicrobial and phytochemical properties [8] [9] . The laboratory studies confirmed the rationale for their use in traditional medical practice in line with other studies elsewhere to evaluate TMs targeted at finding alternative safer drugs to treat modern illnesses and to overcome growing resistance to infective agents. Our continuing research is now focused on selective studies and value addition through identifying active principles and action mechanisms [10] [11] . The HIV/AIDS virus still remain a threat to mankind and the current report presents the toxicity and antiviral properties of 14 plant species used in traditional medical practice which could have a bearing on the development of potential antiviral and anticancer drugs. This report highlights some of the key findings of the preliminary laboratory studies on toxicity and antiviral activities in the pilot study [12] .

2. MATERIALS AND METHODS

2.1. Ethnobotanical Surveys and Plant Materials

Ethno-botanical surveys on traditional medicinal plants used for treating common ailments, those most traded and threatened were carried out in five districts: Bulilima, Chimanimani, Chipinge, Mangwe and Matobo and the plants identified and authenticated by the National Herbarium and Botanic Gardens, Harare, as previously reported [8] . The plant materials roots, leaves, twigs, tuber, bark or whole, were dried, finely ground and stored.

2.2. Preparation of Extracts for Bioactive Tests

Selected plant parts were powdered (30 g), macerated in methanol (200 ml) [13] , filtered and the filtrate evaporated off under reduced pressure, freeze dried and stored at −20°C. Part of the lyophilised extracts were dissolved in DMSO (10 mg/ml), filtered under aseptic conditions and stored for further bioactive use at −20°C.

2.3. Brine Shrimp Bioactivity Testing

Bioactivity testing using the Brine Shrimp (Artemia salina) Lethality Test (BSLT) method was carried out in triplicate with different concentrations of the sterilized plant extracts in brine solution (10 μg/ml - 1000 μg/ml) and the percentage lethality of the nauplii determined [14] . The concentration of the extract that kills 50% of the shrimps, "Lethal Concentration$_{50}$" (LC_{50}) values were recorded using Nerium oleander as the positive control by GraphPad Prism 5.0 linear regression and Pearson's two tailed analysis for 95% confidence limits (95% CI) [15] .

2.4. Cytotoxicity and Antiviral Activity of the Plant Extracts

The lyophilized methanol extracts were dissolved in DMSO to a concentration of 10 mg/ml. Vero cells (African green monkey kidney) (Highveld Ltd., South Africa) were grown and maintained in essential growth medium supplemented with 5% foetal calf serum and incubated under 5% carbon dioxide at 37°C. The confluent cells were removed and trypsinated with phosphate buffered saline solution to achieve a cell concentration of 1 - 2 × 10^4 cells per well for VERO cells.

Herpes simplex virus 2 (HSV-2) (Highveld Ltd., South Africa) was propagated on Vero cells and the recovered virus suspension (1 ml) diluted by infecting Vero confluent monolayers grown on microtitre plates with 0.1 ml of serial tenfold dilutions of the virus suspension quadruplicated and observed for 7 days for cytopathic effect (CPE). The HSV-2 virus titre was obtained from the 50% tissue culture infections dose per ml ($TCID_{50/ml}$) [16] .

Cytotoxicity of the different plant extracts that could cause non-specific cytopathic effect (CPE) on confluent Vero cells in growth medium was determined by using 100 ml of serial two fold dilution of the plant extracts onto confluent Vero cells on microtitre plates in quadruplicate. The 50% cytotoxicity concentration (CC_{50}) was defined as the plant extract concentration causing 50% CPE compared to uninfected cells [16] . The maximum nontoxic dilution (MNTD) was taken as the next dilution after the $TCID_{50}$ of the plant extract and used for the antiviral assays.

2.5. Antiviral Assays

Antiviral assays were carried out using the Neutralization test (NT) to determine non-cytotoxic concentration (ID_{50}) that inhibits/protects 50% of the monolayer cells against destruction by the virus compared to uninfected cells using the Spearman-Karber formula [17] and by the End point titration technique (EPTT) [18] . In the EPPT assay, confluent monolayer Vero cells were grown on 96-well microtitre plates with the medium removed from the wells. 0.05 ml (MNTD) of the plant extract and 0.05 ml of growth maintenance medium followed 1 h later by 0.05 ml of tenfold HSV-2 serial diluted virus suspension and the mixture incubated at 37°C. Controls consisted of Vero cells infected with virus and not treated with plant extract and virus uninfected and untreated cells. All tests were compared with a positive control, acyclor (Sigma). The antiviral activity was determined as the reduction factor (RF) of the viral titre: the ratio of virus titre in the absence and presence of the plant extract. A promising antiviral has a RF $\geq 10^3$ [13] [18] .

3. RESULTS AND DISCUSSION

3.1. Brine Shrimp Bioactivity Testing

The bioactivity of the plant extracts using the Brine shrimp lethality test (BSLT) indicated that plants in this category were relatively toxic with 15 of the 26 extracts recording $LC_{50} < 500$ µg/ml (Table 1). Only 6 extracts had $LC_{50} > 1000$ µg/ml. Potentially very toxic plants were T. sericea (LC_{50} = 66.7 µg/ml) and K. africana (LC_{50} = 117.4 µg/ml) showing greater toxicity than the positive control N. oleander (LC_{50} = 141.7 µg/ml). The BSLT bioassay is widely used as a basic screening test for toxicity in prospecting for suspected biological activity in crude and isolated plant extracts from traditional folklore [4] [14] [15] [19] - [22] .

3.2. Cytotoxicity and Antiviral Activity

Cytotoxicity effects for the extracts on Vero cells showed all 26 extracts had $CC_{50} < 500$ ug/ml, 19 with $CC_{50} < 100$ ug/ml and 12 with $CC_{50} < 50$ ug/ml (Table 1) indicating potential antitumour and/or antiviral activity. Antiviral

Table 1. Plants chosen for toxicity and antiviral tests.

No	Species, family name, voucher number, status	Plant part	BSLT LC_{50} (ug/ml)	CC_{50} (50% CPE) (ug/ml)	NT-ID_{50} (ug/ml)	TI: CC_{50}/ID_{50}	EPTT (RF^b)
1	*Cassia abbreviate* Oliv., Fabaceae (2721): Cm. Status: Cm	lf	454.93 ± 18.60	156.25	20.83	7.5	10^2
		rt	445.72 ± 22.15	156.25	10.41	15.0	10^3
		bk	1319.37 ± 356.63	39.06	NA	-	1
2	*Dichrastachys cinerea* (L.) Wight and Arn. Mimosaceae (32): Cm.	lf	539.39 ± 78.24	78.13	10.41	7.5	10^4
		rt	4304.59 ± 685.69	312.50	83.33	3.7	10^2
3	*Elaedendron matabelicum* (Loes). Steedman Celastraceae (2121): Th., Endemic	rt	1012.31 ± 217.69	78.13	NA	-	1
4	*Elephantorrhiza goetzei* Harms. Fabaceae (7136): Th.	rt	356.55 ± 24.55	156.25	83.33	1.9	10^2
5	*Flacourtia indica* (Burm. f.) Merr. Flacourticaceae (57/62): Th.	lf	281.81 ± 26.14	78.13	83.33	0.9	10^2
		rt	467.31 ± 39.01	156.25	125.00	1.3	10^3
6	*Gymnosporia senegalensis* (Lam.) Loes. Celastraceae (15): Cm.	lf	789.37 ± 104.06	39.06	10.41	3.8	10^3
		rt	2185.61 ± 872.25	78.13	20.83	3.8	10^3
		tw	754.70 ± 182.57	19.53	15.63	1.2	10^3
7	*Hypoxis hemerocallidea* Fisch. & Ave'-Lall. Hypoxidaceae (MTDV06): Th.	tb	735.34 ± 89.39	156.25	10.41	15.0	10^3
8	*Khaya anthotheca* (Welw.) *DC.* Meliaceae (892): Cm.	bk	482.19 ± 43.49	39.06	31.25	1.2	10^2
9	*Kigelia africana* (Lam.) Benth. Bignoniaceae (5990): Cm.	bk	262.20 ± 25.07	39.06	31.25	1.2	10^3
		ft	117.41 ± 30.27	39.06	31.25	1.2	10^4
		rt	501.35 ± 34.88	39.06	NA	-	1
10	*Rhuschirindensis* Baker f. Anacardiaceae (103/67): Cm.	lf	1023.26 ± 161.69	312.50	NA	-	1
		rt	316.60 ± 30.07	78.13	20.83	3.8	10^2
11	*Sclerocarya birrea* (A. Rich.) Hochst. Anacardiaceae (3114): Th, Cm.	bk	1112.37 ± 210.04	39.06	20.83	1.9	10^3
12	*Securidaca longepedunculata* Fresen. Polygalaceae (264/59): V Th.	rt	351.89 ± 35.70	78.13	20.83	3.8	10^3
13	*Terminalia sericea Burch. Ex DC.* Combretaceae (5): Th., endemic	lf	66.66 ± 49.31	39.06	31.25	1.2	10^2
		rt	295.33 ± 37.19	39.06	20.83	1.9	10^3
14	*Warburgia salutaris* (Bertol.f.) Chiov. Canellaceae (CPDV06): V Th.	lf	351.41 ± 29.58	78.13	NA	-	1
		rt	426.10 ± 55.55	39.06	31.25	1.2	10^3
		bk	359.66 ± 14.33	19.53	NA	-	1
15	*Nerium oleander (ref-positive control)*	lf	141.67 ± 68.15		Acylor-1.50		

Plant part investigated: lf = leaf; rt = root; tb = tuber; bk = bark: Environmental status S: Cm = common; Th = threatened; VTh = very threatened; BSLT. Brine shrimp lethality test LC_{50}. lethal concentration that kills 50% of the shrimps with corresponding 95% confidence intervals (95% CI); CC_{50} (50% CPE), plant extract cytotoxicity concentration that kills 50% tissue cells; NT. neutralisation test: NA = no activity: ID_{50}. non-cytotoxic concentration that inhibits/protects 50% uninfected cells, NA = no activity: TI: therapeutic index = CC_{50}/ID_{50} EPPT. End point titration test. RF^b anti-viral reduction factor: ratio of viral titre of control in absence of extract over viral titre in presence of extract.

screening tests indicated some plant extracts possess potential activities to protect the cells against HSV-2 induced destruction with EPTT reduction factors (RFs) ≥ 10^3 (Table 1). D. cinerea (RF = 10^4) and K. Africana (RF = 10^4) were particularly active, however, most plants in the group had RFs ≥ 10^3which activities were also mirrored in the neutralization test (NT) ID_{50}s where acyclovir the positive control had ID_{50} = 1.5. G. senegalensis, T. sericea and W. salutaris previously noted for their high antimicrobial activities [8] were also in the group of potential antiviral plants with high cytotoxicities (CC_{50} < 50 µg/ml) and high reduction factors (RF = 10^3) for some of their extracts. However, in addition to having a high reduction factor, a plant extract should also have a high therapeutic index (TI)/high sensitivity value in order to prevent cell destruction and be considered for further potential antiviral investigation. From this group (Table 1), Cassia abbreviata had TI values of 7 and 15 for leaf and root extract respectively, Dichrostachys cinerea TI = 7.5 for leaf extract and Hypoxis rooperi had TI = 15 for the tuber. Kigelia fruit showed a high reduction factor 10^4 but low therapeutic efficacy 1.2 in line with other reports where its anti-tumour and other biological properties are highlighted [21] - [28] . Plant extracts Gymnosporia senegalensis (Celastraceae), Warburgia salutaris (Canellaceae) and Terminalia sericea (Combretaceae) are popular plants in traditional practice. Recent studies are focusing on the toxicity, antiviral and antitumor properties of crude and isolates against different viral species especially HIV-1, HSV-1, HSV-2, anti- Dengue [29] - [38] . In general, all the plants in this study displayed both antiinfective and toxic properties (Table 1) whilst reports from phytochemical studies reflect flavonoids and condensed tannins largely responsible for the observed antiviral and anticancer activities [25] - [28] [38] [39] . Isolated fractions and derivatives can have greater toxicity and/or activity than the crude extracts, the case of artemisinin from Artemisia annua in the treatment of malaria [39] and the isolated phytoconstituents from Kigelia and Hypoxis species [24] - [27] .

4. CONCLUSION

Sub-Saharan Africa still suffers the great burden for HIV/AIDS with millions infected and affected. Though prevalence has fallen, availability of antiviral drugs poses a continuing challenge. The use of TMPs is often plagued with issues of safety and/or toxicity. The antimicrobial and antiviral activities demonstrated by the plant extracts are further proof to support claims by TPs of their ability to treat more serious illness including HIV/ AIDS and opportunistic infections [5] . Studies carried out from cited literature on traditional medicinal plants elsewhere have also demonstrated their potential antimicrobial and antiviral activities. Whilst the laboratory studies and biological results confirm traditional folkloric uses, their application in modern medicine has enormous challenges with respect to the standardisation of crude extracts, isolation and characterisation of key active principles and the adoption of clinical protocols gave the wide range of related plant species in different countries. This may not be helped by the lack of sponsorship for such research gave the poor state of the economies of most developing countries. In some cases, the active principles and their derivatives have proven more active than their crude extracts: artemisinin and derivatives (from Chinese Artemisia annua) [39] in the treatment of malaria; sitosterols, hypoxoside and rooperol (from African Hypoxis spp.) in the treatment of prostate cancer [25] . Likewise extracts from D. cinerea and C. abbreviata (Fabaceae), Kigelia africana (Bignoniaceae) and T. sericea (Combretaceae) should be studied for active principles, derivatives and followed by pre-clinical trials possible due to the wide spectrum of activities for isolates. The wide geographical variables make this some task but regions can adopt standards consistent with their findings from the wild where they are endemic or from propagation efforts where these are applicable. Our studies focused on local plants and reported wide scientific data which should attract more research. Such increased research activity has the potential to promote traditional medical practice by value addition to the medicinal plants used to treat patients, improve quality standards and the formulation of traditional remedies, bring awareness to the environmental status of TMPs and create opportunities for further development of TMPs into modern medicines.

ACKNOWLEDGEMENTS

The authors would like to thank GEF/UNDP through the Ministry of Environment and Tourism (MET) for supporting two postgraduate students on the project, the local communities and traditional practitioners from the districts involved, the School of Pharmacy and the Research Board of the University of Zimbabwe and other national institutions for making this work possible through many linkages and collaborations.

CITE THIS PAPER

Deniz Iklim Viol, Lameck Shoriwa Chagonda, Sylvester Rodgers Moyo, Ali Hikmet Mericli (2016) Toxicity and Antiviral Activities of Some Medicinal Plants Used by Traditional Medical Practitioners in Zimbabwe. *American Journal of Plant Sciences*,**07**,1538-1544.

REFERENCES

1. Chavunduka, G.L. (1997) Traditional Medicine in Modern Zimbabwe, Harare. University of Zimbabwe Publications, Harare.
2. Breyer-Brandwijk, M.G. and Watt, J.M. (1962) The Medicinal and Poisonous Plants of Southern and Eastern Africa. 2nd Edition, E. and S. Livingstone Ltd., Edinburgh.
3. Gelfand, M., Mavi, M., Drummond, R.B. and Ndemera, N. (1998) Traditional Medical Practitioners in Zimbabwe. Mambo Press, Gweru.
4. Hostettmann, K., Marston, A., Ndjoko, K. and Wolfender, J.L. (2000) The Potential of African Plants as a Source of Drugs. Current Organic Chemistry, 4, 973-1010.
5. Mumba, E.M., Ed. (2007) Introduction to the NEPAD/Southern African Network for Biosciences. Proceedings of the First NEPAD/SANBIO Workshop on Scientific

Validation of Traditional Medicine for Affordable Treatments of HIV/AIDS and Opportunistic Infections, Lusaka, 15-16 March 2007, 9-11.
6. Munodawafa, T., Chagonda, L.S., Viol, I., Muchuweti, M. and Moyo, R.S. (2010) Total Phenolic Content and Antioxidant Activity of Some Zimbabwean Traditional Medicinal Plants. Recent Progress in Medicinal Plants, Drug Plants III, 29, 364-373.
7. Viol, D.I., Chagonda, L.S., Munodawafa, T., Muchuweti, M.M., Moyo, S.R. and Mericli, A.H. (2013) Antioxidant Activity and Total Phenolic Contents of some Traditional Medicinal Plants from Zimbabwe. Journal of Biologically Active Products from Nature, 3, 345-352.
8. Viol, D.I., Chagonda, L.S., Moyo, S.R. and Mericli, A.H. (2013) Antibacterial, Antifungal and Phytochemical Activities of Some Medicinal Plants from Zimbabwe. Journal of Biologically Active Products from Nature, 3, 273-284.
9. Munodawafa, T., Chagonda, L.S. and Moyo, S.R. (2013) Antimicrobial and Phytochemical Screening of Some Zimbabwean Medicinal Plants. Journal of Biologically Active Products from Nature, 3, 323-330.
10. Munodawafa, T., Tagwireyi, D., Gadaga, L., Gomo, E., Bierman, F. and Chagonda, L.S. (2015) Acute and Subacute Oral Toxicity of Hydroethanolic Root Extract of Annona stenophyla Engl. and Diels in Sprague Dawley Rats. Journal of Biologically Active Products from Nature, 5, 349-356.
11. Taderera, T., Chagonda, L.S., Gomo, E. and Shai, L.J. (2015) Inhibitory Activity of α-Glucosidase and α-Amylase by Annona stenophyllaroot Extract as Mechanism for Hypoglycaemic Control of DM. International Journal of Pharmacy, Photon, 106, 436-444.
12. Viol, D.I. (2013) Screening of Traditional Medicinal Plants from Zimbabwe for Phytochemistry, Antioxidant, Antimicrobial, Antiviral and Toxicological Activities. http://ir.uz.ac.zw/jspui/bitstream/10646/1105/1/Viol_thesis.pdf
13. Betancur-Galvis, L.A., Morales, G.E., Forero, J.E. and Roldan, J. (2002) Cytotoxic and Antiviral Activities of Colombian Medicinal Plant Extracts of the Euphorbia Genus. Memorial Institute Oswaldo Cruz, 97, 541-546.
14. McLaughlin, J.L., Chang, C.J. and Smith, D.L. (1991) Bench-Top Bioassays for the Discovery of Bioactive Natural Products: An Update. In: Rhaman, A.U., Ed., Studies in Natural Products Chemistry, Elsevier, Amsterdam, 383-409.
15. Krishnaraju, A.V., Tayi Rao, V.N., Sundararaju, D., Vanisree, M., Tsay, H.S. and Subbaraju, G.V. (2006) Biological Screening of Medicinal Plants Collected from Eastern Ghats of India Using Artemiasalina (Brine Shrimp Test). International Journal of Applied Science and Engineering, 2, 115-125.
16. Reed, L.J. and Muench, H. (1938) A Simple Method of Estimating Fifty Percent Endpoints. American Journal of Hygiene, 27, 493-497.
17. Villegas, P. and Purchase, H.G. (1983) Titration of Biological Suspensions. In: Hitchner, S.B., Domermuth, C.H., Purchase, H.G. and Williams, J.E., Eds., Isolation and Identification of Avian Pathogens, 2nd Edition, The American Association of Avian Pathologists, 124-128.
18. Vlietinck, A.J., van Hoof, L., Totte, J., Lasure, A., van den Berghe, D., Rwangabo, P.C. and Mvukiyumwami, J. (1995) Screening of Hundred Rwandese Medicinal Plants for Antimicrobial and Antiviral Properties. Journal of Ethnopharmacology, 46, 31-47.
19. Mongalo, N.I. and Mafoko, B.J. (2013) Cassia abbreviate Oliv. A Review of Its Ethnomedicinal Uses, Toxicology, Phytochemistry, Possible Propagation Techniques and Pharmacology. African Journal of Pharmacy and Pharmacology, 7, 2901-2906.
20. Gadir, S.A. (2012) Assessment of Bioactivity of Some Sudanese Medicinal Plants Using Brine Shrimp (Artemisia salina) Lethality Assay. Journal of Chemical and Pharmaceutical Research, 4, 5145-5148.
21. Maregesi, S.M., Pieters, L., Ngassapa, O.D., Apers, S., Vingerhoets, R., Cos, P., Vanden Berge, D.A. and Vlietinck, A.J. (2008) Screening of Some Tanzanian Medicinal Plants from Bunda District for Antibacterial, Antifungal and Antiviral Activities. Journal of Ethnopharmacology, 119, 58-66.

22. Moshi, M.J., Mbwambo, Z.H., Kapingu, M.C., Mhozya, V.H. and Marwa, C. (2006) Antimicrobial and Brine Shrimp Lethality of Extracts of Terminalia mollis Laws. African Journal of Traditional Complementary and Alternative Medicines, 3, 59-69.
23. Mathabe, M.C., Nikolova, R.V., Lall, H. and Nyazema, N.Z. (2006) Antibacterial Activities of Medicinal Plants Used for the Treatment of Diarrhoea in Limpopo Province, South Africa. Journal of Ethnopharmacology, 105, 286-293.
24. Drewes, S.E., Elliot, E., Khan, F., Dlamini, J.T.B. and Gcumisa, M.S.S. (2008) Hypoxis hemerocallidea—Not Merely a Cure for Benign Prostate Hyperplasia. Journal of Ethnopharmacology, 119, 593-598.
25. Ncube, B., Ndlala, A.R., Okem, A. and van Staden, J. (2013) Hypoxis (Hypoxidaceae) in African Traditional Medicine. Journal of Ethnopharmacology, 150, 818-827.
26. Amusan, O.O.G., Sukati, N.A., Dlamini, P.S. and Sibandze, F.G. (2007) Some Swazi Phytomedicines and Their Constituents. African Journal of Biotechnology, 6, 267-272.
27. Arkhipov, A., Sirdaarta, J., Raya, P., McDonnell, P.A. and Cook, I.E. (2014) An Examination of the Antibacterial, Antifungal, Anti-Gardial and Anticancer Properties of Kigelia africana Fruit Extracts. Pharmacognosy Communications, 4, 62-76.
28. Neondo, J.O., Mbithe, C.M., Njenga, P.K. and Muthuri, W. (2012) Phytochemical Characterisation, Antibacterial Screening and Toxicity Evaluation of Dichrostachys cinerea. International Journal of Medicinal Plant Research, 1, 32-37.
29. Bessong, P.O., Obi, C.L., Igumbor, E., Andreola, M.L. and Litvak, S. (2004) In Vitro Activity of Three Selected South African Medicinal Plants against HIV Type 1 Reverse Transcriptase. African Journal of Biotechnology, 3, 555-559.
30. Bessong, P.O., Obi, C.L., Andreola, M.L., Rojas, L.B., Pouysegu, L., Igumbor, E., Meyer, J.J.M., Quideau, S. and Litvak, S. (2005) Evaluation of Selected South African Medicinal Plants for Inhibitory Properties against Human Immunodeficiency Virus Type 1 Reverse Transcriptase and Integrase. Journal of Ethnopharmacology, 99, 83-91.
31. Müller, V., Chavez, H., Reginatto, F.H., Zucolotto, S.M., Niero, R., Navarro, D., Yunes, R.A., Schenkel, E.P., Barardi, C.R.M., Zanetti, C.R. and Simoes, C.M.O. (2007) Evaluation of Antiviral Activity of South American Plant Extracts against Herpes Simplex Virus Type 1 and Rabies Virus. Phytotherapy Research, 21, 970-974.
32. Rajhandari, M., Mentel, R., Jha, P.K., Chaudhary, R.P., Bhattarai, S., Gevali, M.B., Karmachya, N., Hipper, M. and Lindequist, U. (2007) Antiviral Activity of Some Plants Used in Nepalese Traditional Medicine. Evidence Based Complementary and Alternative Medicine, 6, 517-522.
33. Abad, M.J., Guerra, J.A., Bermejo, P., Irurzun, A. and Carrasco, L. (2000) Search for Antiviral Activity in Higher Plants. Phytotherapy Research, 14, 604-667.
34. Verma, H., Patil, P.R., Kolhapure, R.M. and Gopalkrishna, V. (2008) Antiviral Activity of the Indian Medicinal Plant Extract, Swertia chirata against Herpes Simplex Viruses: A Study by in Vitro and Molecular Approach. Indiana Journal of Medical Microbiology, 26, 322-326.
35. Spindola, K.C.W., Simas, N.K., Salles, T.S., Meneses, M.D.F., Sato, A., Ferreira, D., Romao, W. and Kuster, R.M. (2014) Anti-Mayaro Virus Activity of Cassia australis Extracts (Fabaceae, Leguminosa). Parasites & Vectors, 7, 537.
36. Nandi, C.S., Sharon, A., Sarkar, M.C., Chakabarti, S. and Chattopadhyay, D. (2013) An Indole Alkaloid from Tribal Folklore Inhibits Immediate Early Event in HSV-2 Infected Cells with Therapeutic Efficacy in Vaginally Infected Mice. PLoS ONE, 8, e77937.
37. Tang, L.I.C., Ling, A.P.K., Koh, R.Y., Chye, S.M. and Voon, K.G.L. (2012) Screening of Anti-Dengue Activity in Methanolic Extracts of Medicinal Plants. BMC Complementary and Alternative Medicine, 12, 3.
38. Swain, S.S. and Dubey, J. (2013) Anti-Dengue Medicinal Plants: A Mini-Review. Research and Reviews: Journal of Pharmacognosy and Phytochemistry, 1, 5-9.

39. De Ridder, S., van der Kooy, F. and Verpoorte, R. (2008) Artemisia annua as Self Reliant Treatment in Developing Countries. Journal of Ethnopharmacology, 120, 302-314.

CHAPTER 6

Aromatic Plants from Western Balkans: A Potential Source of Bioactive Natural Compounds

Ana M. Džamić[1]* and Jelena S. Matejić[2]

[1] *Faculty of Biology, Institute of Botany and Botanical Garden "Jevremovac", University of Belgrade, Belgrade, Serbia*

[2] *Faculty of Medicine, University of Niš, Niš, Serbia*

ABSTRACT

Documentation of traditionally used aromatic and medical plants has been carried out in many European countries over the last several years. Over the last decade, the Western Balkans has become the area of a huge number of ethnobiological field studies. Many of those focused on Balkans ethnobotany are linked to the long and ongoing history of gathering and trading local wild aromatic and medicinal plants from this territory into Western European markets. But only less than a half percent of these have been studied for their chemical composition and medicinal value. The most investigated aromatic species in this area belongs to the few biggest families: Asteraceae, Apiaceae, Lamiaceae and Rosaceae. Medicinal value of plants lies in some chemical substances that produce physiological action on the human body, which leads to positive effect on health. Essential oils are secondary metabolites which are the most examined, as well as various plant extracts. Isolation and identification of the compounds in combination with its biological screening can considerably contribute to plant studies. Also, application of new activities and novel techniques for susceptibility testing provide better knowledge of wild growing medicinal plants as potential sources of biological agents and justified their traditional uses.

Keywords: Western Balkans, ethnobiological studies, wild growing plants, essential oil, extract

1. INTRODUCTION

World is endowed with wealth of medicinal and aromatic plants. Plants have always been the principal form of traditional medicine in many undeveloped countries, and presently, they are becoming popular throughout the developed world, where people strive to stay healthy because of influence of chronic stress and pollution. One definition says that aromatic plants are the local heritage of global importance [1]. In many countries of Balkans, people practiced traditional medicine which is based on the use of plants. The way of use is determined with culture, philosophy and personal attitude. This practice is usually different than conventional medicine. Long period of using traditional medicine has demonstrated that it is safe and effective [2]. Medicinal plants also play an important role in the lives of rural people, particularly in remote parts of developing countries with a few health facilities. It is estimated that around 70,000 plant species have been used at one time or another for medicinal purposes. At the present time, when there is a tendency to respect nature and natural processes more and more, the value of local knowledge on the stability of ecosystem dynamics *in sensu lato* deserves the most attention of scientists [3]. Various pathological conditions of humans that could not be fully treated by conventional pharmaceutical agents are numerous [4]. For this reason, there is a growing tendency in use of herbal preparations. However, scientific researches need to provide additional evidence of their safety and efficacy. Medicinal plants contain active substances which may have active effects on human body and its health. Different diseases could be treated with different plants, which may have 2–3 up to 30–40, sometimes even more, different active substances. Any wrong use can cause various complications, due to the established cytotoxicity for many plant compounds [5]. For all these reasons, medicinal and aromatic plants are in focus of many scientific groups which doing experiments and trying to find new useful compounds.

2. PLANT DIVERSITY OF WESTERN BALKANS COUNTRIES

It is estimated that at least 265,000 species of seed plants [6] exist on earth. Only less than a half percent of these have been studied exhaustively for their chemical composition and medicinal value [7]. With about 6340 different vascular plant species reported, the Balkans, compared to 10,500 species accepted in the Flora Europaea, is one of the most important biodiversity centres of Europe [8]. Region of Serbia is rich in medicinal herbs; about 700 species exist on its territory which is 10.7% of total flora with 3662 taxa. Abundance of medicinal herbs placed Serbia in 158 centres of biodiversity in the world [9]. Despite the inconspicuous extension of the territory of Bosnia and Herzegovina (Western Balkan Peninsula; Southeast Europe), the country holds about 3600 species of vascular plants [10]. In Flora Croatica, around 5000 species and subspecies are registered. Among them, 1144 species and subspecies (21% of total flora) are used for different purpose. The greatest number (25%) of these plants is used in traditional as well in official medicine [11]. Croatia is also one of the centres of endemism in south-eastern Europe. No less than 5.65% of the total numbers of known plant species are endemic. Flora of Montenegro consists about 2632 species of vascular plants, as well as of numerous lower infraspecies taxa as varieties and forms [12]. New data suggest 104 taxa more in second supplement to Rohlena`s Conspectus [13]. The flora of the Republic of Macedonia is among the richest floras not only in respect to the Balkan Peninsula, but also in the context of the whole European continent. According to recent data, 210 families, 920 genera and 3700 species comprise the flora of higher plants, angiosperms being the richest group with about 3200 species. Endemic taxons represent a special characteristic and value of the Macedonian flora. Among them, 114 flowering plant species are Macedonian endemics [14].

3. ETHNOBOTANICAL STUDIES

The knowledge of plants, which are used by local people in different geographical areas, is documented in ethnobotanical studies. People in rural areas with traditional knowledge about plants, which are used as a food, spices, flavouring and for medicinal purposes, transmitted that information from one generation to the next. This knowledge is associated to plant identification, conservation and uses. An extensive ethnobotanical work yielded interesting results that confirmed many assumptions and estimates about plant uses [15].

A group of scientists investigated the local traditions of using wild food plants around Lake Vrana (northern Dalmatia, Zadar region) [16]. This research includes studies and interviews of 43 inhabitants of six traditional villages north of Lake Vrana. On average, 12 species were listed, which in total produced an inventory of 55 food plants. The most common medicinal species were *Rosa canina* L., *Salvia officinalis* L., *Achillea millefolium* L., *Helichrysum italicum* (Roth) G. Don and *Mentha* spp.

Ethnobotanical study on medicinal use of wild and cultivated plants in middle, south and west Bosnia and Herzegovina was presented with 228 wild and cultivated species (belonged to 50 families and 124 genera) and 730 different preparations for the use in human therapy. Species of the genus *Achillea* L., *Hypericum* L., *Mentha* L., *Teucrium* L., *Thymus* L. and *Urtica* L. were particularly highly recommended, by almost 90% of the informants that used these species in preparations. Medicinal plants are often used for treatment of different illness such as urogenital tract—16.0%, respiratory system—16.0%, gastrointestinal tract—14.4%, skin ailments—11.5%, blood system—8.8%, nervous system—7.1%, cardiovascular system—6.6%, rheumatism—5.9%; metabolism disorders—3.8%, senses disorders—2.1%, influenzal infections—2.1%, lever and gall disorders—1.9%, parasitic induced ailments—1.6%, inflammations—1.0%, musculoskeletal system—0.8%, endocrinological—0.3% [10].

Rosaceae—16, Brassicaceae—5, Alliaceae—4 and Apiaceae—4. On this region were identified 128 plants and 2 fungi. Most of them are used in ethnomedicine but some of them in ethnoveterinary medicine. Some plants have Asteraceae—17, Usually used plants from the region of Suva Planina Mountain are from families of Lamiaceae—20, 'other' purposes. While the most common conditions treated with medicinal plants are respiratory (79), urogenital (53), gastrointestinal (51), skin (43) and those relating to the circulatory system (35) [17]. On Kopaonik Mountain (the central part of Serbia), 83 wild plant species reported by informants to have medicinal properties have been recorded. The most often used plants belong to six families: Asteraceae—15.8%, Lamiaceae—15.8%, Rosaceae—7.3%, Malvaceae—3.7%, Apiaceae—3.7% and Plantaginaceae—3.7%, within 69 genera and 41 families. In all other families, authors detected the uses of only two or one species (2.4 and 1.2%). Based on investigations, they concluded that tested plants had the highest influence on gastrointestinal ailments—50%, skin injures and problems—25.6%, respiratory—20.5%, urinary-genital—20.5%, cardiovascular problems—19.2%, antiseptic—15.4% and anti-infective effect—14.1% [18]. Botanical remedies in the South-western Serbia (Zlatibor district) comprise 69 species belonging to 36 families. The predominant botanical families were Rosaceae (16%), Lamiaceae (13%) and Asteraceae (12%). From 69 mentioned species, 23 species are included in European Pharmacopoeia 7.0 [19]. Those plants are used for treatment of gastrointestinal—33.1%, respiratory diseases—29.6%, dermatologic diseases—14.8%, urinary tract ailments—6.1%, circulatory system—5.1%, nervous system and psyche—5.1%, nutritional endocrine glands and metabolism—4.3%, skeletal, muscle and connective tissues problems—1.0%, gynecological complications—0.5%, disease of the sensorial system—0.3% [20].

On Prokletije Mountains (Montenegro), the most often used are Rosaceae—11 species which makes 11.7%, Asteraceae—10 species or 10.6% and Lamiaceae—7 species which make 7.4%. Ninety-four species are reported, 35 species in Ph. Eur. 6.0, which makes 37.2%, 12 species in national pharmacopoeias or 12.8%. In most cases, aerial parts were used—43.6%. People usually used those plants for treating gastrointestinal—57.4% and respiratory diseases—41.5% [21].

In one of the most remote and poorest areas of Europe, the village of Theth, which is located in the upper Shala Valley in the Northern Albanian Alps, 79 botanical taxa known and used by the local population. This area is characterised by total absence of medical assistance, but villagers of Theth used a small number of medicinal herbs for minor ailments. They usually used aerial parts of *Origanum vulgare* L., also *Hypericum maculatum* Crantz, as well as aerial part of *Agrimonia eupatoria* L., and they used roots of *Gentiana lutea* L. [22]. Among of ethnobiological study of Pešter plateau (South-western Serbia), where two communities of Serbians and Albanians have been lived together from three centuries ago when they immigrated to the area, 62 taxa in 129 plant-based remedies recorded as well as 204 plant uses. *Chenopodium-bonus hendricus* (L.)Rchb., *Gentiana lutea* L., *O. vulgare* L., *Hypericum* spp., *R. canina* L., *Urtica dioica* L., which are mostly used in the same way and for the same folk medical purposes, may be viewed as the medicinal plants whose cultural salience, measured through the lens of quotations elicited during the free listing exercises, appears remarkable in both communities [23].

A medico-ethnobotanical study was conducted among Albanians, Macedonians and Gorani in 41 villages located in the Sharr Mountains in Western Macedonia. Seventy-six mainly wild taxa (belonging to 34 families) were found to represent the remaining folk medical heritage of the area. The most frequently cited families were Lamiaceae (15.7%), Asteraceae (14.4%), Rosaceae (5.2%), Malvaceae (5.2%) and Fabaceae (5.2%). The large majority of the recorded plants are used in form of teas, and mainly for minor dysfunctions of the respiratory system (46%). According to results of investigations, leaves of *Ballota nigra* L. as a tea is used as a digestive, leaves and fruits of *Cornus sanguine* L. as a tea against stomachaches, topically applied leaves of *Chenopodium urbicum* L. for treating hemorrhoids, aerial parts of *Convolvulus arvensis* L. as a tea against hypertension. This could be interesting for further phytopharmacological investigations. Similar uses were reported by Gorani and Albanians, because those two communities lived together over the century in the same area shared faith, while Orthodox Macedonians reported different uses of these ethnic groups [24].

Comparison with traditional therapy in neighbouring countries showed that there exists considerable similarity with respect to plants used and the way of usage [25, 18, 26]. Results obtained in the area of Suva Planina Mountain comparing with results from other investigated areas of the Western Balkans showed similarities with Zlatibor region—37.2% and Kopaonik Mountain—32.3%. Compare the wider regions showed similarities with Bosnia and Herzegovina—40.9% and Bulgaria—40.6% [17]. It has also been observed that the inhabitants of mountain regions have longer traditional therapeutic tradition (more than six generations) compared to the rest of the population [10].

An important point in research is paying attention on analysis of endemic and rare species, because they are potential source of new active compounds. However, strict rule and regulations for protection of its localities are necessary because of threat of its extermination. If phytochemical analyses show that those plants represent valuable sources of new compounds, it could be necessary to find adequate way to its cultivations and preservation. The new activities and novel techniques for susceptibility testing provide field for research of well-known medicinal plants as well as unknown plants.

4. COMPOSITION AND BIOLOGICAL ACTIVITY OF ESSENTIAL OILS AND EXTRACTS OF WILD GROWING LAMIACEAE SPECIES

The knowledge of plants which in ethnobotanical studies of Western Balkans countries *O. vulgare* is recognised as a plant with a range of health benefits. This uses are confirmed by scientific research. Essential oil of this, well-known species, was characterised with dominance of carvacrol (64.1%) and linalool (17.6%) and possesses strong antifungal activity against wild-type strains [27]. The cytotoxic, antioxidant and antibacterial activities are attributed mostly to the presence of the isomeric phenolic constituents, carvacrol and thymol [28]. The dominance of 1.8-cineole (44.3%) was found in essential oils of *Rosmarinus officinalis* L. as well as in sample of wild growing *Hyssopus officinalis* L. subsp. *pilifer* (Pant.) Murb (36.4%) (**Table 1**) [27, 29]. The results showed that antifungal activity offered by 1.8-cineole was incomplete and exhibited moderate activity at the highest concentration tested. Otherwise, this compound was arguably the most effective compounds regarding fumigant activities against storage pest *Cryptolestes ferrugineus* Stephens followed by camphor and carvacrol [30]. Phenolic acids were identified in post-distillation waste extract of *H. officinalis* subsp. *pilifer* which possessed valuable radical scavenging activities (**Table 1**). The antioxidant activity of phenolic acids is related to the number and position of hydroxyl groups [29]. *Stachys anisochila* Vis. et Panč. is Balkans endemic species, distributed in Bulgaria, Serbia and Albania. Essential oil composition of *S. anisochila* collected from Stara Planina, Serbia, shows α-pinene (7.6%), α-copaene (6.2%), β-pinene (5.3%) and β-caryophyllene (4.5%) as the main compounds. Many species of genus *Stachys* L. possess strong antioxidant and antimicrobial activity [31].

TABLE 1. Main components and biological activity of essential oils from wild growing plants from Western Balkans—Lamiaceae family.

Plant	Origin	Essential oil/main compounds	Activity	References
Hyssopus officinalis L. subsp. *pilifer* (Pant.) Murb	Sicevo gorge (Serbia)	1.8-cineole (36.43%), β-pinene (19.55%), isopinocamphone (15.32%) pinocamphone (6.39%)	Antifungal, antioxidant	[29]
Micromeria dalmatica Benth.	Mt. Lovcen (Montenegro)	piperitenone (41.46%), pulegone (19.02%), piperitenone oxide (14.49%), D-limonene (6.23%) and p-menthone (5.06%)	Antimicrobial	[35]
Satureja montana L.	Mt Biokovo (Croatia)	Carvacrol (63.4%), thymol (19.4%)	Antioxidant	[34]
Satureja montana L. ssp. *montana*	Budva Cetinje MtOrijen (Montenegro)	Thymol (24.7%), carvacrol (15.2%), linalool (15.4%). Carvacrol (24.5%), linalool (17.9%), cis-sabinene hydrate (14.61%), terpinen-4-ol (10.6%). Linalool (32.6%), cis-sabinene hydrate (23.0%), terpinen-4-ol (11%), nerolidol (9.4%)	Antimicrobial, antioxidant, anticholinesterase activity	[39]
Satureja horvatii Šilić	Mt Orijen (Montenegro)	p-cymene (33.1%), thymol (26.1%) and thymol methyl ether (15.1%)	Antimicrobial	[32]
Satureja cuneifolia Ten.	Mt Biokovo (Croatia)	Carvacrol (17.7%), spathulenol(13.2%), caryophyllene oxide (9.5%), α-cadinol (7.1%), amorpha-4,9-dien-2-ol (6.7%)	Antioxidant	[34]
Thymus praecox Opiz ssp. *polytrichus*	Mt Pasjača (Serbia)	trans-nerolidol (24.2%), germacrene-D (16.0%), thymol (9.6%)	Antimicrobial, antioxidant	[38, 40]
Thymus longicaulis C. Presl	Jasenice (Croatia)	Thymol (46.3%), γ-terpinene (16.2%), thymyl methyl ether (11.4%) and p-cymene (9.4%)	Antimicrobial	[41]

Nine species of genus *Satureja* L. have been registered in the area of the Central and Western Balkans. The essential oils isolated from various *Satureja* species have shown antibacterial, fungicidal, antiviral and antioxidant activities. Essential oil of *Satureja horvatii* Šilić (**Table 1**) shows significant antimicrobial effect, and this oil successfully inhibited the development of *Lysteria monocytogenes* in pork meat and also can be a useful source of natural antioxidants [32]. Oxygenated monoterpene hydrocarbons were the major compounds: carvacrol, thymol, carvacrol methyl ether and β-linalool in essential oil of *S. montana* L. subsp. *pisidica* (Wettst.) Silic from two localities (mountains Korab and Galičica, Macedonia). Both oils showed excellent antimicrobial activity and good cytotoxic potential [33]. Antioxidant activity of essential oils and water wastes after distillation process of *S. montana* L. and *S. cuneifolia* Ten. was demonstrated (**Table 1**). The extract from waste water after hydrodistillation of these plant species was rich in thymoquinone (38.7%) and (E)-coniferyl alcohol (18.1%), respectively [34]. *Micromeria dalmatica* Benth. is an endemic species of Balkan Peninsula. In essential oil of *M. dalmatica*, the dominant compounds were piperitenone, pulegone and piperitenone oxide (**Table 1**). This oil possess high antimicrobial efficacy against food spoilage microorganisms, and significant activity of this oil in pork meat against *Salmonella typhimurium* enhances the possibility to use it in food preservation. Also, the essential oil of *S. horvatii* and oregano has been shown to be more or less effective against spoilage microbiota in meat products [35].

Collection of *Sideritis raeseri* Boiss & Heldr. subsp. *raeseri* has a long tradition in local communities in the Ohrid-Prespa region. Recently, identified 17 compounds could be classified into flavonoid glycosides or hydroxycynnamic acid derivatives [36]. They found that hypolaetin derivatives, known for their anti-inflammatory activity, were more abundant in *S. raeseri* subsp. *raeseri* grown in National Park Galičica (Macedonia) in comparison with *S. scardica* Griseb grown near Galičica.

Since ancient times, plants from the genus *Thymus* L. are known according to their therapeutic benefit. Recently, scientists discovered its new medicinal properties. They reported their chemical composition, biological activities and occurrence of *Thymus* species from Balkans [37]. Essential oil of wild growing *T. praecox* Opiz showed high radical scavenging and antimicrobial activities (**Table 1**). The major component, which is identified in supercritical extracts, was thymol, with the strongest antimicrobial activity. Hexane/ethanol extract possessed strongest antibacterial potential, while supercritical extracts were more efficient antifungal agents. In 1,1-diphenyl-2-picrylhydrazyl (DPPH), radical scavenging assay hexane/ethanol extract of *T. praecox* showed the best antioxidant activity, which is connected with the high level of phenolics antioxidants [38].

5. COMPOSITION AND BIOLOGICAL ACTIVITY OF ESSENTIAL OILS AND EXTRACTS OF WILD GROWING APIACEAE SPECIES

Many plants from "spices" family, Apiaceae, are phytochemicaly characterised, and they are used in official medicine, but for much of them data are still missing or scant. *Eryngium palmatum* Vis. et Pancic is an endemic perennial herb, distributed in the central part of the Balkan Peninsula (Serbia, Bulgaria, Macedonia and Albania). Essential oil of *E. palmatum*, collected from Ozren Mountain (Serbia), was characterised by high presence of sesquicineole (21.3%), caryophyllene oxide (16.0%), spathulenol (16.0%) and sabinene (5.5%). Related species, *Eryngium serbicum* Pancic, endemic in Serbia, in essential oil of aerial parts contains germacrene D (19.7%), β-elemene (10.0%) and spathulenol (6.9%). The main compounds in both studied species consisted of sesquiterpenes [42]. In general, sesquiterpenes possess promising anti-inflammatory, antiparasitic and anticarcinogenic activities, but some of them can cause serious toxicity. The essential oils of different *Seseli* L. species show antimicrobial and antifungal activity, while extracts exhibit anti-inflammatory and antinociceptive properties. Essential oil from aerial parts in the flowering stage and rosette in the vegetative stage of *Seseli annum* L. collected on mountain Rtanj (Serbia), as dominant compounds contain sesquiterpenes β-selinene, germacrene D, caryophyllene oxide, germacrene A (0–21.4, 3.4–19.1, 1.2–18.1,0–14.6%, respectively) and in lowest percentage α-selinene, β-elemene, E-caryophyllene, α-pinene, vetiselinenol, isodaucene (0–12.4, 0–11.4, 3.2–10.3, 0.6–7.5, 0–6.3, 0–5.4%, respectively) [43]. Also, different extracts from three *Seseli* taxa (*Seseli pallasii* Besser, *S. libanotis* (L.) Koch ssp. *libanotis* and *S. libanotis* (L.) Koch ssp. *intermedium* (Rupr.) P. W. Ball, growing wild in Serbia) possessed antioxidant and antimicrobial activity [44]. Correlation from the total phenolic and flavonoid content and in vitro antioxidant and antimicrobial activity of methanol and water extracts was presented from *Tordylium maximum* L. (Ozren Mountain, Serbia) [45]. Essential oil isolated from the aerial parts and fruits of *Cachrys cristata* DC., a rare and critically endangered species in the flora of Serbia (**Table 2**), were rich in sesquiterpenes (45.7%) and oxygenated sesquiterpenes (32.9%), while fruit oil consisted of a higher percentage of sesquiterpenes (48.3%) and oxygenated sesquiterpenes (36.7%) [46]. In fruit and aerial part was found germacrene D which is known for its effect on insects and present a potential source of antibacterial and antifungal agents. It cannot be said that only germacrene D is responsible for high activity because essential oils present mixture of different components which have synergetic effect. Various extracts of traditionally used medicinal plants prepared with different extraction method and solvent of different polarity possessed valuable antimicrobial and antioxidant activity. Good activity is related to high concentration of total phenolic and total flavonoid content. Recently, in the study of four *Peucedanum* L. species (*P. officinale* L., *P. longifolium* W. et K., *P. aegopodioides* (Boiss.) Vandas and *P. alsaticum* L. from Serbia), their various extracts were found significant correlation coefficient between radical scavenging DPPH and 2,2-azino-bis(3-ethylbenz-thiazoline-6-sulfonic acid) (ABTS) radical scavenging activities with the total phenolic content/total flavonoids content. Therefore, total phenolic components can be used as an indicator in assessing the antioxidant activity of fruits and vegetables, including *Peucedanum* species [47]. Many *Heracleum* L. species are edible, and medicinal plants *Heracleum sphondylium* L. are used as food or food additive in many countries. Essential oil and extracts of these plants were presented in **Table 2**. Aerial parts of *H. sphondylium* showed antioxidant and antimicrobial effects [48]. Essential oil of *Heracleumorphanidis* Boiss consisted mainly of fatty acid derivate (80.8%) was effective antimicrobial agent (**Table 2**). The methanol extracts from aerial parts and roots of *H. orphanidis* were characterised by furanocoumarins, but stronger antioxidant potential was detected by extracts from aerial parts. *H. orphanidis* showed considerable biological potential considering free radical and various pathogenic strains, including wild type of *Pseudomonas aeruginosa*. One of a new strategy for the treatment of bacterial infections is inhibition of bacterial quorum sensing activity. The higher plants represent native source rich in anti-quorum compounds as novel virulence inhibitors [49].

6. COMPOSITION AND BIOLOGICAL ACTIVITY OF ESSENTIAL OILS AND EXTRACTS OF WILD GROWING ASTERACEAE SPECIES

Considering chemical composition of *Artemisia absinthium* L. essential oil, different chemotypes have been established in Croatia β-thujone/(Z)-epoxy ocimene chemotype, while in samples of *A. absinthium* from Serbia, pure β-thujone chemotype and β-thujone/cis-epoxy ocimen chemotype and sabinene/α-phellandrene/sabinyl acetate chemotype were found. In previous study were analysed chemical composition and biological activity of the essential oil of *A. absinthium* (**Table 3**) [54]. This oil showed high antimicrobial and significant antioxidant activity. Undiluted *A. absinthium* essential oil shown negative skin irritant reaction on all 30 tested volunteers.Essential oil in oral application did not cause mortality in the treated mice. Negative effects were neurological, muscle and gastrointestinal problems. In examinations with *Drosophila melanogaster*, the result showed that the essential oil was toxic for insect larvae in development.

Table 2. Main components and biological activity of essential oils from wild growing plants from Western Balkans—Apiaceae family.

Plant	Origin	Essential oil/Main compounds	Activity	References
Cachrys cristata DC.	Mt Rujan (Serbia)	Aerial parts: phytol (13.1%), germacrene D (12.9%), β-caryophyllene (9.7%), β-bourbonene (8.5%) Fruit: suberosin (19.7%), germacrene D (12.3%), germacrene B (10.0%)	Antimicrobial, antioxidant	[46, 50]
Echinophora sibthorpiana Guss.	Stip (Macedonia)	Aerial parts: methyl eugenol (60.4%), p-cymene (11.2%), α-phellandrene (10.2%), α-Phellandrene epoxide (6.9%)	Antimicrobial, antioxidant	[51]
Ferulago macedonica Micevski & E. Mayer	Negotino city (Macedonia)	Inflorescence: α-pinene (43.1%) and sabinene (26.7%), limonene (6.5%) Aerial part: α-pinene (22.8%), sabinene (15.5%), terpinen-4-ol (9.6%), cis-chrysanthenyl acetate (9.5%) and p-cymene (9.1%)	Antimicrobial, antioxidant	[52]
Heracleum orphanidis Boiss.	Mt Pelister (Macedonia)	n-octanol (39.6), octylhexanoate (17.6%), n-octyl acetate (14.1%).	Antimicrobial, antioxidant, antiquorum sensing	[47]
Heracleum sphondylium L.	Mt Kopaonik (Serbia)	ar-curcumene (13.4%), β-sesquiphellandrene (11.9%), caryophyllene oxide (9.2%)	Antimicrobial, antioxidant	[48]
Laserpitium latifolium L.	Mt Staraplanina (Serbia)	Sabinene (47.8%), α-pinene (24.9%), β-pinene (7.1%), terpinen-4-ol (5.4%)	Antimicrobial, antioxidant	[53]

Table 3. Main components and biological activity of essential oils from wild growing plants from Western Balkans—Asteraceae family.

Pant	Origin	Essential oil/Main compounds	Activity	References
Artemisia absinthium L.	Near the city of Niš (Srbia)	Sabinene (24.5%), sabinyl acetate (13.6%), α-phellandrene (10.3%)	Antimicrobial, antioxidant, insect repellent	[54]
Tanacetum parthenium (L.) Sch.Bip.	Mt Staraplanina (Serbia)	Camphor (51.4%), camphene (7.3%)	Antimicrobial, antioxidant	[53]
Achillea grandifolia Friv.	Jerma Gorge (Serbia)	Camphor (45.4%), 1,8-cineole (16.4%), α-thujone (15.1%), borneol (8.1%)	Antimicrobial, antioxidant	[53]
Achilea crithmifolia Waldst. & Kit.	Mt Staraplanina (Serbia)	Artemisia ketone (31.7%), camphor (25.4%), 1,8-cineole (14.8%)	Antimicrobial, antioxidant	[53]
Helichrysum italicum (Roth) G. Don	Bar (Montenegro)	γ-curcumene (22.5%), α-pinene (15.9%) and neryl acetate (7.8%), β-selinene (6.9%)	Antifungal	[55]

The chemical analysis of essential oils of *Tanacetum parthenium* (L.) Sch. Bip., *Achillea grandifolia* Friv. and *Achillea crithmifolia* Waldst. & Kit. (collected on the territory of Southeast Serbia) was contained the monoterpenes. The major class of components showed camphor in *T. parthenium*, *A. grandifolia* and *A. crithmifolia* and 1,8—cineole in *A. grandifolia* and *A. crithmifolia* (**Table 3**). The antiradical and antioxidant activities were determined using ABTS and DPPH radical scavenging methods. The essential oil of *A. grandifolia* showed the highest antioxidant activity. The antimicrobial activity was tested against 16 multi-resistant pathogenic bacteria isolated from human source material. The essential oils of plant species used in the present study can be considered as potential source of antimicrobial substances and may contribute to solution of bacterial multiresistance [53].

7. COMPOSITION AND BIOLOGICAL ACTIVITY OF ESSENTIAL OILS AND EXTRACTS OF WILD GROWING ROSACEAE SPECIES

Investigation was carried out on air-dried flowers, leaves, stem-bark and wood of *Prunus mahaleb* L. (wild growing trees near Sinj, sout Croatia). The main component that was isolated from bark volatiles is coumarin with 34.1%, while in the wood is hexadecanoic acid with 46.0%. In the wood, eicosane was also detected with 12.9%. Hexadecanoic acid was isolated in the leaves and bark (17.8 and 9.3%, respectively). As the main compound in the flowers are isolated n-alkanes, heneicosane with 22.1% and octacosane with 13.0%. Phytol was isolated in the leaves with 5.1%. In addition to the main components, three fatty acids—dodecanoic, tetradecanoic and linoleic acids—were presented in all samples [56].

Geum rhodopeum Stoj. & Stef. is representative of the East Moesian endemic taxa of the Balkans floristic subregion. Aerial parts of *G. rhodopeum* were collected at Prestojceva mahala (Cemernik Mountain, Serbia). The major component of aerial parts from this essential oil was α-bisabolol with 12.7% constituting of the oil. It is documented that α-bisabolol may be a useful therapeutic candidate for the treatment of skin inflammation. Significant compounds were also myrtenal (9.5%), 1-isopropyl-4,8-dimethylspiro[4.5]dec-8-en-7-one (7.7%), palmitic acid (6.4%), myrtanal (5.8%) and 1-octen-3-ol (5.3%) [57]. The major component of aerial part from *G. coccineum* Sibth. et Sm. essential oil (collected from natural populations of Jablanica Mountain, Macedonia) was phytol constituting 24.3% of the oil [58]. According to the literature data, dewberry (*Rubus caesius* var. *aquáticus* Weihe. & Nees.) leaves have been used in traditional medicine due to their anti-inflammatory, antiviral and antimicrobial, antiproliferative activity against cancer cells and antitumor and wound-healing properties. Leaf methanol, ethanol, acetone and water extracts from dewberry collected near city of Zaječar (Serbia) were examined for its antioxidant activity using DPPH, ABTS, ferric reducing power (FRAP) assay, total reducing power (TRP) methods. Acetone extract was with the highest antioxidant activity against DPPH and ABTS•+ radicals as well as for total reducing capacity. Obtained results showed that phenolic compounds are major contributors to the antioxidant properties of *R. caesius* var. *aquaticus*) leaves [59].

8. CONCLUSION

According to numerous references, many aromatic plants from Western Balkans represented valuable sources of potential new bioactive compounds. Considerable efforts were made for investigation of essential oils and extracts and discovering new, natural antimicrobials, antioxidants or cytotoxic agents. The results encourage the application of the plants for further evaluations of other possible bioactivities and detection of active pure compounds as constituents of the essential oils and extracts. Studies to date have identified a number of plant compounds and explained its mechanism of action in organisms. All presented data additionally validate the use of well-known aromatic plants in new treatments, as well as endemic or rare plants which could be only scientifically investigated. High protection of their natural localities is necessary, and new potential active compounds from those plants could be used only with their possible cultivation. The beneficial effect of the aromatic plants from Western Balkans that are recognised in traditional knowledge could be useful for conventional medicine or other aspects for improving life quality.

9. ACKNOWLEDGEMENTS

The authors are grateful to the Ministry of Education, Science and Technological Development of the Republic of Serbia for financial support (Grant No. 173029).

REFERENCES

1. Thirumagal A, Sivasubramanian S. Bibliometric analysis medicinal and aromatic plants research. Library Philosophy and Practice. Academic OneFile; 2014. http://digitalcommons.unl.edu/libphilprac/1129
2. World Health Organization (WHO). General guidelines for methodologies on research and evaluation of traditional medicine. World Health Organization: Geneva; 2000. 1 p. (document reference WHO/EDM/TRM/2000.1)
3. Cunningham AB. Applied ethnobotany: people, wild plant use and conservation. Earthscan: London; 2001.
4. Redzić S. The ecological aspect of ethnobotany and ethnopharmacology of population in Bosnia and Herzegovina. Collegium Antropologicum. 2007;31:869–890.
5. Miler D. Natural medicine. 1st ed. Metaphysica: Belgrade; 2009.
6. Stevens PF. (2001 onwards). Angiosperm Phylogeny Website [Internet]. 2012. Available from: http://www.mobot.org/MOBOT/research/APweb/. [Accessed: 2016-15-09]
7. Cox PA, Balick MJ. The ethnobotanical approach to drug discovery. Scientific American. 1994;270(6):82–87.
8. Turrill WB. The plant-life of the Balkan Peninsula. Oxford: Leipzig; 1929.
9. Sarić M. The medicinal plants of the Federal Republic of Serbia. SAAS: Belgrade; 1989.
10. Sarić-Kundalić B, Dobes C, Klatte-Asselmeyer V, Saukel J. Ethnobotanical study on medicinal use of wild and cultivated plants in middle, south and west Bosnia and Herzegovina. Journal of Ethnopharmacology. 2010;131:33–55.
11. Nikolić T. Sistematska botanika: Raznolikost i evolucija biljnog svijeta. [Systematic botany: Diversity and evolution of plant life]. 1th ed. Alfa d.d: Zagreb; 2013. 20 p.
12. Rohlena J. Conspectus Florae Montenegrinae. Preslia; 1942. 20p.
13. Stešević D, Petrović D, Bubanja N, Vuksanović S, Biberdžić V. Contribution to the flora of Montenegro (Supplementum to the material for Vascular flora of Montenegro). Natura Montenegrina. 2008;7(3):463–480.

14. Niketić M, Tomović G, Melovski L, Stevanović V, Matevski V. New species for the vascular flora of Republic of Macedonia and their distribution in the Balkan Peninsula. Botanica Serbica. 2014;38(1):57–67.
15. Barata AM, Rocha F, Lopes V, Bettencourt E, Figueiredo AC. Medicinal and Aromatic Plants—Portugal. Eolss Publishers: Oxford; 2014.
16. Łuczaj Ł, Fressel N, Perković S. Wild food plants used in the villages of the Lake Vrana Nature Park (northern Dalmatia, Croatia). Acta Societatis Botanicorum Poloniae. 2013;82(4):275–281.
17. Jarić S, Mačukanović-Jocić M, Djurdjević L, Mitrović M, Kostić O, Karadžić B, Pavlović P. An ethnobotanical survey of traditionally used plants on Suva planina mountain (south-eastern Serbia). Journal of Ethnopharmacology. 2015;175:93–108.
18. Jarić S, Popović Z, Mačukanović-Jocić M, Djurdjević L, Mijatović M, Karadžić B, Mitrović M, Pavlović P. An ethnobotanical study on the usage of wild medicinal herbs from Kopaonik Mountain (Central Serbia). Journal of Ethnopharmacology. 2007;111:160–175.
19. European Pharmacopoeia 7.0. European directorate for the quality of medicines and healthcare. Council of Europe EDQM: Strasbourg, France; 2011.
20. Šavikin K, Zdunić G, Menković N, Živković J, Ćujić N, Tereščenko M, Bigović D. Ethnobotanical study on traditional use of medicinal plants in South-Western Serbia, Zlatibor district. Journal of Ethnopharmacology. 2013;146:803–810.
21. Menković N, Šavikin K, Tasić S, Zdunić G, Stešević D, Milosavljević S, Vincek, D. Ethnobotanical study on traditional uses of wild medicinal plants in Prokletije Mountains (Montenegro). Journal of Ethnopharmacology. 2011;133(1):97–107.
22. Pieroni A. Local plant resources in the ethnobotany of Theth, a village in the Northern Albanian Alps. Genetic Resources and Crop Evolution. 2008;55:1197–1214.
23. Pieroni A, Giusti MA, Quave CL. Cross-cultural ethnobiology in the Western Balkans: medical ethnobotany and ethnozoology among Albanians and Serbs in the Pešter Plateau, Sandžak, South-Western Serbia. Human Ecology: An Interdisciplinary Journal. 2011:39(3):333–349.
24. Rexhepi B, Mustafa B, Hajdari A, Rushidi-Rexhepi J, Quave CL, Pieroni A. Traditional medicinal plant knowledge among Albanians, Macedonians and Gorani in the Sharr Mountains (Republic of Macedonia). Genetic Resources and Crop Evolution. 2013;60:2055–2080.
25. Gursky Z. Zlatna knjiga ljekovitog bilja. Nakladni zavod matice Hrvatske: Zagreb; 1978.
26. Pieroni A, Giusti ME. The remedies of the folk medicine of the Croatians living in Ćićarija, northern Istria. Collegium Antropologicum. 2008;32:623–627.
27. Stupar M, Ljaljević-Grbić M, Džamić A, Unković N, Ristić M, Jelikić A, Vukojević J. Antifungal activity of selected essential oils and biocide benzalkonium chloride against the fungi isolated from cultural heritage object. South African Journal of Botany. 2014;93:118–124.
28. Coccimiglio J, Alipour M, Jiang Z-H, Gottardo C, Suntres Z. Antioxidant, antibacterial, and cytotoxic activities of the ethanolic *Origanum vulgare* extract and its major constituents. Oxidative Medicine and Cellular Longevity. 2016;2016:1–8.
29. Džamić AM, Soković MD, Novaković M, Jadranin M, Ristić MS, Tešević V, Marin PD. Composition, antifungal and antioxidant properties of *Hyssopus officinalis* L. subsp. *pilifer* (Pant.) Murb.essential oil and deodorized extracts. Industrial Crops and Products. 2013;51:401–407.
30. Rozman V, Kalinovic I, Liška A. Bioactivity of 1,8-cineole, camphor and carvacrol against rusty grain beetle (*Cryptolestes ferrugineus* Steph.) on stored wheat. Proceedings of the 9th International Working Conference on Stored Product Protection; 15–18 October 2006; Campinas, São Paulo, Brazil. Brazilian Post-harvest Association—ABRAPOS, Passo Fundo, RS, Brazil, 2006. pp. 688–694.
31. Grujic-Jovanovic S, Dzamic A, Ristic MS, Marin PD. Essential oil composition of *Stachys anisochila*. Chemistry of Natural Compounds. 2011;47(5):823–825.

32. Bukvički D, Stojković D, Soković M, Vannini L, Montanari C, Pejin B, Savić A, Veljić M, Grujić S, Marin PD. *Satureja horvatii* essential oil: in vitro antimicrobial and antiradical properties and in situ control of *Listeria monocytogenes* in pork meat. Meat Science.2014;96(3):1355–1360.
33. Kundaković T, Stanojković T, Kolundzija B, Marković S, Sukilović B, Milenković M, Lakušić B. Cytotoxicity and antimicrobial activity of the essential oil from *Satureja montana* subsp. *pisidica* (Lamiceae). Natural Product Communications. 2014;9(4):569–572.
34. Ćavar S, Šolić EM, Maksimović M. Chemical composition and antioxidant activity of two *Satureja* species from Mt. Biokovo. Botanica Serbica. 2013;37(2):159–165.
35. Bukvicki D, Stojkovic D, Sokovic M, Nikolic M, Vannini L, Montanari C, Marin PD. Potential application of *Micromeria dalmatica* essential oil as a protective agent in a food system. LWT-Food Science and Technology. 2015;63:262–267.
36. Menković N, Gođevac D, Šavikin K, Zdunić G, Milosavljević S, Bojadži A, Avramoski O. Bioactive compounds of endemic species *Sideritis raeseri* subsp. *raeseri* grown in National Park Galičica. Records of Natural Products. 2013;7(3):161–168.
37. Zeljković ČS, Maksimović M. Chemical composition and bioactivity of essential oil from *Thymus* species in Balkan Peninsula. Phytochemical Review. 2015;14:335–352.
38. Petrović NV, Petrović SS, Ristić MS, Džamić AM, Ćirić AD. Chemical composition and antimicrobial activity of *Thymus praecox* Opiz ssp. *plytrychus* essential oil from Serbia. Records of Natural Products. 2016;10(5):633–638.
39. Capetanos C, Saroglou V, Marin P, Simić A, Skaltsa H. Essential oil analyses of two endemic *Eryngium* species from Serbia. Journal of Serbian Chemical Society. 2007;72:961–965.
40. Kovacevic NN, Marcetic MD, Lakusic DV, Lakusic BS. Composition of the essential oils of different parts of *Seseli annuum* L. (Apiaceae). Journal of Essential Oil Bearing Plants. 2016;19(3):671–677.
41. Matejić J, Džamić A, Mihajilov-Krstev T, Ranđelović V, Krivošej Z, Marin PD. Total phenolic content, flavonoid concentration, antioxidant and antimicrobial activity of methanol extracts from three *Seseli* L. taxa. Central European Journal of Biology. 2012;7(6):1116–1122.
42. Matejić J, Džamić A, Mihajilov-Krstev T, Ranđelović V, Krivošej Z, Marin PD. Total phenolic and flavonoid content, antioxidant and antimicrobial activity of extracts from *Tordylium maximum*. Journal of Applied Pharmaceutical Science. 2013;3(1):055–059.
43. Matejić J, Džamić A, Ristić M, Ranđelović V, Marin P. Essential oil composition of *Cachrys cristata*—a rare and endangered species in the flora of Serbia. Natural Products Communications. 2012;7:1–2.
44. Matejić J, Džamić A, Ćirić A, Ranđelović V, Krivošej Z, Marin P. Antioxidant and antimicrobial activities of extracts of four *Peucedanum* L. species. Digest Journal of Nanomaterials and Biostructures. 2013;8(2):655–665.
45. Matejic JS, Dzamic AM, Mihajilov-Krstev T, Ristic MS, Randjelovic VN, Krivošej ZĐ, Marin PD. Chemical composition, antioxidant and antimicrobial properties of essential oil and extracts from *Heracleum sphondylium* L. Journal of Essential Oil Bearing Plants. 2016;19(4):944–953.
46. Mileski KS, Ćirić AD, Trifunović SS, Ristić MS, Soković MD, Matevski VS, Tešević VV, Jadranin MB, Marin PD, Džamić AM. *Heracleum orphanidis*: chemical characterisation, and comparative evaluation of antioxidant and antimicrobial activities with specific interest in the influence on *Pseudomonas aeruginosa* PAO1. Food and Function. 2016;7:4061–4074.
47. Mihajilov-Krstev TM, Jovanović B, Jović JL, Ilić BS, Miladinović DL, Matejić JS, Rajković JS, Djordjević L, Cvetković VJ, Zlatković BK. Antimicrobial, antioxidative, and insect repellent effects of *Artemisia absinthium* essential oil. Planta Medica. 2014;80(18):1698–1705.
48. Stanković N, Mihajilov-Krstev T, Zlatković B, Matejić J, StankovJovanović V, Kocić B, Čomić L. Comparative study of composition and antioxidant and antimicrobial

activities of essential oils of selected aromatic plants from Balkan Peninsula. PlantaMedica. 2016;82(7):650–661.

49. Mastelić J, Jerković I, Mesić M. Volatile constituents from flowers, leaves, bark and wood of *Prunus mahaleb* L. Flavour and Fragrance Journal. 2006;21:306–313.

50. Miladinovic DL, Ilic BS, Matejic JS, Randjelovic VN, Nikolic DM, Mihajilov-Krstev TM, Mladenovic IO. Chemical composition of the essential oil of *Geum rhodopeum*. Chemistry of Natural Compounds. 2014;50(5): 926–928.

51. Miladinovic DL, Ilic BS, Matejic JS, Randjelovic VN, Nikolic DM. Chemical composition of the essential oil of *Geum coccineum*. Chemistry of Natural Compounds. 2015;51(4): 785–786.

52. Veličković I, Grujić S, Džamić A, Krivošej Z, Marin PD. *In vitro* antioxidant activity of dewberry (*Rubus caesius* L. var. *aquáticus*Weihe. and Nees.) leaf extracts. Archives of Biological Sciences. 2015;67(4):1323–1330.

53. Matejić JS, Džamić AM, Mihajilov-Krstev TM, Ranđelović VN, Mileski KS, Marin PD. Total phenolic and flavonoid contents and biological activities of *Cachrys cristata* Dc. extracts. Archives of Biological Science. 2014;66(3):1117–1123.

54. Mileski K, Džamić A, Ćirić A, Grujić S, Ristić M, Matevski V, Marin PD. Radical scavenging and antimicrobial activity of essential oil and extracts of *Echinophora sibthorpiana* Guss. from Macedonia. Archives of Biological Sciences. 2014;66(1):401–413.

55. Mileski K, Džamić A, Ćirić A, Ristić M, Grujić S, Matevski V, Marin PD. Composition, antimicrobial and antioxidant properties of endemic species *Ferulago macedonica* Micevski and E. Mayer. Records of Natural Products. 2015;9(2):208–223.

56. Mihajilov-Krstev T, Radnović D, Kitić D, Stankov-Jovanović V, Mitić V, Stojanović-Radić Z, Zlatković B. Chemical composition, antimicrobial, antioxidative and anticholinesterase activity of *Satureja montana* L. ssp *montana* essential oil. Central European Journal of Biology. 2014;9(7):668–677.

57. Petrović NV, Petrović SS, Džamić AM, Ćirić AD, Ristić MS, Milovanović SL, Petrović SD. Chemical composition, antioxidant and antimicrobial activity of *Thymus praecox* supercritical extracts. Journal of Supercritical Fluids. 2016;110:117–125.

58. Vladimir-Knežević S, Kosalec I, Babac M, Petrović M, Ralić J, Matica B, Blažeković B. Antimicrobial activity of *Thymus longicaulis* C. Presl essential oil against respiratory pathogens. Central European Journal of Biology. 2012;7(6):1109–1115.

59. Stupar M, Ljaljević-Grbić M, Džamić A, Unković N, Ristić M, Vukojević J. Antifungal activity *Helichrysum italicum* (Roth) G. Don (Asteraceae) essential oil against the fungi isolated from cultural heritage objects. Archives of Biological Sciences.2014;66(4):1539–1545.

CHAPTER 7

Gas Exchange and Growth of Medicinal Plant Subjected to Salinity and Application of Biofertilizers

Sergiane Beatriz da Silva Mesquita[1], José Adriano da Silva[2], Rafael Santiago da Costa[1], Móises Wilkson Nunes dos Santos[1], Claudivan Feitosa de Lacerda[2], Aiala Vieira Amorim[1], Antonio Marcos Esmeraldo Bezerra[2]

[1]*Universidade da Integração Internacional da Lusofonia Afro-Brasileira (UNILAB), Campus da Liberdade, Redenção, Brasil*

[2]*Universidade Federal do Ceará (UFC), Fortaleza, Brasil*

ABSTRACT

The objective of this study was to evaluate the use of biofertilizers and saline waters on gas exchange and growth of medicinal plant Plectrantus amboinicus. The experiment was conducted in the period February to May 2013 in a greenhouse. The experimental design was completely randomized in a 2 × 4 factorial arrangement, with two levels of salinity of irrigation water (ECw: 0.7 and 3.1 dS m^{-1}) and four levels of bovine liquid biofertilizer applied to the soil, corresponding to 0%, 10%, 20% and 30% of the soil volume, with five replications. The experiment lasted 60 days, counted from the beginning of the treatments. The stomatal conductance (gs), photosynthesis (A), transpiration (E), intrinsic water use efficiency (WUEi) were performed at the end of the experiment, and the height, number of leaves and stem diameter at the beginning and at the end. Generally plants subjected to salinity of irrigation water of 3.1 dS m^{-1} had the lowest values of gas exchange. Moreover, the application of biofertilizers and the interaction between this and salinity did not affect any growth variable studied except the stem length in the final phase which was influenced by salinity at 5% probability by F test. The average values of this variable were 57.22 cm and 69.65 cm when applied water ECw: 0.7 to 3.1 dS m^{-1}, respectively. The application of biofertilizers can reduce the effect of salinity on the final plant height of Plectrantus amboinicus, especially when the plants were fertilized with a dose of 20% of biofertilizers.

Keywords: Medicinal Plants, Salt Stress, Fertilizers

1. INTRODUCTION

High concentration of salts is an important characteristic of the soils of semi-arid and arid regions of the world and is a continuing threat to crop production [1] -[3] . It has been estimated that high soil salinity brought about by mismanaged irrigation systems undermines the yield productivity of at least one-third of the world's irrigation lands [4] . Recent trends and future projections suggest that the need to produce more food and fiber for the expanding population will lead to an increase in the use of salt-prone water and land resources for crop-production systems, and this will be met by using salt-tolerant crops or halophytes [5] [6] . Biosaline agriculture is now becoming a reliable strategy for using saline environment [7] .

As [8] there are several effects caused by salt stress in plants: the osmotic effect, the toxic effect of mineral elements (chlorides, boron, sodium) and the indirect effects, which occur when high concentrations of sodium or other cations in solution interfere with the physical condition of the soil or the availability of certain elements, affecting the growth and development of plants indirectly. The excess salts from irrigation water cause a reduction in photosynthetic rates, stomatal conductance and transpiration of plants. In addition to the gas exchange variables [9] found declines in chlorophyll levels, which result from imbalances in the physiological and biochemical activities promoted by the content of salts, above the limit tolerated by crops. However, saline soils can be utilized by growing salt tolerant crops. For example, A. majus is a potential medicinal crop, it could be grown on salt-affected lands if it possesses high degree of salt tolerance [10] .

One of the alternatives that can minimize the deleterious effects of salinity and other adverse effects of the environment is the use of organic substances. The organic material functions as an organic soil amendment, since it can reduce the exchangeable sodium percentage (ESP) due to the release of CO_2 and the production of organic acids during their decomposition, increase water retention and acts as a source of calcium and magnesium rather than the sodium [11] . An example of this was observed by [12] , to detect the mitigating action of bovine biofertilizer in plants of paluma guava (Psidium guajava) subjected to salinity of irrigation water. However, little is known about the use of biofertilizers on medicinal plants.

Of the potential medicinal plants being cultivated these days, Plectranthus amboinicus (Lour) Spreng, belonging to family Lamiacea, is a traditional medicinal herb which is also known as Karpuravalli, Omavalli in Tamil, Patta ajavayin, Patharcur in Hindi, Country borage in English [13] . This species is a large succulent aromatic perennial herb which together with Plectranthus barbatus have the widest geographical range of family Lamiacea occurring beyond Africa and Asia continents into the Americas [13] . The species Plectranthus amboinicus has a variety of properties. The leaves have been used in malarial fever, hepatopathy, renal and vesical calculi, cough, chronic asthma, hiccough, bronchitis, anthelmintic, colic and convulsions [13] , are often rubbed into the hair and body after bathing [14] , are mixed with sugar and used as an intoxicant [15] , used as insect repellants [16] , in food stuffings [17] , for flavouring and marinating beef and chicken [18] -[22] , to mask odor of strong smells associated with goat, fish and shellfish [14] . Also this plant is offered to the spirits when a house is being built [14] .

Based on the foregoing, the aim of the present study was to evaluate the use of biofertilizers and salt on gas exchange and development of medicinal plants of the Plectranthus amboinicus.

2. MATERIALS AND METHODS

2.1. The Experiment Site

The experiment was conducted in a greenhouse owned by the Center for Teaching and Research in Urban Agriculture (NEPAU), Federal University of Ceará-UFC, located in Fortaleza-CE, latitude 3°44' S, longitude 38°33' W, in altitude of 20 m during the period from February to May 2013. The climate is Aw (rainy tropical) according to Köeppen classification. The average values of temperature and relative humidity inside the greenhouse were 30.02°C and 79.5%, respectively.

2.2. Plant Material

For this study we used the species (Plectranthus amboinicus (Lour) Spreng). The seedlings of this species were produced by vegetative propagation (cuttings) and placed in polyethylene bags containing a mixture of substrate as aloof + earthworm humus, in a 2:1 ratio, where they will remain for a period of 30 days. Subsequently, seedlings were selected for uniformity, height and diameter of the pile, and transplanted to plastic pots with a capacity of 7.5 L, containing the same substrate used for the training of seedling. The characteristics of the substrate were pH in H_2O (1:2.5) = 6.7; Ca^{2+} = 5.1 $cmol_c/dm^3$; Mg^{2+} = 3.6 $cmol_c/dm^3$; Na = 1.39 $cmol_c/dm^3$; K^+ = 2.15 $cmol_c/dm^3$; SB = 12.25 $cmol_c/dm^3$; H + Al = 1.65 $cmol_c/dm^3$; Al^{3+} = 0.00; V% = 88.13; P = 493.84 mg/dm^3; C = 1.43%; M.O. = 2.46%.

Before application of treatments the seedlings were kept for 10 days in greenhouse with 50% brightness, in order to restore the stress suffered at transplanting and started to issue new roots and leaves. The experiment lasted 60 days, counted from the beginning of the application of the treatments, the plants were kept in a greenhouse with 50% brightness.

2.3. Experimental Design and Treatments

The experiment was conducted according to a completely randomized design in a 2 × 4 factorial arrangement, with two levels of salinity of irrigation water ECw: 0.7 and 3.1 dS m^{-1} and four levels of biofertilizers applied liquid bovine to the ground, corresponding to (0, 10, 20 and 30%) of the volume of the soil, with five replications.

Water sources used were from two wells, one with electrical conductivity of 0.7 dS m^{-1} and the other with ECw of 3.1 dS m^{-1}. Irrigation was performed every two days and the amount of water applied was estimated with the objective of soil to reach field capacity, adding a leaching fraction of 0.15 for the water percole the bottom of the vessels in order to avoid excessive accumulation of salts. The application of water was manual, localized way to prevent direct effects on the leaves.

The biofertilizer was prepared by anaerobic fermentation of manure in black plastic pots containing fresh water at a ratio of 50% (volume/volume = v/v) for a period of thirty to sixty days. The results of three chemical analyzes of biofertilizers performed throughout the experiment are shown in Table1

2.4. Variables Evaluated

At the end of the experiment were performed on fully expanded leaf measurements of stomatal conductance (gs), net photosynthetic rate (A) and transpiration rate (E), using an infrared gas analyzer (IRGA) model Li-6400XT (Licor, USA). The measurements were conducted between 08:00 and 12:00h under natural conditions of air temperature, and CO_2 concentration, and photosynthetic active radiation of 1600 μmol m^{-2} s^{-1}. From the data obtained was estimated intrinsic water use efficiency (A/gs). During the measurements of gas exchange is also estimated the IRC, using laptop SPAD 502 (Minolta).

Height was determined with the aid of a graduated ruler, the number of leaves was determined by manual counting and stem diameter using a caliper.

2.5. Statistical Analyzes

The results were statistically analyzed using the "ASSISTAT 7.6 BETA" program. The data of analyzed variables were subjected to analysis of variance and subsequently as significant by the F test, submitted to Tukey's

Table 1 . Chemical attributes of biofertilizer used in the experiment.

Elements	1st Application	2nd Application	3rd Application	units
N	0.7	1.1	1.0	$g{\cdot}kg^{-1}$
P	0.1	0.3	0.3	$g{\cdot}kg^{-1}$
P_2O_5	0.2	0.7	0.7	$g{\cdot}kg^{-1}$
K	1.5	1.7	1.5	$g{\cdot}kg^{-1}$
K_2O	1.8	2.1	1.8	$g{\cdot}kg^{-1}$
Ca	2.0	1.6	1.5	$g{\cdot}kg^{-1}$
Mg	0.5	0.8	0.6	$g{\cdot}kg^{-1}$
Fe	28.1	69.8	69.1	$mg{\cdot}kg^{-1}$
Cu	0.6	0.3	0.9	$mg{\cdot}kg^{-1}$
Zn	6.7	17.9	20.4	$mg{\cdot}kg^{-1}$
Mn	5.3	12.1	14.0	$mg{\cdot}kg^{-1}$

$p < 0.05$. In regression analysis, the equations that best fit the data were selected based on the significance of the regression coefficients at 1% and 5% probability by F test and greater coefficient of determination.

3. RESULTS

3.1. Gas Exchange

In the analysis of variance shown in Table 2, it can be observed that with the exception of chlorophyll content all gas exchange variables were influenced by salinity. The biofertilizer and biofertilizer/salinity interaction did not affect the chlorophyll content, and intrinsic water use efficiency (WUEi), while stomatal conductance (gs), photosynthesis (A) and transpiration (E), were influenced by these two factors, the significance level of 1% probability by F test.

Observe in Figure 1 that generally plants subjected to salinity of irrigation water from 3.1 dS m^{-1}had the lowest values of gas exchange. However, it appears that plants irrigated with irrigation conductivity of 0.7 dS m^{-1} water suffered an abrupt decrease when applied 30% of biofertilizer, which did not occur in plants under higher salinity.

3.2. Growth

In the analysis of variance presented in Table 3, it can be observed that no variable was influenced by salinity, biofertilizer or the interaction between these two factors, with exception of the height at the end of the experiment which was influenced by salinity at a significance level of 1% probability by F test. This shows the impossibility of detecting the benefit of the use of biofertilizers in alleviating the effect of salinity on plants of the Plectranthus amboinicus, at least at the levels used in the present study.

It is observed in Figure 2 that the plants Plectrantus amboinicus collected at the end of the experiment, independent of the electrical conductivity of the irrigation water, had lower height values when doses were fertilized with 30% of biofertilizers. The highest value for this variable was observed in plants grown with 3.1 ECw water and fertilized with 20% of biofertilizers.

4. DISCUSSION

Similar to that observed in this study, other authors have observed that salinity reduces gs, A and E in most plant species [23] -[25] . Salt-induced reduction of A can be caused by stomatal limitation with stomatal closure [26] [27]

Table 2. Values summarized the analysis of variance for stomatal conductance (gs), rates of net photosynthesis (A), transpiration (E), chlorophyll, intrinsic water use efficiency (WUEi) in plants of Plectrantus amboinicus under different levels of biofertilizers and water salinity irrigation.

Sources of variation	DL	Medium Square				
		gs	*A*	E	Chlorophyll	WUEi
Salinity (A)	1	0.11^{**}	129.70^{**}	55.13^{**}	0.11^{ns}	4532.85^{**}
Biofertilizer (B)	3	0.01^{**}	34.76^{**}	8.27^{**}	2.20^{ns}	46.17^{ns}
Int. (A × B)	3	0.02^{**}	45.90^{**}	13.44^{**}	10.75^{ns}	207.30^{ns}
Treatments	7	0.03^{**}	53.10^{**}	17.18^{**}	5.56^{ns}	756.18^{**}
Residue	32	0.0027	4.64	1.48	16.32	210.16
CV%	-	46.40	37.20	36.11	13.88	25.58
Salinity 0.7 dS m^{-1}	-	0.16a	7.59a	4.55a	29.16a	46.03b
Salinity 3.1 dS m^{-1}	-	0.06b	3.99b	2.20b	29.05a	67.32a
Biofertilizer 0%	-	0.13a	7.04a	3.76a	29.27a	59.35a
Biofertilizer 10%	-	0.14a	6.96a	4.06a	29.57a	56.64a
Biofertilizer 20%	-	0.12a	6.07a	3.65a	28.46a	56.62a
Biofertilizer 30%	-	0.05b	3.07b	2.04b	29.13a	54.09a

**, * and ns—significant at 1% and 5% of probability and not significant by F test, respectively. CV—coefficient of variation in percent. DF—Degrees of liberty. Means followed by the same lower case letter in columns do not differ by Tukey test ($p < 0.01$ and $p < 0.05$).

Table 3. Values summarized analysis of variance for the height, number of leaves and stem diameter of plants of Plectrantus amboinicus under different levels of biofertilizers and salinity of irrigation water.

Sources of variation	DL	Medium Square					
		Height		Number of leaves		Stem diameter	
		initial	end	initial	end	initial	end
Salinity (A)	1	2.50^{ns}	1543.80^{*}	0.02^{ns}	90.00^{ns}	0.12^{ns}	1.52^{ns}
Biofertilizer (B)	3	11.84^{ns}	251.02^{ns}	1.49^{ns}	35.00^{ns}	0.48^{ns}	0.63^{ns}
Int. (A × B)	3	11.61^{ns}	166.83^{ns}	1.49^{ns}	490.20^{ns}	1.19^{ns}	1.20^{ns}
Treatments	7	10.41^{ns}	399.62^{ns}	1.28^{ns}	237.94^{ns}	0.73^{ns}	1.00^{ns}
Residue	32	14.51	212.03	2.07	245.00	0.80	0.82
CV%	-	20.24	22.95	22.60	19.49	13.61	11.62
Salinity 0.7 dS m^{-1}	-	18.57a	57.22b	6.40a	81.90a	6.64a	7.61a
Salinity 3.1 dS m^{-1}	-	19.07a	69.65a	6.35a	78.90a	6.53a	8.00a
Biofertilizer 0%	-	19.85a	60.85a	6.10a	82.20a	6.61a	7.44a
Biofertilizer 10%	-	17.55a	63.00a	6.00a	78.50a	6.61a	7.94a
Biofertilizer 20%	-	18.30a	70.60a	6.80a	81.80a	6.83a	7.87a
Biofertilizer 30%	-	19.60a	59.30a	6.60a	79.10a	6.29a	7.99a

**, * and ns—significant at 1% and 5% of probability and not significant by F test, respectively. CV—coefficient of variation in percent. DF—Degrees of liberty. Means followed by the same lower case letter in columns do not differ by Tukey test ($p < 0.01$ and $p < 0.05$).

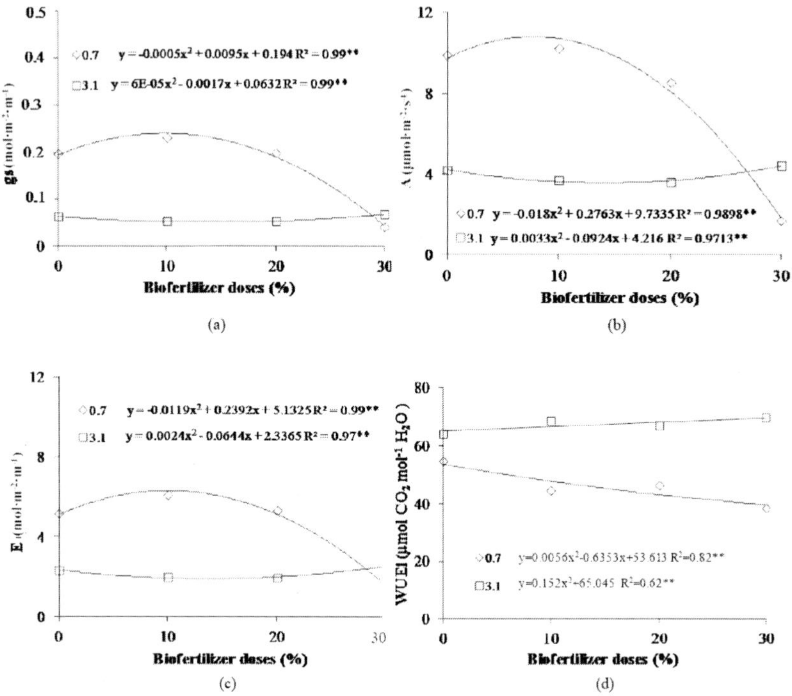

Figure 1. Stomatal conductance (gs), rates of net photosynthesis (A), transpiration rate (E), intrinsic water use efficiency (WUEi) in plants of Plectrantus amboinicus under two levels of irrigation water salinity (ECw: 0.7 and 3.1 dS m^{-1}) and four doses of biofertilizer.

nonstomatal limitation or both limitations with stomatal closure at low tissue salt concentration, and a disturbance of photosynthetic activity at high tissue salt content [28] [29] . However, some researchers not find any significant effect of salt stress on gas exchange in herbal A. majus [10] , Hibiscus cannabinus [30] , Hordeum vulgare [31] , Trifolium repens [32] , Triticum aestivum [33][34] , and Olea europea [35] .

According [2] , inhibition of growth in plants grown under salinity may be due to osmotic effects of salts, reducing the availability of water in the soil solution and/or the excess of ions absorbed by the plant metabolism leading to toxic effects, and this phase that clearly separates species and genotypes that differ in tolerance or sensitivity to salinity. However, in the present work to fertilizing with biofertilizer at least up to 20%, seems to have alleviated the effect of salinity stress on plant height at the end of the experiment, whereas the values of this variable were higher when irrigated with saline water higher electrical conductivity.

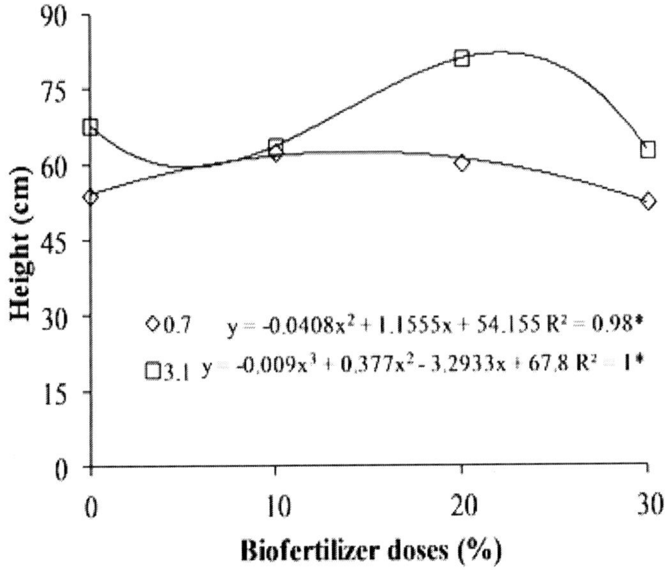

Figure 2. Plant height of Plectrantus amboinicus under two levels of salinity of irrigation water (ECw: 0.7 and 3.1 dS m^{-1}) and four doses of biofertilizer.

Excess salts can alter the physiological and biochemical functions of plants, resulting in disturbances in water relations, changes in the absorption and use of essential plant nutrients, affecting their final growth and yield. However, according to [36] , the intensity of salt stress is dependent on many factors, among them, the plant species, cultivar, developmental stage, saline medium composition, intensity and duration of stress, as well as the climatic conditions and irrigation management. This may explain in part the fact that salinity did not influence the studied plants.

Regarding the relationship beneficial for biofertilizer with salinity, some studies have shown that the use of biofertilizer in saline environments can partially mitigate the impact of salinity on the growth of plants [37] . The importance of the use of liquid biofertilizers, as simple or enriched microbial fermentation, is the quantitative elements in the diversity of chelated and made available for biological activity as an enzyme activator and plant metabolism nutrients.

5. CONCLUSIONS

The stomatal conductance, rates of net photosynthesis and transpiration showed lower values when the plants of Plectrantus amboinicus were irrigated with ECw: 3.1 dS m^{-1}. However, it can be seen that the interaction of biofertilizers to 30% with saline water benefited gas exchange of these plants.

The application of biofertilizers can reduce the effect of salinity on the final plant height of Plectrantus amboinicus, especially when the plants were fertilized with a dose of 20% of biofertilizers. This positive effect can be associated with better nutrition of the plant.

ACKNOWLEDGMENTS

Sergiane Beatriz da Silva Mesquita, José Adriano da Silva, Rafael Santiago da Costa, Móises Wilkson Nunes dos Santos, Claudivan Feitosa de Lacerda, Aiala Vieira Amorim, Antonio Marcos Esmeraldo Bezerra The authors thank CAPES and INCTSal funding this research and Unilab the granting of a research grant.

REFERENCES

1. Ashraf, M. (2002) Salt Tolerance of Cotton: Some New Advances. Critical Reviews in Plant Sciences, 21, 1-30.
2. Munns, R. (2002) Comparative Physiology of Salt and Water Stress. Plant, Cell and Environment, 25, 239-250.
3. Tester, M. and Davenport, R. (2003) Na+ Tolerance and Na+ Transport in Higher Plants. Annals of Botany, 91, 1-25.
4. Reeve, R.C. and Fireman, M. (1967) Salt Problems in Relation to Irrigation. In: Hagan, R.M., Haise, H.R. and Edminster, T.W., Eds., Irrigation of Agricultural Lands, American Society of Agronomy, Madison, 988-1008.
5. Khan, M.A., Ansari, R., Ali, H., Gul, B. and Nielsen, B.L. (2009) Panicum turgidum, a Potentially Sustainable Cattle Feed Alternative to Maize for Saline Areas. Agriculture, Ecosystems and Environment, 129, 542-546.
6. Yensen, N.P. and Biel, K.Y. (2006) Soil Remediation via Salt-Conduction and the Hypotheses of Halosynthesis and photoprotection, Tasks for Vegetation Science Series 40. Ecophysiology of High Salinity Tolerant Plants, 313-344.
7. Kafi, M., Asadi, H. and Ganjeali, A. (2010) Possible Utilization of High-Salinity Waters and Application of Low Amounts of Water for Production of the Halophyte Kochia scoparia as Alternative Fodder in Saline Agroecosystems. Agricultural Water Management, 97, 139-147.
8. Larcher, W. (2003) Physiological Plant Ecology: Ecophysiology and Stress Physiology of functional groups. 4th Edition, Springer, New York, 513.

9. Munns, R. and Tester, M. (2008) Mechanisms of Salinity Tolerance. Annual Reviews Plant Biology, 59, 651-681.
10. Ashraf, M., Mukhtar, N., Rehman, S. and Rha, E.S. (2004) Salt-Induced Changes in Photosynthetic Activity and Growth in a Potential Medicinal Plant Bishop's Weed (Ammi majus L.), Photosynthetica, 42, 543-550.
11. Baalousha, M., Motelica-Heino, M. and Coustumer, P.L. (2006) Conformation and Size of Humic Substances: Effects of Major Cation Concentration and Type, pH, Salinity, and Residence Time. Colloids and Surfaces A: Physicochemical and Engineering Aspects, 272, 48-55.
12. Cavalcante, L.F., Vieira, M.S., Santos, A. F., Oliveira, W.M. and Nascimento, J.A.M. (2010) água salina e esterco bovino líquido na formação de mudas de goiabeira cultivar paluma. Revista Brasileira de Fruticultura, 32, 251-261.
13. Kaliappan, N.D. and Viswanathan, P.K. (2008) Pharmacognostical Studies on the Leaves of Plectranthus amboinicus (Lour) Spreng. International Journal of Green Pharmacy, 2, 182-184.
14. Morton, J.F. (2010) Country Borage (Coleus amboinicus Lour.): A Potent Flavoring and Medicinal Plant. Journal of Herbs, Spices and Medicinal Plants, 1, 77-90.
15. Jain, S.K. and Lata, S. (1996) Amazonian Uses of Some Plants Growing in India. Indigenous Knowledge and Development Monitor, 4, 21-23.
16. Omolo, M.O., Okinyo, D., Ndiege, I.O., Lwande, W. and Hassanali, A. (2004) Repellency of Essential Oils of Some Kenyan Plants against Anopheles gambiae. Phytochemisty, 65, 2797-2802.
17. Purseglove, J.W. (1991) Tropical Crops. Dicotyledons. Longman Scientific and Technical. John Wiley and Sons, Inc. New York.
18. Epling (1981) P. amboinicus (Lour.) Spreng. Herbarium Specimen held at K Collected from the Pacific (Epling 18080).
19. Kuebel, K.R. and Tucker, A.O. (1988) Vietnamese Culinary Herbs in the United States. Economic Botany, 42, 413-419.
20. Bodner, C.C. and Gereau, R.E. (1988) A Contribution to Bontoc Ethnobotany. Economic Botany, 42, 307-369.
21. Craig, Mayenda (1990) P. amboinicus (Lour.) Spreng. Herbarium Specimen held at K. Collected from the Pacific (Craig & Mayenda 25).
22. Brown, D. (1997) Grenada: Isle of Spices. Herbs, 22, 6-7.
23. Gibberd, M.R., Turner, N.C. and Storey, R. (2002) Influence of Saline Irrigation on Growth, Ion Accumulation and Partitioning, and Leaf Gas Exchange of Carrot (Daucus carota L.) Annals of Botany, 90, 715-724.
24. Tezara, W., Mitchell, V., Driscoll, S.P., Lawlor, D.W. (2002) Effects of Water Deficit and Its Interaction with CO2 Supply on the Biochemistry and Physiology of Photosynthesis in Sunflower. Journal of Experimental Botany, 53, 1781-1791.
25. Ashraf, M. and Shahbaz, M. (2003) Assessment of Genotypic Variation in Salt Tolerance of early CIMMYT Hexaploid Wheat Germplasm Using Photosynthetic Capacity and Water Relations as Selection Criteria. Photosynthetica, 41, 273-280.
26. Pascale, S.D. and Barbieri, G. (1995) Effect of Soil Salinity from Long-Term Irrigation with Saline-Sodic Water on Yield and Quality of Winter Vegetable Crops. Scientia Horticulturae, 64, 145-147.
27. Goldstein, G., Drake, D.R., Alpha, C., Melcher, P., Heraux, J. and Azocar, A. (1996) Growth and Photosynthetic Responses of Scaevola sericea, a Hawaiian Coastal Shrub, to Substrate Salinity and Salt Spray. International Journal of Plant Sciences, 157, 171-179.
28. Downton, W.J.S., Loveys, B.R. and Grant, W.J.R. (1990) Salinity Effects on the Stomatal Behaviour of Grapevine. New Phytologist, 116, 499-503.
29. Yeo, A.R., Lee, K.S., Izard, P., Boursier, P.J. and Flowers, T.J. (1991) Short and Long Term Effects of Salinity on Leaf Growth in Rice (Oryza sativa L.). Journal of Experimental Botany, 42, 881-889.

30. Curtis, P.S. and Läuchli, A. (1986) The Role of Leaf Area Development and Photosynthetic Capacity in Determining Growth of Kenaf under Moderate Salt Stress. Australian Journal of Plant Physiology, 13, 553-565.
31. Rawson, H.M., Richards, R.A. and Munns, R. (1988) An Examination of Selection Criteria for Salt Tolerance in Wheat, Barley and Triticale Genotypes. Australian Journal of Agricultural Research, 39, 759-772.
32. Rogers, M.E. and Noble, C.L. (1992) Variation in Growth and Ion Accumulation between Two Selected Populations of Trifolium repens L. Differing in Salt Tolerance. Plant and Soil, 146, 131-136.
33. Hawkins, H.J. and Lewis, O.A.M. (1993) Combination Effect of NaCl Salinity, Nitrogen form and Calcium Concentration on the Growth and Ionic Content and Gaseous Properties of Triticum aestivum L. cv. Gamtoos. New Phytologist, 124, 161-170.
34. Ashraf, M. and O'Leary, J.W. (1996) Responses of Some Newly Evolved Salt-Tolerant Genotypes of Spring Wheat to Salt Stress: II. Water Relations and Gas Exchange. Acta Botanica Neerlandica, 45, 29-39.
35. Loreto, F., Centritto, M. and Chartzoulakis, K. (2003) Photosynthetic Limitations in Olive Cultivars with Different Sensitivity to Salt Stress. Plant, Cell and Environment, 26, 595-601.
36. Yeo, A.R. (1998) Predicting the Interaction between the Effects of Salinity and Climate Change on Crop Plants. Science Horticultural, 78, 159-174.
37. Sousa, G.G., Marinho, A.B., Albuqueruqe, A.H.P., Viana, T.V.A. and Azevedo, B.M. (2012) Crescimento inicial do milho sob diferentes concentraç ões de biofertilizante bovino irrigado com águas salinas. Revista Ciência Agronômica, 43, 237-245.

CHAPTER 8

Antioxidant and Antimicrobial Properties of Five Medicinal Libyan Plants Extracts

Rabia Alghazeer[1*], Hussein El-Saltani[1], Nabeel Saleh[1], Asma Al-Najjar[2], Fatma Hebail[1]

*[1]Chemistry Department, Faculty of Sciences, Tripoli University, Tripoli, Libya; *Corresponding Author: Rabia_alghazeer@yahoo.com*

[2]Soil Department, Faculty of Agriculture, Tripoli University, Tripoli, Libya

ABSTRACT

Five Libyan medicinal plants Thapsia garganica, Hammada scoparia, Euphorbia serrata, Hyoscyamus albus and Retama rateam were selected to evaluate their biological activities. Their total phenolic and flavanoid contents were assessed. The antioxidant activity was estimated using 2, 2-diphenyl-1-picrylhydrazyl (DPPH) as free radical scavenger. Their crude extracts showed reducing potential proportional to their concentration. The correlation coefficient (R^2) between antioxidant activity and their total phenolics and flavanoids content is 0.77 and 0.98 respectively. Crude aqueous, methanolic as well as alkaloids extracts of the five plants were tested against a number of G +ve and G –ve sensitive resistant (e.g. MRSA) bacteria beside some fungal species. The aqueous extracts displayed weak antibacterial activity whereas methanolic extracts were profoundly effective against both G +ve and G –ve bacteria. The extracts of E. serrata and H. scoparia were highly effective against E. coli in particular. The alkaloid-rich extracts of H. albus and H. scoparia induced remarkable bacteriostatic and fungistatic effects. The bioactive ingredients of H. scoparia, E. serrata and R. rateam extracts are shown to be potential sources of natural antioxidant and antimicrobial ingredients favoring their possible use in industrial pharmacology on large scale.

Keywords: Medicinal Plants; Flavonoids; Free Radical Scavenging Capacity; Antioxidant; Phenolics; Antibacterial; Antifungal

1. INTRODUCTION

Medicinal plants MP have been used in folk medicine in Libyan rural areas at relatively cheaper expenses than modern medicine. They have been widely used as diuretics, topical anti-inflammants, and haemostatics [1]. Plants generally produce several secondary metabolites like phenols, flavonoids, quinones, tannins, alkaloids, saponinsand sterols which are important sources of biocides and many other pharmaceutical drugs [2-8]. Medicinal plants are important in pharmacological research and drug development [9]. The most important secondary plant metabolites are phenols and flavonoids [10,11], they have distinctive biological activity as natural antioxidants which exceed many other synthetic ones [12].

Antioxidants are used to preserve food quality mainly because they arrest oxidative deterioration of lipids [13, 14]. Moreover, the exploration of naturally occurring antimicrobials for food preservation receives increasing attention due to consumer awareness of natural food products and a growing concern of microbial resistance towards conventional preservatives [15]. These factors have inspired the widespread screening of new plant species for possible medicinal and antioxidant properties, while the isolation and characterization of diverse phytochemicals [16,17].

The relationship between oxidative stress and antioxidants is very often explained in a very simple manner. First, reactive oxygen species (ROS) are reacted together as one functional entity producing destruction compounds such as secondary lipid peroxidation products. However, there are many different ROS that have separate and essential roles in normal physiology and are required for a variety of normal processes. Toxic reactive oxygen (ROS) can be produced by many sources such as ultra violet radiation (UV) and heat from sunlight. These species are substantially reactive and interact with a number of cellular molecules and metabolites thereby leading to a number of destructive processes causing cell damage and neurodegenerative diseases [18]. The aim of this study was to evaluate in vitro antioxidant as well as the antibacterial and antifungal properties of crude methanolic and alkaloid extracts of some Libyan medicinal plants.

2. MATERIALS AND METHODS

2.1. Plant Collection and Materials

Shade-dried leaves of Euphorbia serrata, Retama raetem, Thapsia garganica, Hammada scoparia and Hyoscyamus albus that commonly used in folk medicine in Libyan countryside with reference to their utilized parts were subjected to extraction. All plants were collected from different regions of Libya between spring and early summer of 2009, then classified and authenticated by taxonomists, Department of Botany, Faculty of Science, Tripoli University (Tripoli, Libya).

2.2. Microorganisms

The bacterial (Bacillus subtilis, Staphylococcus aureus, Escherichia coli, Pseudomonas aeruginosa and Salmonella typhi) and fungal (Rizochtonia solani, Trichoderma spp, Sclerotium spp and Aspergillus niger) species were kindly provided by the Microbiological Laboratory, Faculty of Pharmacy, Dr. Najat Elghariany Department of Plant protection, Faculty of Agriculture, (Tripoli University) and Mohamed Elghazally Department of Microbiology Biotechnology Research Center (Twaisha), while resistant bacteria (clinically isolated) were kindly provided from the Clinical Microbiology Laboratory, Tripoli Medical Center, (Tripoli, Libya).

2.3. Preparation of Plant Crude Extracts

Collected plants were dried at room temperature (25°C ± 2°C). The dried leaves were powdered with the aid of a clean electric blender and passed through a reasonable mesh sieve, then preserved in an air-tight dark colored glass container. Milled plant material (20 g) was soaked for 6 hours in 100 ml of distilled water in tightly sealed vessels at room temperature. The same weight of plant materials were extracted with boiling water for 10 - 15 min. The methanolic extracts were prepared by extracting the dry leaves powder (20 g) three times each with 200 ml of methanol. All crude extracts were filtered through Whatman filter paper and kept at 4°C until tested.

2.4. Phytochemical Screening

Phytochemical screening for major constituent (alkaloids, anthraquinones, coumarins, flavonoids, saponins, tannins, and terpenes) was undertaken using standard qualitative methods that has been described previously [19-23].

2.5. Alkaloids Extraction

Powdered plant materials (50 g) were extracted several times with methanol (300 ml). Methanol extraction was continued until the plant material gave a negative result for alkaloids (Mayer's test). The obtained methanolic extract was evaporated under reduced pressure at 40°C, to minimize any possible thermal degradation of the alkaloids and other thermolabile compounds. The crude alkaloid mixture was then separated from neutral and acidic materials, and water soluble ingredients. The initial extraction was carried out with aqueous acetic acid. After which, the dichloromethane extraction was carried out and basification. The aqueous solution subjected to further dichloromethane extraction thereafter [24].

2.6. Determination of Total Polyphenol Content

The phenolic containing extracts were obtained using a method described by Marinova (2005) [25]. The phenolic contents were determined with the Folin-Ciocalteu method [26]. The tenfold diluted extract was left to react with the Folin-Ciocalteu reagent (1 ml), and then the mixture was neutralized with Na_2CO_3. The mixture was thereafter incubated at room temperature for 90 min. Finally, the absorbance was measured using spectrophotometer at 750 nm. The total phenolic content was expressed as mg gallic acid equivalents (GAE)/g sample.

2.7. Determination of Total Flavonoids Content

Flavonoid content of the crude extracts was estimated according to the modified method of Marinova (2005) [25]. The diluted standard solutions or plant extract (0.5 ml) were separately mixed with 1.5 ml of 80% ethanol, 0.1 ml of 10% aluminum chloride (Merck, Germany), 0.1 ml of 1M potassium acetate (Kemika, Croatia) and 2.8 ml of distilled water. After incubation at room temperature for 30 minutes, the absorbance of the reaction mixture was measured at 415 nm. Rutin (Sigma, Germany) was used for calibration curve (25, 50 and 100 $\mu g \cdot ml^{-1}$ in 80% ethanol).

2.8. Determination of Antioxidant Activity

2.8.1. Reducing Power

The antioxidant activity was investigated using the ferric reducing antioxidant power (FRAP) assay for plant extracts was investigated according to the method of Oyaizu (1986) [27]. Plant extract of 1, 10, 50 or 100 mg in 1 ml of ethanol was mixed with 2.5 ml of phosphate buffer (2 M, pH 6.6) and 2.5 ml of potassium ferricyanide (10 g/L) then the mixture was incubated at 50°C for 20 min. 1.5 ml of trichloroacetic acid (100 g/L) was added to the mixture, which was then centrifuged at 3000 rpm for 10 min. Finally, 0.5 ml of the supernatant solution was mixed with 1.0 ml of distilled water and 0.5 ml of $FeCl_3$ (1.0 g/L) and the absorbance was measured at 665 nm. Increased absorbance of the reaction mixture indicates increasing of reducing power of plant tissue ingredients.

2.8.2. DPPH Radical Scavenging Activity

The free radical scavenging activity quantitatively tested using 2,2-diphenyl-1-picrylhydrazyl (DPPH) based on Wong (2006) method [28]. The initial absorbance of DPPH in methanol was measured at 515 nm until the absorbance remains constant. A total of 40 µl of extract was added to 3 ml of 0.1 mM methalonic DPPH solution. The mixture was incubated at room temperature for 30 min. after that the absorbance at 515 nm was measured. The percentage of inhibition was calculated using the following formula:

$$\text{Inhibition}\ (\%) = \left[\left(A_{515}\text{of control} - A_{515}\text{of sample}\right)/A_{515}\text{of control}\right] \times 100$$

The IC_{50} was calculated as the amount of antioxidants present in the sample necessary to reduce the initial DPPH concentration by 50%. All determinations were performed in triplicates.

2.9. Antibacterial and Antifungal Assessment of Plant Extracts

The antimicrobial activity of the crude and alkaloid extracts was determined by the "hole-plate diffusion methods" [29]. Each test organism was maintained on nutrient agar slant and was recovered for testing by growth in nutrient broth (Biolab, Difco) for 14 h at 37°C before streaking. Cultures were routinely adjusted to a suspension of 1×10^6 to 2×10^6 CFU/ml using pre-made calibration curve representing viable cell count ($X \times 10^6$) against OD 660 nm (Y). 150 µl aliquots of the extract were placed into wells of 8mm diameter. The plates were kept for 1h at 4°C for allowing better diffusion of the extract into the agar. Subsequently, plates were incubated at 37°C for 18 h. Vancomycin and Tetracycline (30 µg each) were used as positive control while sterile water or 70% methanol was used as negative control. Diameters of inhibition zones (DIZ) were measured in mm and the results were recorded as the mean of triplicate experiments.

For fungi, Petri dishes containing 25 ml of Sabaruod dextrose agar (SDA) medium were evenly seeded with 0.5 ml of adjusted fungal inoculum. According to CLSI M38-A2 (for filamentous fungi), 0.5McFarland standard was used as a reference to visually adjust the turbidity of the fungal suspension [30]. Holes of 8 mm diameter were punched and filled with 150 µl of plant extract, and the plates were incubated at 28°C ± 1°C for 48 hours [31].

2.10. Determination of Minimum Inhibitory Concentration (MIC)

Minimum Inhibitory Concentration (MIC) was carried out using agar two fold serial dilution assays [32]. The concentrations (mg/ml) of extracts were prepared; reference antibiotics and the solvent were also assayed. The MIC was recorded as the lowest concentration of the sample that inhibits visible growth giving clear zones of inhibition after 24 h incubation.

2.11. Statistical Analysis

Data were expressed as means ± standard deviations (SD) of triplicate determinations and then analyzed by SPSS V.16 (Statistical Program for Social Sciences, SPSS Corporation, Chicago, IL) using one way analysis of variance (ANOVA). P values < 0.05 were considered to be significant. The Pearsons correlation analysis was performed between antioxidant activity, total phenolic and flavonoids contents.

3. RESULTS AND DISCUSSION

3.1. Phytochemical Screening

Five Libyan medicinal plants were rottenly screened (as described in methodology) for their phytochemical constituents. **Table 1** shows the contents of the phytochemicals obtained from the extraction of the five plants rated from (+ve) for faint to (++++ve) for dense turbidity. Results demonstrated that, most of them are either rich or of moderate content of alkaloids and flavonoids which is the case in most of the families to which these plants belong [33]. Tannins are in almost all investigated plants except H. albus. All are of poor levels of saponins except H. albus (**Table 1**).

3.2. Total Alkaloids Content

On the dry weight basis, **Figure 1** shows the relatively high percentage of alkaloids content in both T. garganica (1.25%) and H. albus (1%), while, H. scoparia and E. serrata are of moderate content (0.63%, 0.83% respectively), however, R. raetem was comparatively poor (<0.2%).

3.3. Total Phenolics and Flavonoid Content of Tested Medicinal Plants

Table 2 summarizes the contents of total phenolics, and flavanoid contents of the different plant extracts. Phenolics present in medicinal plants have received considerable attention because of their potential antioxidant activity [34]. Methanol and ethanol were proven as effective solvents to extract antioxidant phenolic compounds [35]. H. scoparia showed the highest total phenolic content (119.76 ± 15.76 mg GAE/g DW), while T. garganica had the lowest value (28.53 ± 3.82 mg GAE/g DW). The amount of polyphenolic compounds in E. serrata was significantly higher ($P < 0.05$) compared to their amounts in H. albus.

Table 1. Preliminary phytochemical screening of extracts of the five tested Libyan medicinal plants.

Phytochemical Compounds	*Retama raetem* (*Fabaceae*)	*Hammada scoparia* (*Chenopodiaceae*)	*Thapsia garganica* (*Apiaceae*)	*Euphorbia serrata* (*Euphorbiaceae*)	*Hyoscyamus albus* (*Rhamnacea*)
Alkaloids	++++ve	+++ve	+++ve	++ve	+++ve
Flavonoids	+++ve	+++ve	++ve	+++ve	+++ve
Tannins	++ve	+++ve	+++ve	+++ve	+++ve
Terpenes	+ve	++ve	+++ve	++ve	++ve
Coumarins	++ve	+ve	+ve	++ve	-ve
Saponins	+ve	+++ve	+++ve	+ve	+ve
Anthraquinones	++ve	++ve	++ve	++ve	++ve

+++ve, rich, ++ve, moderate, +ve, poor, -ve, absent.

Table 2. Total polyphenol and flavonoid contents of methanolic extracts of the tested plants.

Extract	Polyphenols content (mg GAE g DW)	Flavonoids content (mg Rutin g DW)
H. scoparia	119.76 ± 15.76	81.07 ± 6.14
R. raetem	89.35 ± 2.1	45.23 ± 3.64
H. albus	48.54 ± 7.82	27.39 ± 0.87
T. garganica	28.53 ± 3.82	13.35 ± 1.06
E. serrata	65.93 ± 9.45	36.29 ± 4.76

DW: dry weight; Results were recorded as (mean ± SD); *mg GAE/g DW: milligram gallic acid equivalent per gram dry weight; *mg Rutin/ g DW: milligram Rutin equivalent per gram dry weight.

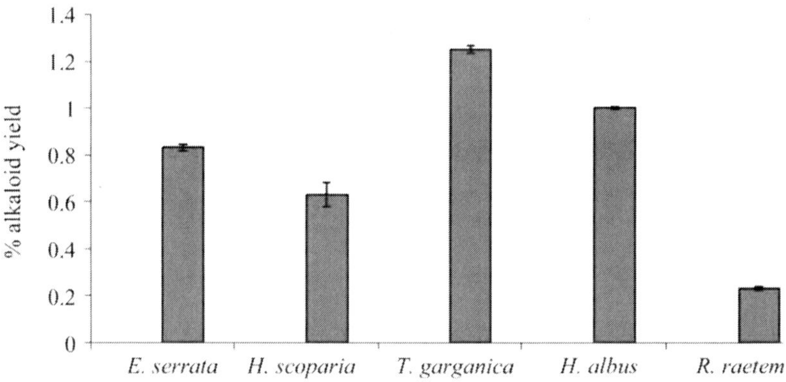

Figure 1. Percentage of alkaloid yield (mg/g plant dry weight) extracted from the five tested medicinal plants.

Flavonoids is one of the most diverse and widespread groups of natural products which are probably the most important natural phenolics. Several flavonoids have been reported to quench active oxygen species and inhibit in vitro oxidation of low-density lipoproteins [36, 37]. A variety of plant flavonoids and alkaloids have also been shown to be anticarcinogenic in several animal models [38,39]. From analysis, it can be concluded that most of these plants are rich in flavonoids except T. garganica (**Table 2**). The amount of flavonoids in the tested plants varies from 13.35 to 81.07 mg/g rutin equivalent of the crude extract. The highest flavonoid content was found in H. scoparia which was in agreement with pervious study [40], while the lowest amount was obtained in T. garganica [41]. The flavonoid contents obtained from E. serrata extract was remarkably higher than its content in R. raetem and H. albus ($P < 0.05$).

3.4. Antioxidant Activity

The antioxidant activity of the investigated botanical methanolic extracts was initially determined using the reducing power and DPPH assays.

3.4.1. Reducing Power of the Tested Plant Extracts

Determination of reducing potential can reflect some aspects of antioxidant activity in the extracts (**Figure 2**). The intensity of color resulting from reduction of ferric ions depends on the reducing potential of the compounds presented in the extracts. The intensity of the color reflects the absorption which is parallel to antioxidant activity [42]. Reducing potential was of concentrationdependent up to 1 mg/ml pattern and ranged from 0.78 in T. garganica to 1.73 in H. scoparia. The reducing power of H. scoparia extract was significantly higher as compared to H. albus, T. garganica and R. raetem extracts (P < 0.05). The obtained results are comparable with those of the total phenolic and flavonoid contents (**Table 2**), indicating that these plants represent reliable source of antioxidants.

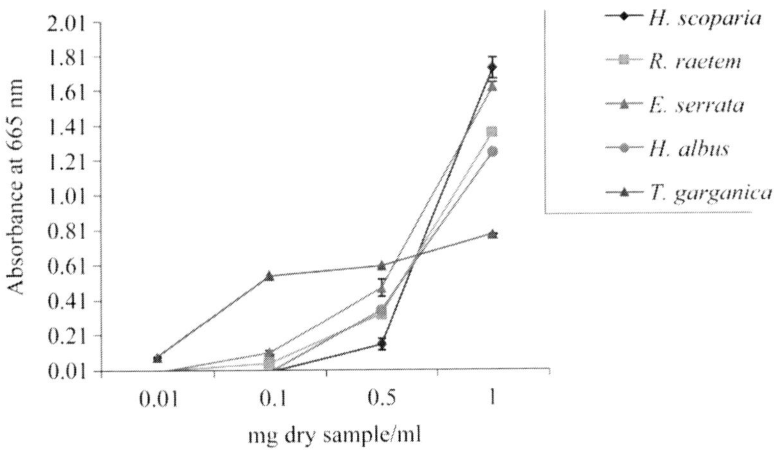

Figure 2. Reducing potential of methanolic plant extracts determined by FRAP activity assay. Data are mean ± SD.

3.4.2. Free Radicals Scavenging Activity

2,2-diphenyl-1-picrylhydrazyl (DPPH) is a commercial oxidizing stable free radical, which is readily reduced by antioxidants. Increased reduction of DPPH is related to the high scavenging activity given by particular sample [43]. The IC_{50} was calculated, and expressed as the amount of antioxidant exists in the sample necessary to decrease the initial DPPH concentration by 50%. The lower the IC_{50} value, the higher is the antioxidant activity.

Although all tested plant extracts were of observed antioxidant activity, the methanol extract of H. scoparia was able to reduce the DPPH concentration with an IC_{50} of 8 ± 1.5 µg/ml, which was stronger than that of control antioxidant, ascorbic acid, (IC_{50} = 10 ± 1.6 µg/ml) (**Figure 3**). The antioxidant activity of E. serrata and R. raetem extracts were less than H. scorpia extract, where IC_{50} of E. serrata and R. raetem extracts was five-folds higher than of that H. scorpia extract ($P < 0.05$). T. Garganica and H. albus extracts was showed the lowest antioxidant activity in comparison to tested extracts. The low antioxidant activity in DPPH assay is related to the amount of the total phenolic compounds of the plant [44]. Since, T. garganica has no considerable amounts of phenolic compound in the methanolic extract; yet it could be considered as moderate antioxidant which can be used for different purposes (**Table 2**) which is true also for E. serrata and R. raetem.

The correlation coefficient between antioxidant activity and total phenolic contents of the studied plants is $R^2 = 0.77$ (**Figure 4**), whereas the correlation coefficient between the total flavonoids and antioxidant activity is equal to 0.98 (**Figure 4**). These results revealed that 77% of the antioxidant capacity is due to the contribution of the phenolic compounds.

The distinguishable relationship between the antioxidant activity and the total phenolics may be explained in various ways, in fact, the total phenolics content does not incorporate all the antioxidants. In addition, the synergism between the antioxidants in the mixture makes the antioxidant activity not only dependant on the concentration, but also on the structure and the interaction between the contained antioxidants. This can be explained because some plants like H. albus with respectable amount of phenolic compounds (48 mg/g) give low antioxidant activity compared with T. garganica extract (**Table 2**). The results confirm that there is a direct correlation between total phenolic content and antioxidant capacity of the medicinal plants [45].

3.5. Antimicrobial and Antifungal Activities

Many medicinal plants are considered to be potential antimicrobial crude drugs as well as a source of novel compounds such as flavonoids and alkaloids with antimicrobial activity [46,47]. This expectation that some naturally occurring plant ingredients can kill antibioticresistant strains of bacteria such as Bacillus cereus, Escherichia coli, Micrococcus luteus and Staphylococcus aureus has been confirmed [48].

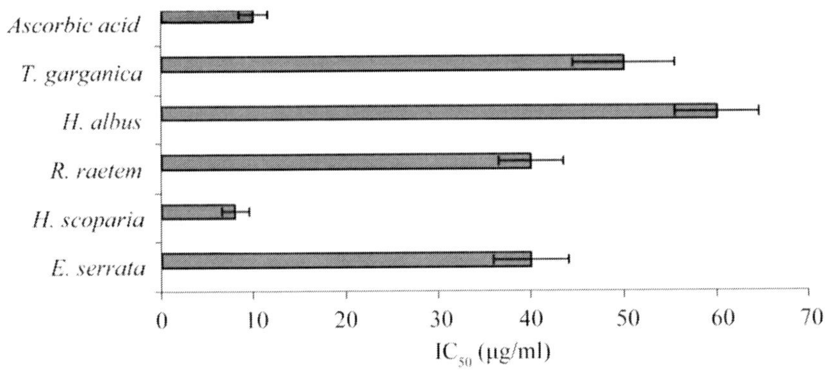

Figure 3. IC_{50} (µg/ml) values of plant extracts for free radical scavenging activity by DPPH radical (A lower IC_{50} value indicates higher antioxidant activity; ascorbic acid was used as positive control).

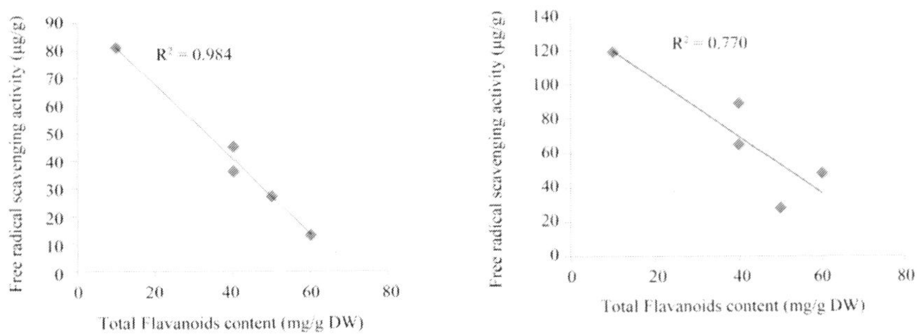

Figure 4. Linear correlation of free radical activity with respect to the total polyphenols and flavanoids content of the five tested Libyan plants.

The antimicrobial activities of medicinal plant extracts against microorganisms examined in the present study and their potency was qualitatively and quantitatively assessed by the presence or absence of inhibition zones, zone diameter and MIC values. The results obtained from testing the extracts presented in **Table 3** indicated moderate or high antimicrobial activity against some microorganisms. High antibacterial activity against most strains was attained by E. serrata and H. scoparia most probably due to their high content of phenolics and flavonoids, while the high activity of H. albus might be attributed to its high content of alkaloids The antimicrobial effects of aqueous extracts (ambient or hot) of five medicinal plant species are shown in Table 3. It can be concluded that; the activity of cold water extracts was mostly of weak to very weak against both G +ve and G –ve bacteria except H. albus and E. serrata. However, the antibacterial potential was increased notably using hot aqueous or methanol plant extracts.

The five plant extracts, in most cases, displayed remarkable antibacterial potential against all tested bacteria (**Table 3**). H. albus extract showed high antibacterial activity exceeds the reference antibiotics vancomycin and tetracycline (at 30 µg/disc) since it gives DIZ range between 26 - 32 mm on G +ve and 18 - 24 cm on G-ve bacteria. High antibacterial activity was also obtained from extracts of E. serrata and H. scoparia, which inhibited the growth of E. coli by 27 mm and 20 mm, respectively. The biological activity of H. scoparia has been reported by other study [49].

Since the antibacterial activity tests demonstrated that, both crude methanolic and/or hot water extracts are of good bactericidal properties against susceptible bacteria (with variations between species), it was advisable to test them against antibiotic-resistant clinical isolates species, including the MRSA. It was found that extract of H. scoparia, E. serrata and H. albus inhibited growth of some tested antibiotic-resistant microbes such as P. aeruginosa and MRSA. It was noteworthy that the MRSA growth was strongly inhibited by methanolic extract of E. serrata followed by H. scoparia and H. albus (**Table 4**).

Table 3. In vitro antibacterial activity of the aqueous and methanolic plant extracts against narrow spectrum of susceptible bacteria Well diameter: 8 mm; W: water at 250°C; HW: hot water [70°C - 80°C].

Extract	Solvent	Inhibition Zone (mm)			
		Gram positive		Gram negative	
		S. aur	*B. sub*	*E. coli*	*S. typhi*
E. serrata	*W*	**17**	8	**20**	14
	HW	**15**	12	**12**	**18**
	MeOH	20	25	27	22
T. garganica	*W*	8	10	10	10
	HW	12	10	11	14
	MeOH	16	18	16	13
H. scoparia	*W*	9	8	8	8
	HW	14	**17**	**19**	14
	MeOH	20	14	20	18
R. raetem	*W*	8	8	9	8
	HW	12	8	11	10
	MeOH	19	16	14	14
H. albus	*W*	**17**	**19**	15	10
	HW	**17**	**17**	20	18
	MeOH	32	26	18	24
Vancomycin (30 µg)		**10**	**15**	**24**	**21**
Tetracyeline (30 µg)		**26**	**24**	**27**	**25**

S. aur: *Staphylococcus aure*, *E. coli*: *Escherichia coli*, *B. sub*: *Bacillus subtilis*, *S. typhi*: *Salmonella typhi*.

Table 4. In vitro antimicrobial activity of the methanolic plant extracts against resistant bacteria.

Methanolic extract	*E. agg*	*P. Syr*	*S. spp*	*S. mar*	*P. aer.*	*ESBL**	*MRSA*
E. serrata	8	8	8	8	19	13	35
T. garganica	8	8	8	8	10	25	15
H. scoparia	14	22	15	15	13	17	25
R. raetem	8	8	8	8	24	12	10
H. albus	8	9	8	18	21	30	24

E. agg: *Enterobacter agglomerans*; *P. syr*: *Pseudomonas syrenga*; *S. spp*: *Salmonella spp*; *S. mar*: *Serratia marcescens*; *P. aer*: *Pseudomonas aeruginosa*; *MRSA*: *Methicilline resistant staphylococcus aur*; *ESBL**: *Extended-spectrum β-Lactamases*.

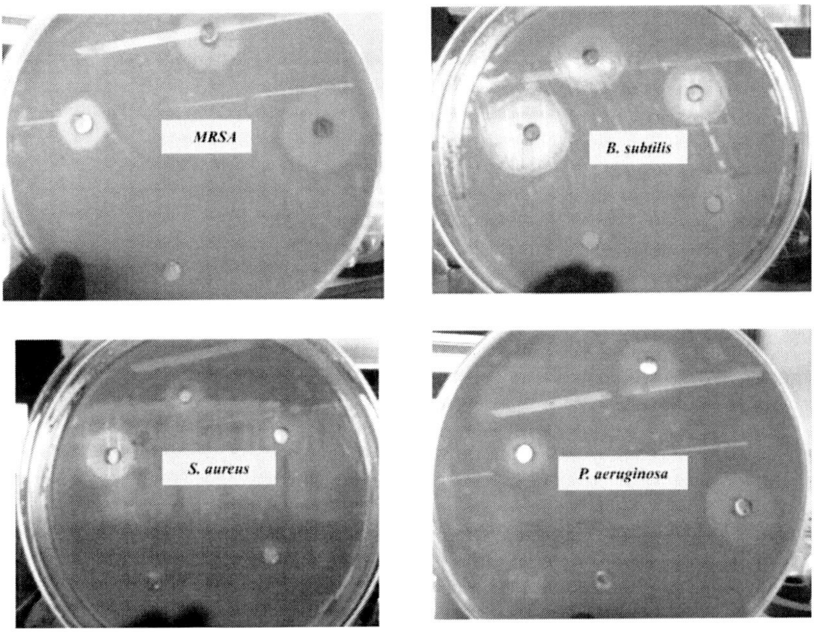

Figure 5. In vitro antibacterial activity (MIC) of methanolic extract of H. scoparia incubated with MRSA, B. subtilis, S. aureus and P. aeruginosa for 24 h.

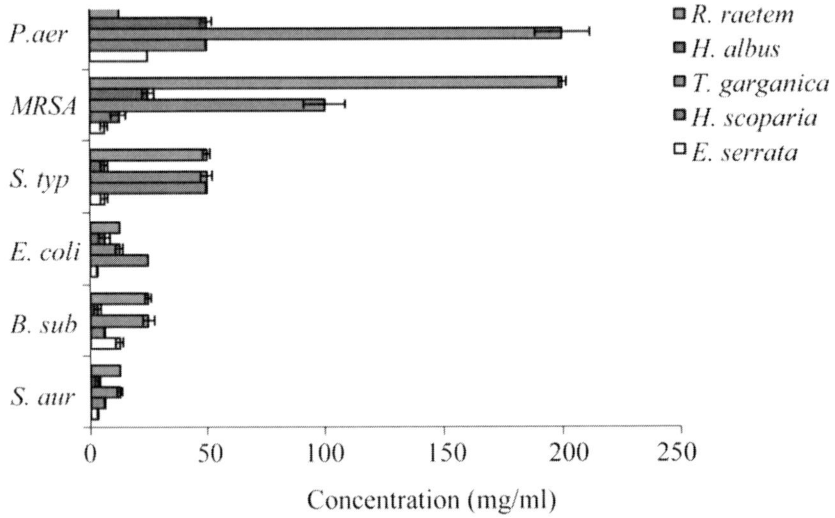

Figure 6. The in vitro antimicrobial activity of methanolic crude extracts of tested plants expressed as minimum inhibitory concentration (MIC) (mg/ml) against some bacteria. S. aur: Staphylococcus aureus; B. sub: Bacillus subtilis, E. coli: Escherichia coli; MRSA: Methicillin-resistant Stapylococcus aureus; P. aer: Pseudomonas aeruginosa (resistant); S. typhi: Salmonella typhi.

Seeking quantitative expression of the antibacterial activity, the MICs of methanolic extracts of tested MPs were determined (**Figure 5**). The lowest MIC values were recorded with H. albus extract on G +ve and G –ve susceptible bacterial species with values between 3.13 - 6.25 mg/ml (**Figure 6**). These values were 8 times higher on resistant strains of the same species (e.g. MRSA). The methanol extract of E. serrata showed high antimicrobial activity against Gram-positive bacteria than the other extracts with MICs of 3.13 - 6.25 mg/ml against S. aureus, E. coli, S. typhi, and MRSA. The R. raetem extract was more effective as antimicrobial agent with relatively low MIC against G –ve such as P. aeruginosa in comparison with G +ve like MRSA. Methanol extracts of R. raetem flowers revealed high antimicrobial activity against Grampositive bacteria [47] which was in good agreement with our results. The high resistance of tested Gram-negative bacteria to external agents has been reported earlier, and it is mainly attributed to the presence of lipopoysaccharides in their outer cell membranes, which make them inherently resistant to antibiotics, detergent, and hydrophilic dyes [50]. The reason for higher sensitivity of the Gram-positive bacteria than Gramnegative bacteria could be attributed to the differences between their cell wall compositions. The Gram-positive bacteria contain an outer peptidoglycan layer instead of LPS, which is an ineffective permeability barrier [51,52]. Without outer membrane, the cell wall of Gram-positive ones can be permeated more easily and extracts can disturb the cytoplasmic membrane, the proton motive force (PMF), electron flow, active transport and coagulation of cell contents [1]. Therefore, the structural difference of bacterial cell wall plays an important role in their susceptibility. Collectively, H. albus followed by H. scoparia and E. serrata extracts were of higher potential as antimicrobials, (on basis of their MICs) regardless of the sensitivity of the test organisms (Susceptible or Resistant).

Although alkaloids are typical natural products of plants, it also produced by bacteria, fungi and animals produce alkaloids. It has been estimated that more than 20% of all plant species have alkaloids, especially members of the families Ranunculaceae, Berberidaceae, Cactaceae, Apocynaceae, Rutaceae, Solanaceae, Boraginaceae, and Leguminosae [53].

The antibacterial efficacy of the alkaloid extracts of the five investigated medicinal plants against susceptible and antibiotic-resistant bacteria of both species was investigated; the obtained results are shown in **Table 5**. The antimicrobial results showed that the H. albus alkaloid extract was effectively arrested growth of susceptible G +ve (DIZ = 41 - 43 mm), G –ve (DIZ = 34 - 35 mm) including resistant bacteria, especially MRSA (DIZ = 32 mm). These results showed high antimicrobial activity in comparison with the positive control. This can be partly attributed to the presence of higher alkaloidal content, as reported in other study [54].

Like the alkaloid extract of H. albus, E. serrata (DIZ = 21 - 41 mm) and H. scoparia (DIZ = 22 - 37 mm) alkaloid extracts showed similar results with remarkable effect against tested bacteria. Whereas, R. raetem (DIZ = 15 - 33 mm) and T. garganica exhibited moderate effect (DIZ = 19 - 24 mm). E. serrata alkaloid extract was very effective on MRSA (DIZ ~41 mm) compared with other plant species tested.

In terms of MIC values, the quantitative differences between the efficacies of alkaloid extracts of the five plants were determined and the results are illustrated in **Figure 7**. These results indicate that H. albus extract was of the lowest MIC value (0.156 mg/ml) on both G +ve strains (B. subtilis and S. aureus). Alkaloid extract of H. scoparia showed the same effect at doubled MIC values (0.313 mg/ml) E. serrata alkaloid extract potentially affected S. aureus, E. coli, B. subtilis and S. typhi at MIC values of 0.156, 0.313, 0.625 and 0.625 mg/ml respectively (**Figure 7**). Although, it has been reported that other species of Rhamnacea family possess moderate antimicrobial and antiviral activities [55], the alkaloids extract of H. albus showed remarkable activity on S. aureus, E. coli and S. typhi with MIC values 0.156, 0.313 and 0.313 mg/ml respectively. It is most likely that the toxicity of plant members of this family is mostly related to the high concentration of the alkaloid in their extracts.

Antifungal activity of five botanical extracts was assayed against Rhizoctonia sp, Trichoderma sp, Sclerotium sp, Aspergillus and the obtained data are presented in **Figure 8**. The crude plant extracts of R. raetem and H. scoparia were exhibited significant reduction in the growth of four fungi which was in parallel with pervious study [48] (**Figure 8**(a)). It has been reported that raetem extract showed an antibacterial, antifungal and anticancer activity [56]. Alkaloids extract of H. scoparia and H. albus proved to be effective against tested fungi especially on Trichoderma spp, Sclerotium spp, Aspergillus niger.

H. scoparia demonstrated strong growth inhibition of the malaria parasite in earlier study [57]. In addition, methanolic H. scoparia extract demonstrated richness in alkaloids and showed significant molluscicidal activity [56], while H. albus extract exhibited antiviral activity [54]. The antifungal activity of alkaloids of T. garganica was increased significantly in comparison with crude extract ($P < 0.05$). The alkaloids extract of E. serrata has lower antifungal activity compared with other alkaloid extracts which might be due to its comparatively low alkaloid content (**Figure 8**(b)), this study suggests that methanolic extracts of the screened plants would be helpful in treating diseases in plants caused by Rizochtonia solani, Trichoderma spp, Sclerotium spp and Aspergillus niger.

Table 5. In vitro antimicrobial activity of the alkaloid extract of the five tested Libyan medicinal plants against susceptible and resistant bacteria.

Alkaloids extract	Diameter of Inhibition Zone (mm)*					
	Susceptible Gram +ve		Susceptible Gram -ve		Resistant Bacteria	
	S. aur	*B. sub*	*E. coli*	*S. typ*	*MRSA*	*P. aer*
H. scoparia	31	33	29	30	37	22
R. raetem	25	23	19	21	15	33
H. albus	41	43	34	35	32	30
T. garganica	24	22	20	19	20	24
E. serrata	33	23	39	34	41	21
Vancomycin (30 μg)	10	15	24	21	8	8
Tetracycline (30 μg)	26	24	27	25	8	8

Well diameter: 8 mm; *S. aur*: *Staphylococcus aureus*; *B. sub*: *Bacillus subtilis*; *E. coli*: *Escherichia coli*; *S. typ*: *Salmonella typhi*; *MRSA*: Methicillin-resistant; *P. aer*: *Pseudomonas aeruginosa*.

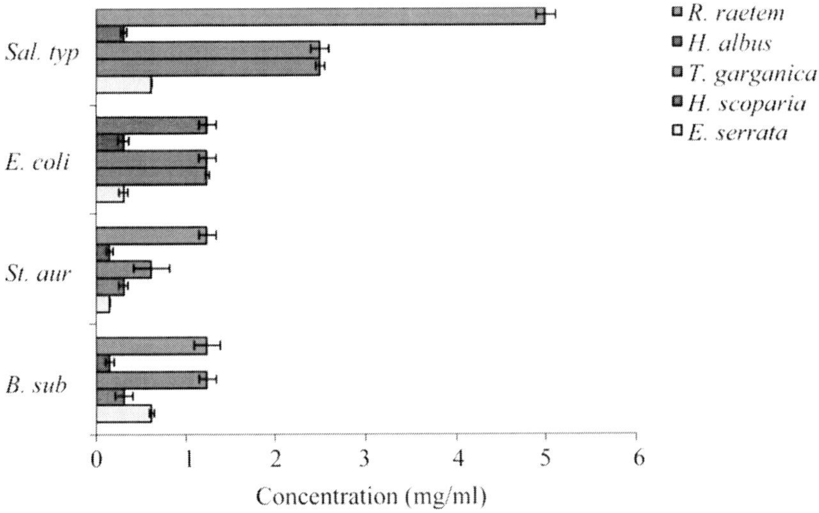

Figure 7. The in vitro MIC values (mg/ml) of alkaloid extracts of tested plants. S. aur: Staphylococcus aureus; B. sub: Bacillus subtilis; E. coli: Escherichia coil; S. typ: Salmonella typhi.

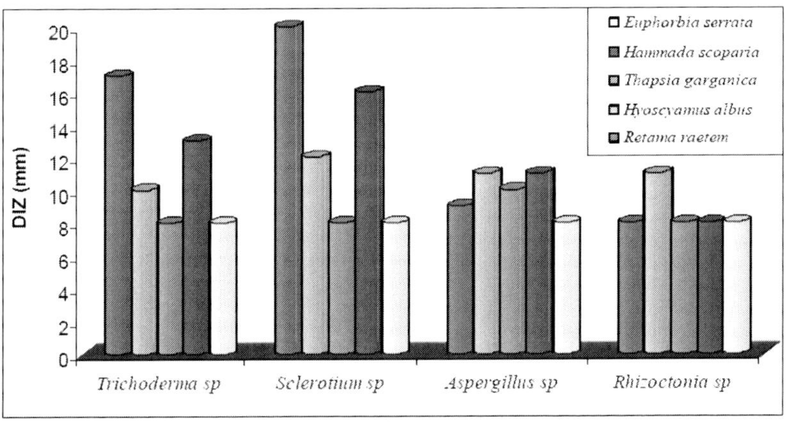

(a)

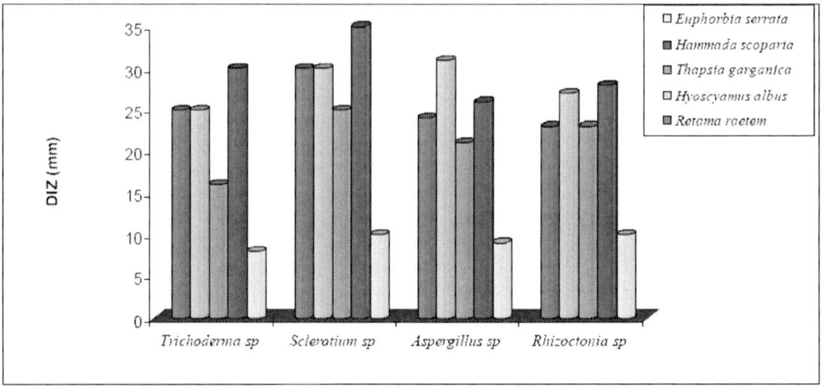

(b)

Figure 8. Antifungal activity of methanolic (a) and alkaloid (b) extracts of five selected medicinal plants.

4. CONCLUSION

Rabia Alghazeer, Hussein El-Saltani, Nabeel Saleh, Asma Al-Najjar, Fatma Hebail Antioxidative and antimicrobial properties of the medicinal plants are proven to be of great interest in both health and industry due to their possible uses as natural food additives to replace synthetic ones. In this respect, the present study was designed to evaluate the in vitro antioxidant properties and antimicrobial activity of five Libyan medicinal plant extracts against various microorganisms. The results presented here can be considered as the first information on the antioxidant and antimicrobial properties of those plants that have traditional uses in Libya. H. scoparia and R. raetem may be useful as excellent antioxidant, antibacterial and antifungal agents while T. garganic can be considered as a good source of antioxidant and antimicrobial agents. Additional studies are required on the mode of action of the active ingredients on pathogenic microorganisms. Despite the significant effect of E. serrata extract against both bacteria and fungi, it belongs to a family that is known to be toxic. Therefore, consumers of any of the investigated plants should also be familiar with science-based information on dosage, contraindications, and efficacy.

REFERENCES

1. Burt, S. (2004) Essential oils: Their antibacterial properties and potential applications in foods. International Journal of Food Microbiology, 94, 223-253.

2. Isman, B.M. (2000) Plant essential oils for pest and disease management. Crop Protection journal, 19, 603-608.
3. Liu, Y.J. and Zhang, K.Q. (2004) Antimicrobial activity of selected Cyathus species. Mycopathologia, 157, 185- 189.
4. Mares, D., Romagnoli. C., Tosi, B., Andreotti. E., Chillemi. G. and Poli, F. (2005) Chicory extracts from Cichorium intybus L. as potential antifungals. Mycopathologia, 160, 85-92.
5. Soylu, E.M., Soylu, S. and Kurt, S. (2006) Antimicrobial activities of the essential oils of various plants against tomato late blight disease agent Phytophthora infestans. Mycopathologia, 161, 119-128.
6. Yazaki, K., Sugiyama, A., Morita, M. and Shitan, N. (2008) Secondary transport as a efficient membrane transport mechanism for plant secondary metabolites. Phytochemistry Review, 7, 513-524.
7. Akhtar, Y., Rankin, K. and Isman, M. (2008) Decreased Response to Feeding Deterrents Following Prolonged Exposure in the Larvae of a Generalist Herbivore, Trichoplusia ni (Lepidoptera: Noctuidae). Phytochemistry Review, 7, 77-88.
8. Naili, M., Alghazeer, R., Saleh., N. and Al-Najjar, A. (2010) Evaluation of antibacterial and antioxidant activities of Artemisia campestris (Astraceae) and Ziziphus lotus (Rhamnacea). Arabian Journal of Chemistry, 3, 73- 134.
9. Li, J.W. and Vederas, J.C. (2009) Drug discovery and natural products: End of an era or an endless frontier? Science, 325, 161-165.
10. Hartmann, T. (2007) From waste products to ecochemicals: 50 years research on plant secondary metabolism. Phytochemistry, 68, 2831-2846.
11. Jenke-Kodama, H., Müller, R. and Dittmann, E. (2008) Evolutionary mechanisms underlying secondary metabolite diversity. Progress in Drug Research, 65, 121-140.
12. Suhaj, M. (2006) Spice antioxidants isolation and their antiradical activity: A review. Journal of Food Composition and Analysis, 19, 531-537.
13. Akinmoladun, A., Ibukun, E., Afor, E., Obuotor, E. and Farombi, E. (2007) Phytochemical constituent and antioxidant activity of extract from the leaves of Ocimum gratissimum. Scientific Research and Essays, 2, 163-166.
14. Alghazeer, R., Gao, H. and Howell, N.K. (2008) Cytotoxicity of oxidised lipids in cultured colonal human intestinal cancer cells (caco-2 cells). Toxicology Letters, 180, 202-211.
15. Nychas, G.J.E. (1995) Natural antimicrobial from plants. In: Gould, G.W., Ed., New Method of Food Preservation, Chapman and Hall, Glasgow, 58-89.
16. Gülçin, I., Oktay, M., Küfrevioğlu, O.I. and Aslan, A. (2002) Determination of antioxidant activity of lichens Cetraria islandica (L) Ach. Journal of Ethnopharmacology, 79, 325-329.
17. Jayaprakasha, G.K., Jaganmohan Rao, L. and Sakariah, K.K. (2006) Antioxidant activities of curcumin, demethoxycurcumin and bisdemethoxycurcumin. Food Chemistry, 98, 720-724.
18. Lindsay, D.G. and Astley, S.B. (2002) European research on the functional effects of dietary antioxidants. Molecular Aspects of Medicine, 23, 1-38.
19. Fadeyi, M.G., Adeoye, A.E. and Olowokodejo, J.D. (1989) Epidermal and phytochemical studies with genus of Boerhavia (Nyetanginaceae). International Journal of Crude Drug Research, 29, 178-184.
20. Odebiyi, A. and Sofowora, A.E. (1990) Phytochemical screening of nigerian medicinal plants. Part III. Lloydia, 41, 234-246.
21. Harborne, J.B. (1992) Phytochemical methods. Chapman and Hall Publications, London, 7-8.
22. Abulude, F.O., Onibon, V.O. and Oluwatoba, F. (2004) Nutrition and nutritional composition of some tree barks. Nigerian Journal of Basic and Applied Sciences, 13, 43- 49.
23. Abulude, F.O. (2007) Phytochemical screening and mineral contents of leaves of some Nigerian woody plants. Research Journal of Phytochemistry, 1, 33-39.

24. Hadi, S. and Bremner, B. (2001) Initial studies on alkaloids from Lombok medicinal plants. Molecules, 6, 117- 129.
25. Marinova, D., Ribarova, F. and Atanassova, M. (2005) Total phenolics and total flavonoids in Bulgarian fruits and vegetables. Journal of Chemical Technology and Biotechnology, 40, 255-260.
26. Oyaizu, M. (1986) Studies on product of browning reaction prepared from glucose amine. Japanese Journal of Nutrition, 44, 307-315.
27. Singleton, V.L., Orthofer, R. and Lamuela-Raventos, R.M. (1999) Analysis of total phenols and oxidization substrates and antioxidants by means of Folin-Ciocalteu reagent. Methods in Enzymology, 299, 152-177.
28. Wong, C., Li, H. and Cheng, F. (2006) A systematic survey of antioxidant activity of 30 Chinese medicinal plants using the ferric reducing antioxidant power assay. Food Chemistry, 97, 705-711.
29. Daud, A., Gallo, A. and Sanchez, A. (2005) Antimicrobial properties of Phrygilanthus acutifolius. Journal of Ethnopharmacology, 99, 193-197.
30. CLSI (2008) Reference method for broth dilution antifungal susceptibility testing of filamentous fungi. Approved Standard M38-A2, Clinical and Laboratory Standards Institute, Wayne.
31. Romero, C., Medina, E., Vargas, J., Brenes, M. and De, Castro (2007) Activity of olive polyphenols against Helicobactor pylori. Journal of Agricultural and Food Chemistry, 55, 680-686.
32. Irith, W., Kai, H. and Robert, E.W.H. (2008) Agar and broth dilution methods to determine the minimal inhibitory concentration (MIC) of antimicrobial substances. Nature Protocols, 3, 163-175.
33. Ayoola, G.A., Sofidiya, T., Odukoya, O. and Coker, A.B. (2006) Phytochemical screening and free radical scavenging activity of some Nigerian medicinal plants. International Journal of Pharmacy and Pharmaceutical Practice, 8, 133-136.
34. Zheng, X., Liu, B., Li, L. and Zhu, Z. (2011) Microwave-assisted extraction and antioxidant activity of total phenolic compounds from pomegranate peel. Journal of Medicinal Plants Research, 15, 1004-1011.
35. Esmaeili, M.A. and Sonboli, A. (2010) Antioxidant, free radical scavenging activities of Salvia brachyantha and its protective effect against oxidative cardiac cell injury. Food and Chemical Toxicology, 48, 846-853.
36. Leake, D. (2001) Flavonoids and the oxidation of lowdensity lipoprotein. Nutrition, 17, 59-63.
37. Aviram, M. and Fuhrman, B. (1998) Polyphenolic flavonoids inhibit macrophage-mediated oxidation of LDL and attenuate atherogenesis. Atherosclerosis, 137, S45-S50.
38. Reddy, L., Odhav, B. and Bhoola, K.D. (2003) Natural products for cancer prevention: A global perspective. Pharmacology & Therapeutics, 99, 1-13.
39. Kandaswami, C., Lee, L.T., Lee, P.P., Hwang, J.J., Ke, F.C., Huang, Y.T. and Lee, M.T. (2005). The antitumor activities of flavonoids. In Vivo, 19, 895-909.
40. Kassem, M., Mosharrafa, S.A., Saleh, N.A. and AbdelWahab, S.M. (2000) Two new flavonoids from Retama raetam. Fitoterapia, 71, 649-654.
41. Djeridane, A., Yousfi, M., Nadjemi, B., Boutassouna, D., Stocker, P. and Vidal, N. (2006) Antioxidant activity of some Algerian medicinal plants extracts containing phenolic compounds. Food Chemistry, 97, 654-660.
42. Zou Y., Lu, Y. and Wei, D. (2004) Antioxidant activity of a flavonoid-rich extract of Hypericum perforatum L. in vitro. Journal of Agricultural and Food Chemistry, 52, 5032-5039.
43. Molyneux, P. (2004) The use of the stable free radical diphenylpicrylhydrazyl (DPPH) for estimating antioxidant activity. Songklanakarin Journal of Science and Technology, 26, 211-219.
44. Katalinic, V., Milos, M., Kulisic, T. and Jukic, M. (2006) Screening of 70 medicinal plant extracts for antioxidant capacity and total phenols. Food Chemistry, 94, 550-557.

45. Brighente, I.M.C., Dias, M., Verdi, L.G. and Pizzolatti, M.G. (2007) Antioxidant activity and total phenolic content of some Brazilian species. Pharmaceutical Biology, 45, 156-161.
46. Rastogi, R.P. and Mehrotra, B.N. (2002) Glossary of Indian medicinal plants. National Institute of Science Communication, New Delhi.
47. Parekh, J., Jadeja, D. and Chanda, S. (2005) Efficacy of Aqueous and Methanol extracts of some medicinal plants for potential antibacterial activity. Turkish Journal of Biology, 29, 203-210.
48. Edziri, H., Mastouri, M., Chéraif, I. and Aouini, M. (2010) Chemical composition and antibacterial, antifungal and antioxidant activities of the flower oil of Retama raetam (Forssk.) Webb from Tunisia. Natural Product Research, 24, 789-796.
49. Abutbul, S., Golan-Goldhirsh, A. Barazani, O. Ofir, R. and Zilberg, D. (2005). Sscreening of the desert plants for use against bacterial pathogens in fish. The Israeli Journal of Aquaculture—Bamidgeh, 57, 71-80.
50. Nikaido, H. (2003) Molecular basis of bacterial outer membrane permeability revisited. Microbiology and Molecular Biology Reviews, 67, 593-656.
51. Matias, V.R. and Beveridge, T.J. (2006) Native cell wall organization shown by cryoelectron microscopy confirms the existence of a periplasmic space in Staphylococcus aureus. Journal of Bacteriology, 188, 1011-1021.
52. Hayder, N., Bouhlel, I., Skandrani, I., Kadri, M., Steiman, R., Guiraud, P., Mariotte, A.M., Ghedira, K., DijouxFranca, M.G. and Chekir-Ghedira, L. (2008) In vitro antioxidant and antigenotoxic potentials of myricetin-3- o-galactoside and myricetin-3-o-rhamnoside from Myrtus communis: modulation of expression of genes involved in cell defense system using cDNA microarray. Toxicology in Vitro, 22, 567-581.
53. Rahman, A. (2000) Studies in natural products chemistry. Bioactive Natural Products (Part B), 21, 3-122.
54. Dawidar, A.M., Abdel-Mogib, M., El-Ghorab, A.H. and Hussien, Kh. (2009) Chemical composition and effect of photooxygenation on biological activities of Egyptian commercial spearmint and dill essential oils. Mansoura Journal of Chemistry, 36, 49-60.
55. Budzanivska, I.G. and Demyanenko, F. (2002) Inhibition of Solanaceae plants by combined effect of increased concentrations of heavy metals and viral infection. Plant Protection Science, 38, 452-454.
56. Merghoub, N., Benbacer, L., Amzazi, S., Morjani, H. and El Mzibri, M. (2009) Cytotoxic effect of some Moroccan medicinal plant extracts on human cervical cell lines. Journal of Medicinal Plants Research, 3, 1045-1050.
57. Mezghani-Jarraya, R., Hammami, H., Ayadi, A. and Damak, M. (2009) Molluscicidal activity of Hammada scoparia (Pomel) Iljin leaf extracts and the principal alkaloids isolated from them against Galba truncatula. Memórias do Instituto Oswaldo Cruz, 104, 1035-1038.

CHAPTER 9

Lesser Known Aromatic Plants in Nigeria

Ngozichukwuka P. Igoli[1*] and John O. Igoli[2]

[1] *Centre for Food Technology and Research, Benue State University, Makurdi, Nigeria*

[2] *University of Agriculture, Makurdi, Nigeria*

ABSTRACT

Herbs and spices are used in all cultures as natural foodstuffs and for medicinal purposes. *Siphonochilus aethiopicus, Monodora myristica* and *Crateva adansonii* are some of the spices which are not commonly used. They improve the taste of food, and through their anti-oxidant, anti-microbial and anti-fungal properties, they could act as food preservatives. There is an accumulation of evidence for the usage of these spices medicinally as anti-inflammatory, anti-plasmodial, anti-sickling, anti-oxidant and chemopreventive agents. There have also been investigations to identify the active constituents of these spices and to verify their pharmacological actions. This article aims at reviewing the available data on these investigations and the basis for usage in several diseases and conditions.

Keywords: *Siphonochilus aethiopicus, Monodora myristica, Crateva adansonii,* herbs, spices

1. INTRODUCTION

Aromatic herbs and spices are widely used in Nigeria for culinary and medicinal purposes. While some are quite common and used worldwide, others like: wild ginger (*Siphonochilus aethiopicus* (Schweinf) B.L Burtt), African nutmeg (*Monodora myristica* (Gaertn) Dunal) and sacred garlic pear (*Crateva adansonii* DC) are not. Herbs and spices make important contributions towards the odour and flavour of foods due to the presence of volatile (essential) oils and fixed oils. They confer new aromas to the foods, and their use for improving the taste of foods is a cultural achievement of all races which also led to cultural exchanges very early in history [1]. Moreover, there is an increasing interest in using the extracts of herbs and spices, for food preservation [2] since as natural foodstuffs, they appeal to many consumers who question the safety of synthetic food additives due to their carcinogenicity or other concerns [3, 4]. Indeed, since prehistoric times, herbs were the basis for nearly all medicinal therapy until synthetic drugs were developed, and even today, herbs are still found in 40% of prescription drugs [5].

Thus, in addition to imparting characteristic pleasant flavours, certain herbs and spices prolong the storage life of foods by preventing rancidity through their anti-oxidant activity or through bacteriostatic or bactericidal activity [6]. Consequently, herbs and spices have medicinal values, anti-oxidant and anti-microbial properties [7], and some do contain potent phytochemicals, which provide significant protection against cancer [8].

Wild ginger is an herb with perennial tuberous roots giving rise to annual leafy stems which grows in sub-Saharan Africa especially in savannah regions or regions with dry season [9]. It belongs to the family Zingiberaceae and has leafy shoots (pseudo-stem) which grow to about 35 cm high after flowering and is common throughout the West African region and elsewhere in tropical Africa [9]. The rhizomes which have a terrific scent of violets and ginger are spindle shaped and are about 5 cm by 1 cm, arranged radially on lateral roots that spread fairly readily underground [9]. The Nigerian variety flowers between April and May after the early rains and the flowers, which appear before the leaves, come up in considerable quantity followed by the leaves and the pseudo-stem [10]. The flowers, just borne above ground level in inflorescences separate from the leafy shoot, are purple with white corolla tube and a yellow flare on the central petal and are 7–10 cm long [9]. The leafy shoot dries up between September and December after which it falls off. The herb is found in the wild and could also be cultivated.

The South African variety is a deciduous aromatic plant, bisexual or female, up to 1 m high and sprout annually from the underground stem in spring [11]. The leaves are glabrous and 30–400 × 50–90 mm in size, light green, lance shaped and borne on the end of stem-like leaf bases [11, 12]. Between October and February, it gives faintly scented flowers that are white to bright pink with yellow markings on lip, white corolla tubes 30–40 mm long and tepal lobes 60–80 mm wide [12]. The tremendously attractive flowers often appear before the leaves in spring, perhaps to allow them to be more visible to pollinators [13]. They may also vary in colour from bright pink, purple-pink, yellow to white with a yellow centre and are delicately scented. About 15 flowers are produced per plant over the flowering season, each lasting a single day [14].

The roots, tubers and rhizomes of wild ginger are used for their aroma and medicinal properties in the West and South of Africa. While they are used as spice by the Igede people of Benue State of Nigeria [15], others mainly use them in traditional medicine for colds, coughs, influenza, hysteria, pain and malaria amongst other ailments [16, 17]. It is also used by Zulu people as a protection against lightning and snakes [18]. Infusions of the rhizome and roots are anti-inflammatory (prostaglandin synthetase inhibition), bronchodilatory, smooth muscle relaxant, mild sedative, anti-candidal and used to treat headache, influenza, mild asthma, sinusitis, sore throat, thrush, epilepsy, hysteria and relieve dysmenorrhoea [11] or administered to horses as prophylactics against horse sickness [19]. It is in such popular demand and coupled with the method of harvesting which involves removal of the entire rhizome that it has become extinct in certain areas [20]. The plant does not set much seed and splitting rhizomes is the best available option for plant propagation [14].

The seeds of *Monodora myristica* (Gaertn.) Dunal either fresh or roasted are used both as a spice in cuisine and medicinally in different parts of Africa and in the Caribbean [21, 22]. It is largely underutilized and has recently been used as popcorn flavouring in an attempt to increase its utilization[23]. It is also responsible for the distinct flavour of the Nigerian delicacy "isi ewu", and it is used in soups and peanut paste [23, 24]. Its aroma which is enhanced by mild roasting is similar to nutmeg; thus, the plant is commonly known as African nutmeg, calabash nutmeg or Jamaican nutmeg; and it is a large tropical tree which can reach 35 m height and 2 m in diameter [25]. It is native to tropical West Africa and further east to Uganda, Kenya and Tanzania where it grows naturally in evergreen forests but has been introduced to Jamaica, other parts of the Caribbean and elsewhere [26]. Its large leaves (35 cm long and 18 cm wide) are purple at first but turn a smooth deep green on the upper side with paler green underneath. They are prominently veined, and the petiole is purplish. The exotic, conspicuous and scented flowers hang down on long stalks and have three calyx lobes and three petals arranged in two whorls [27]. The flowers are followed by large woody fruits filled with brown seeds embedded in an aromatic pulp [25]. The medicinal uses of the seeds are to treat headaches, pains, toothaches, haemorrhoids, stomach ache, relieve constipation and control passive uterine haemorrhage in women immediately after child birth [24, 25, 28] and hypertension [25].

The leaves of sacred garlic pear (*Crateva adansonii*) which are eaten with soups or mixed with cereal are also used medicinally in different parts of Africa and Asia [29, 30]. It belongs to the family Capparaceae and is widely distributed as a small handsome tree of the galleried forest and savannah woodland often found on river banks across Africa [29]. The species is confined to Africa but bears very close affinity to the Asian *Crateva religiosa* G. Forst with which it has been equated by some authorities [29]. It attains a height of 7 m or more with an irregular trunk which is seldom straight and could be cultivated for ornamental purposes due to its dense masses of white flowers borne at the ends of all the shoots [29]. Where it survives bush burning and repeated stripping of its leaves, the tree is often stunted. The wood is strong-smelling when cut and is soft and yellow [29]. In Eastern Nigeria, *C. adansonii* is used medicinally, its leaves being in high demand for the treatment of ear and parasitic infections [31]. The leaves are applied externally to relieve joint pains; the fresh juice from the leaves is used for the relief of ear ache, eye infection and anodyne in toothache [32]. The leaves are also used in fumigations for treating jaundice and yellow fever, applied to the head as a mild counter-irritant for easing headaches, and a steam bath of over the face is used as a remedy for all troubles due to poor vision [29]. The bark is said to be rubefacient and tonic, widely used as a remedy for stomach-troubles and used both internally and externally for treating sterility. It is used in combination with *Flacourtia flavescens* as a treatment for leprosy [29]. Powdered and boiled in oil, it is used as an application for rheumatic condition, and a bark paste is used as a poultice on swellings [29]. The powdered bark is used in rheumatism, itch, epilepsy, stomach troubles and asthma [33]. The powdered leaves and bark are considered to be rubefacient and are used especially on cysts. The root is used as a febrifuge and in several treatments for syphilis. The dried, ground roots are used as an application to swollen parts of the body, while the seeds have unspecified medicinal uses [29].

This review identifies lesser known aromatic plants in Nigeria and current reports on their utilization, constituents and properties. The aim is to provide an insight into the health-promoting potentials of biologically active constituents of: wild ginger (*Siphonochilus aethiopicus* (Schweinf) B.L Burtt), African nutmeg (*Monodora myristica* (Gaertn) Dunal) and sacred garlic pear (*Crateva adansonii* DC). This is because diets rich in plant foods can provide biologically active phytochemicals that promote health [34].

2. CONSTITUENT PHYTOCHEMICALS

The sensory perception of wild ginger depends on a variety of odorants including esters, monoterpenes, sesquiterpenes, aldehydes, pyrazines and thiophenes which are important for the mild and pleasant aroma both in the fresh and in roasted spice as against the hot/pungent flavours of ginger and other Zingiberaceae [35]. Aroma extracts dilution analysis (AEDA) employing the gas chromatography/olfactometry (GC/O) technique was reported for the organoleptic evaluation of these odorants and their odour quality together with quantification from GC-FID/GC-MS profiles [35]. Thus, the sweet/fruity ester flavours, methyl-2-/-3-methyl butanoates and derivatives of the apple flavour were reported to be the most important odorants perceived at the highest dilution of the aroma extract of the fresh spice [35]. These were followed by the monoterpene β-phellandrene which has a terpenish/woody odour and is also important for the aroma of ginger and dill [35–37]. Another sweet/fruity flavour propyl-2-methylbutanoate also an apple flavour follows before the roasty/earthy smelling 2-isopropyl-3-methoxypyrazine and 2-isobutyl-3-methoxypyrazine which are known to have a hot/paprika taste and are also present in paprika pepper and chillies [35–38]. The sesquiterpene curzerenone (sweet/coconut-like) is perceived at the next significant dilution together with the roasty/potato-like methional [35]. In the roasted sample, terpenish/woody β-phellandrene is the most important odorant followed by the roasty/earthy smelling pyrazines before the sweet/fruity flavoured butanoates. The pungent smelling principle 2-acetyl thiophene, which is absent in the fresh sample, is next followed by the sesquiterpene curzerenone (sweet/coconut-like) together with the roasty/potato-like methional [35].

Principal component analysis using GC shows that the constituents of wild ginger are mostly sesquiterpenes as against monoterpenoids or diterpenoids [16, 35, 39]. Five eudesmane sesquiterpenoids as shown in **Figure 1**: 4aαH-3,5α,8aβ-trimethyl-4,4a,8a,9-tetrahydronaphtho[2, 3b]-furan-8-one; 2-hydroxy-4aαH-3,5α,8aβ-trimethyl-4,4a,8a,9-tetrahydro-naphtho[2, 3b]-furan-8(5H)-one; 4aαH-3,5α,8aβ-trimethyl-4,4a,8a,9-tetrahydronaphtho-([2, 3b]-dihydrofuran-2-one)-8-one; 9aβ-hydroxy-4aαH-3,5α,8aβ-trimethyl-4,4a,8a,9-tetrahydronaphtho-([2, 3b]-dihydrofuran-2-one)-8-one and 4aαH-3,5α,8aβ-trimethyl-4,4a,8a-trihydronaphtho-([2, 3b]-dihydrofuran-2-one)-8-one were isolated as the constituents of the South African variety [16, 17]. However, the elemane sesquiterpenoids: curzerenone and *epi*-curzerenone, the germacrane sesquiterpenoids: furanodiene (8,12-epoxy-1(10)E,4Z,7,11-germac ratetra ene); isofuranodiene (8,12-epoxy-1(10)E,4Z,7,11-germacratetraene) and furanod ienone (8,12-epoxy-1(10)E,4Z,7,11-germacratetraen-6-one) together with the labdane diterpenoids: 8(17),12E-labdadiene-15,16-

dial, 15-hydroxy-8(17),12E-labdadiene-16-al, and 16-oxo-8(17), 12E-labdadiene- 15-oic acid (zerumin A) as also shown in **Figure 1** were isolated from the Nigerian variety [35, 39].

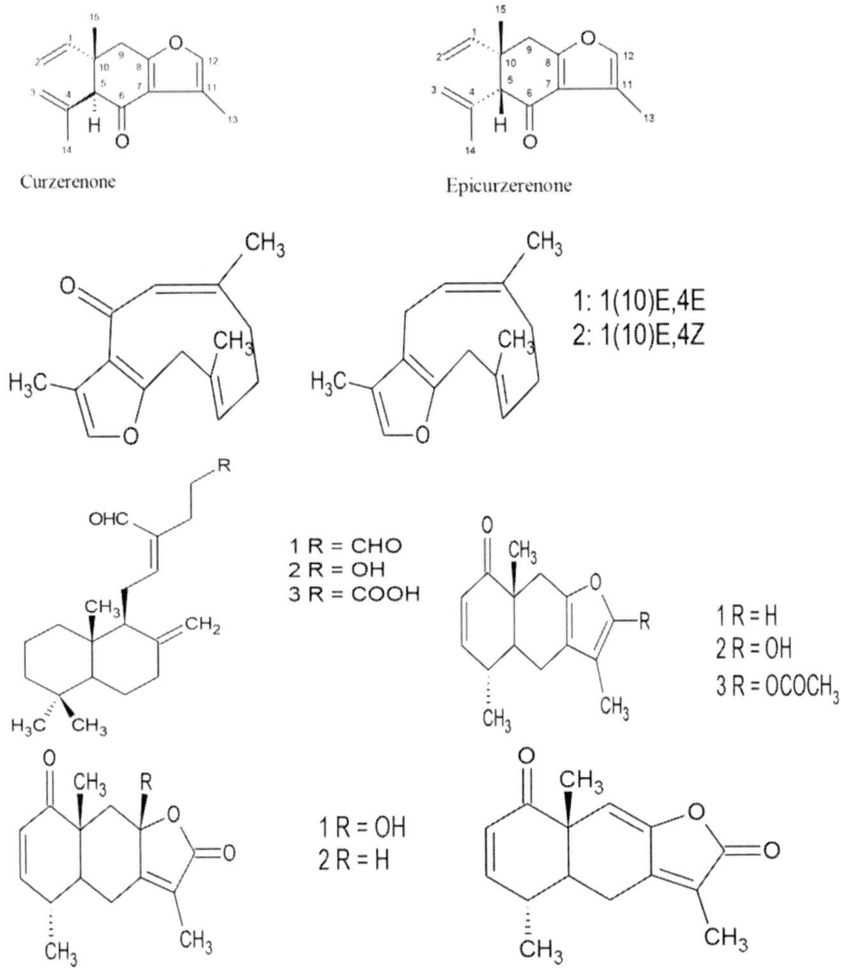

Figure 1. Isolated constituents of *Siphonochilus aethiopicus*.

A yield of 45-6g kg^{-1} essential oils containing 75% monoterpene hydrocarbons; the major compounds being: α-phellandrene (50–4%), α-pinene (5–5%) and myrcene (4–35%) has been reported for the African nutmeg [40]. Few sesquiterpene hydrocarbons (3%) and oxygenated compounds such as germacrene-D-4-ol (9–5%) were also reported as against another report of 25.48% germacrene-D-4-ol [41]. Meanwhile, an essential oil yield of 6.2% (dry weight basis) has also been reported [42].

Prenylated indole alkaloids are considered a chemotaxonomic marker of the genus, and 5-formyl indole and 5-(3-oxo-but-1-enyl) indole as shown in **Figure 2** have been reported from *Monodora myristica* and other species of *Monodora* [21, 43, 44].

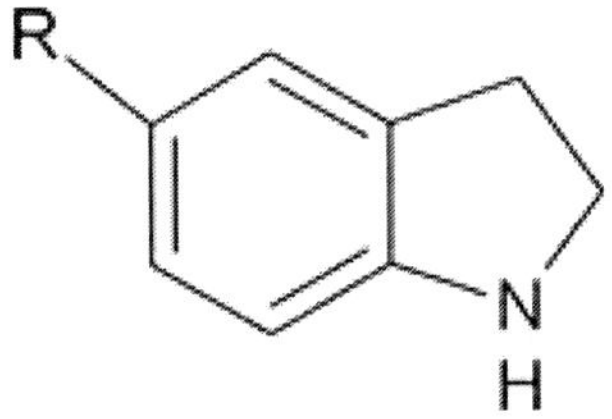

1 R = CHO
2 R = CH=CHCOCH$_3$

Figure 2. Isolated constituents of *Monodora myristica.*

The volatile oils of the sacred garlic pear whole plant reportedly show 43.5 and 41.1% oxygenated monoterpenes and aliphatic compounds, respectively. The major constituents are linalool (30.2%) and nonanal (17.2%), and it contains no sesquiterpene hydrocarbons [45]. The leaves are also known to have a disagreeable smell when crushed [29]. The triterpenes: oleanolic acid and 4-epi-hederagenin were isolated from the 1:1 C_2H_2:MeOH extract of its seed as shown in **Figure 3** [46]. Then, the hexane extract of the leaf yielded the antibiotic aurantiamide acetate while the ethyl acetate extract afforded ethyl pyropheophorbide A, purpurin-18 ethyl ester and pyropheophorbide A as also shown in **Figure 3** [31]. Additionally, the triterpene lupeol was also isolated from a 1:1 dichloromethane/methanol fraction of the leaf as illustrated in **Figure 3** [47].

It has been reported that the leaf and rhizome extracts of *S. aethiopicus* possess anti-microbial and anti-fungal properties [48]. However, the activities of the leaf extracts are much lower than those of the rhizome extracts which inhibited *Bacillus subtilis, Micrococcus kristinae, Bacillus cereus, Staphylococcus aureus, Staphylococcus epidermidis* and *Klebsiella pneumoniae* and showed anti-fungal properties against *Aspergillus flavus, Aspergillus glaucus, Candida albicans, Candida tropicalis, Trichophyton mentagrophytes* and *Trichophyton rubrum* [48, 49]. This is despite the similar chemical composition of the essential oils of the leaf and rhizome [20]. Again, the anti-fungal activities of some of the isolated constituents of the rhizomes and tubers: *epi*-curzerenone and furanodienone against *Candida albicans*, and 8(17),12E-labdadiene-15,16-dial against *Candida tropicalis* and *Candida guilliermondii* have been reported [50, 51]. Moderate activity of the crude rhizome extract and isolated diterpenes: 8(17),12E-labdadiene-15,16-dial and 15-hydroxy-

8(17),12E-labdadiene-16-al against *Mycobacterium tuberculosis* has also been reported [39].

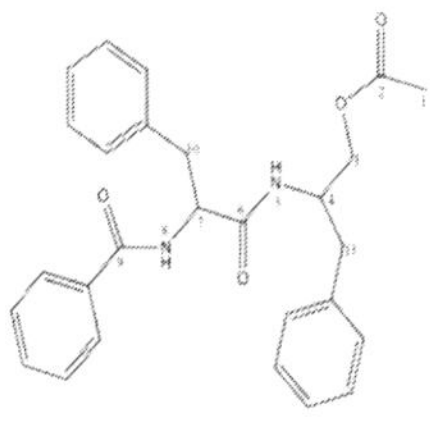

Aurantiamide acetate

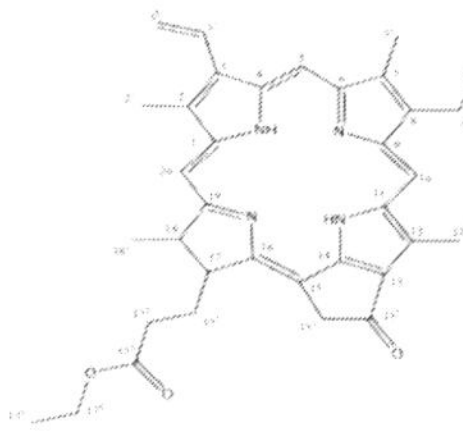

Ethyl Pyropheophorbide

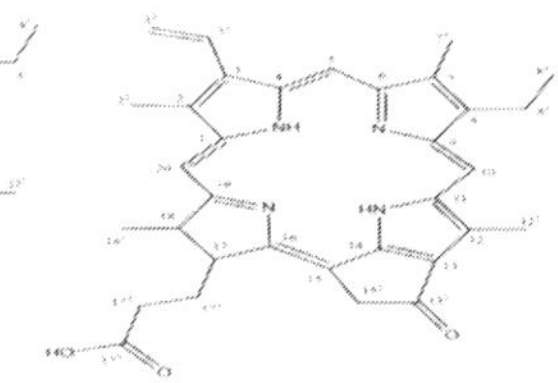

Pyropheophorbide A

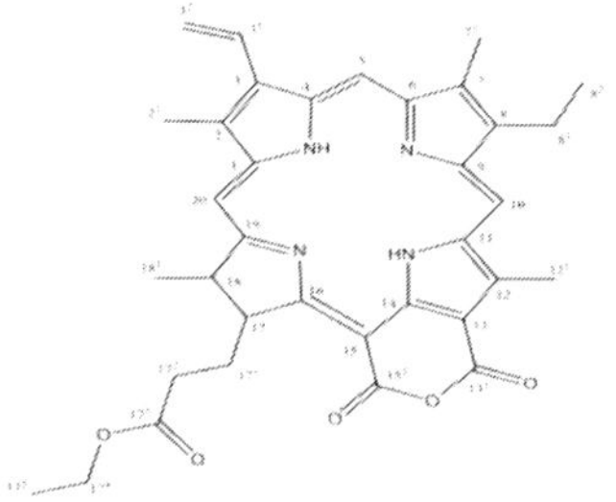

Purpurin-18 ethyl ester

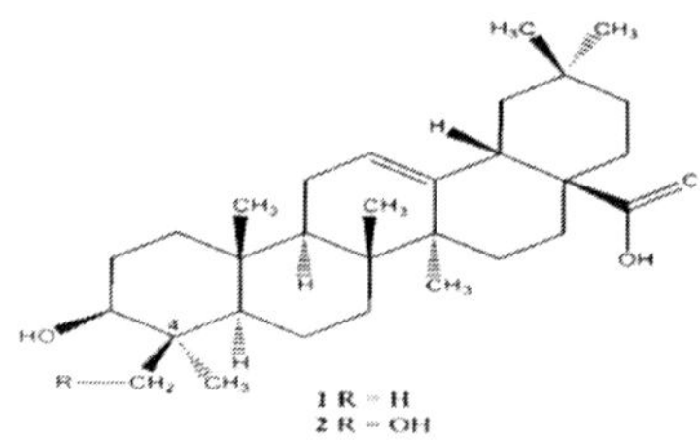

Oleanolic acid (1) and 4-epi-hederagenin (2)

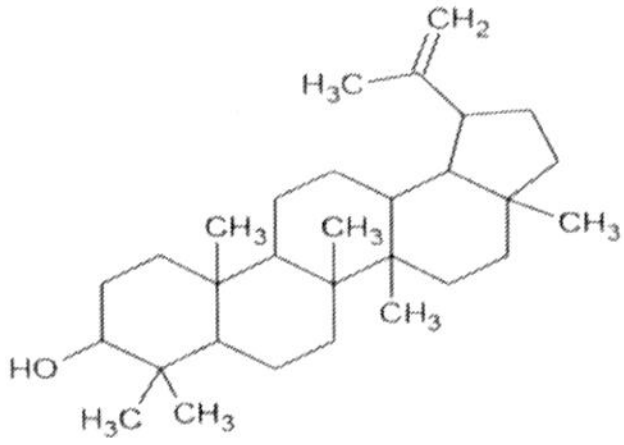

Lupeol

Figure 3. Isolated constituents from *Crateva adansonii* bioactivity.

The in vitro anti-proliferative properties of the essential oils of wild ginger against MCF-7 cancer cells were reported [52]; indeed, it has been suggested that the presence of antiseptic monoterpenoids contributes to its bioactivity [11]. In vitro cytotoxicity determinations of the crude rhizome extract and isolated constituents using five cell lines: SH-SY5Y, Jurkat, L929, Hep G2 and Hs 27 were also carried out [39]. *Epi*-curzerenone and furanodienone were inactive against the five different cell lines tested, while two of the diterpenes reportedly showed specific cytotoxic effects. 8(17),12E-Labdadiene-15,16-dial had moderate effect on the normal cell line Hs 27 and was cytotoxic to SH-SY5Y, the cancerous Jurkat and L929. However, only Jurkat and SH-SY5Y were affected by 15-hydroxy-8(17),12E-labdadiene-16-al [39].

The in vitro and in vivo anti-inflammatory properties of *S. aethiopicus* have also been reported [53]. The rhizome extract and the isolated furanoterpenoid, 4aαH-3,5α,8aβ-trimethyl-4,4a,8a,9-tetrahydronaphtho [2, 3b]-furan-8-one, showed in vitro inhibition of glucocorticoid and histamine H1 receptor binding and phosphodiesterase IV activity [53]. OVA-sensitized and challenged mice showed significantly reduced lung inflammation and the percentage of eosinophils in bronchoalveolar lavage fluid after administration of *S. aethiopicus* extracts but airway hyper reactivity was not influenced supporting anecdotal accounts of effectiveness against asthma, sinusitis, colds and flu [53]. Another report on the anti-inflammatory properties of the extracts of various parts of *S. aethiopicus* showed that the in vitro cyclooxegenase-1 (COX-1) inhibition of the stem and leaf extracts was reportedly higher than that of the rhizome [54]. High inhibition of cyclooxygenase and hence the prostaglandin pathway which should prevent uterine contraction and relieve dysmenorrhoea was again reported for the leaves and tubers of wild ginger [55]. However, in vitro reduction of pre-contracted uterine muscle was not observed [55].

Additionally, the in vitro anti-plasmodial activity for the ethanolic extracts and isolated eudesmane sesquiterpenoids of *S. aethiopicus* rhizomes against the chloroquine-sensitive and chloroquine resistant strains of *Plasmodium falciparum* has also been reported [17]. The substitution of the OH group in the sesquiterpene structure with hydrogen resulted in a threefold increase in activity against the chloroquine-resistant strain and an introduction of a double bond further improved the activity [17]. It is suggested that the anti-plasmodial activity is due to the furan moiety [17, 56]. Further in vitro anti-protozoal property of *S. aethiopicus* was reported against T*rypanosoma brucei brucei* (S427) blood stream forms by the crude rhizome extract, and it increased with the pure components: 8(17),12E-labdadiene-15,16-dial, *epi*-curzerenone and furanodienone [39].

The crude seed extract of *Monodora myristica* and isolated 5-formyl indole and 5-(3-oxo-but-1-enyl)indole reportedly showed no in vitro cytotoxicity against normal PNT2A cells and no anti-trypanosomal activity against *Trypanosoma brucei brucei* (S427) blood stream forms [43]. There was also no in vitro cytotoxicity when the crude seed extract was tested against five different cancerous cell lines [57]. Similarly, no lethality against brine shrimp (*Artemia salina*) and low anti-microbial activity against the *Mycobacterium* species: *M. madagascariense* and *M. indicus pranii* were reported for the stem bark extract of *Monodora carolinae* and constituent 5-formyl indole [58]. However, some of these prenylated indole alkaloids reportedly show interesting in vitro anti-plasmodial properties against the multi-drug-resistant strain K1 of *Plasmodium falciparum* [59].

Again, there are reports on the in vitro anti-oxidant properties and protective potential against free radicals of *M. myristica* seeds which suggest usage in the management of diseases associated with oxidative stress [28, 59, 60]. In vivo studies on the aqueous extracts of the seed and fruit also suggest that anti-oxidant bio-constituents play an important role in the prevention of liver toxicity possibly by inhibiting bioaccumulation of free radicals in animal models and could also reverse liver toxicity induced by high cholesterol diets and exert hypocholesterolemic effects [25, 61].

Also, the crude seed extracts of *M. myristica* reportedly exhibit profound in vitro anti-sickling properties suggesting that the spice/extracts could be used in combination with other foods in the management and prophylactic control of sickle cell crisis [62].

In vitro anti-microbial properties have been reported for *C. adansonii* leaf extracts against *Pseudomonas aeruginosa*, *Escherichia coli*, *Salmonella typhii*, *Staphylococcus aureus*, *Klebsiella pneumoniae*, *Bacillus subtilis*, *Shigella sonei*, *Pasteurella pestis*, *Yersinia enterocolitica* and anti-fungal properties against two fungi: *Aspergillus niger* and *Candida albicans* [63–65]. It has also been suggested that the traditional use of the leaves against several inflammatory diseases such as rheumatism, arthritis and gout is due to xanthine oxidase inhibition [32].

In vitro anti-oxidant properties and in vivo analgesic properties have also been reported for the methanolic extracts of the stem bark [66]. The methanolic extracts of the leaves and constituent lupeol were also reported to show in vitro anti-oxidant properties [47].

Additionally, there is a report on the in vitro anti-trypanosomal activity of the leaf extracts and isolated aurantiamide acetate, ethyl pyropheophorbide A, purpurin-18 ethyl ester and pyropheophorbide A against the African trypanosome *Trypanosoma brucei brucei* (S427) blood stream forms [31]. *In silico* testing of these ligands with the potential biomolecular targets of *T. brucei*: riboflavin kinase, trypanothione reductase, sterol-14α-demethylase, rohedsain and glutathione synthetase revealed multi-functional scaffolds validating the possibility of anti-trypanosomal activity [31].

3. CONCLUSION

Overall, organoleptic studies encourage the increased utilization of wild ginger, African nutmeg and sacred garlic pear to flavour foods. Moreover, a significant number of in vitro and laboratory animal studies support and explain the folk medicinal usage of these herbs and spices. These spices have anti-microbial, anti-oxidant, anti-inflammatory and in some instances anti-plasmodial and anti-cancer actions. As several metabolic diseases and age-related degenerative disorders are closely associated with oxidative processes in the body, further clinical studies on the use of these spices or their constituents as sources of anti-oxidants and anti-inflammatory agents are needed.

REFERENCES

1. Alais C, Linden G. Food Biochemistry. 2nd ed. Maryland: Aspen; 1999. 213p.
2. Smid EJ, Gorris LGM. Natural antimicrobials for food preservation. In: Rahman MS, editor. Handbook of Food Preservation. New York: Marcel Dekker; 1999. pp. 285–308.
3. Ito N, Fukushima S, Hasegawa A, Shibata M, Ogiso T. Carcinogenicity of butylated hydroxy anisole in F344 rats. J. Natl. Cancer Inst. 1983; 70: 343–347. d
4. Shan B, Cai Y-Z, Brooks JD, Harold Corke H. The in vitro antibacterial activity of dietary spice and medicinal herb extracts. Int. J. Food Microbiol. 2007; 117: 112–119.
5. Smith RJ, Winder ML. Medicinal Garden. In: Ober R, editor. The National Herb Garden Guidebook. Springfield: The Herb Society of America; 1996. pp. 61–71.
6. Beuchat LR, Golden DA. Antimicrobials occurring naturally in foods. Food Technol. 1989; 43: 134–142.
7. Rehm S, Espig G. The Cultivated Plants of the Tropics and Subtropics. Berlin: Priese GmbH; 1991. pp. 277–321.
8. Lai PK, Roy J. Antimicrobial and chemopreventive properties of herbs and spices. Curr. Med. Chem. 2004; 11: 1451–1460.
9. Burkill HM. The Useful Plants of West Tropical Africa. 2nd ed. vol. 5. Families S – Z. London: Crown Agents for Overseas Governments; 2000. 320–321p.
10. Igoli NP. Identification of the flavour components of wild ginger (*Siphonochilus aethiopicus* (Schweinf) B.L. Burtt). [thesis]. Nsukka. University of Nigeria; 2009.
11. Manzini TZ. Production of wild ginger (*Siphonochilus aethiopicus*) under protection and indigenous knowledge of the plant from traditional healers. [thesis]. Pretoria. University of Pretoria; 2005. [Internet] Available from: http://hdl.handle.net/2263/27497 [Accessed: 2016-09-06].
12. Kiew KY. Taxonomic Studies in the genus *Kaempferia* (Zingiberaceae). Notes from the Royal Botanic Garden Edinburgh. 1980; 38: 1–12.
13. van Wyk B-E, Gericke N. People's Plants: A Guide to Useful Plants of Southern Africa. Pretoria: Briza Publications; 2007. 352pp.
14. Nichols G. Some notes on the cultivation of Natal ginger (*Siphonochilus aethiopicus*). Veld Flora. 1989; 75: 92–93.
15. Igoli JO, Ogaji OG, Tor-Anyiin TA, Igoli NP. Traditional medicine practices amongst the igede people of nigeria. Part 11. Afr. J. Trad. CAM 2005;2: 134–152.

16. Holzapfel CW, Marais W, Wessels PL, Van Wyk BE. Furanoterpenoids from *Siphonochilus aethiopicus*. Phytochemistry. 2002;59: 405–407.
17. Lategan CA, Campbell WE, Seaman T, Smith PJ. The bioactivity of novel furanoterpenoids isolated from *Siphonochilus aethiopicus*. J. Ethnopharmacol. 2009; 121: 92–97.
18. Hutchings A. A survey and analysis of traditional medicinal plants as used by the Zulu, Xhosa, and Sotho. Bothalia. 1989; 19: 111–123.
19. Watt JM, Breyer-Brandwijk MG. The medicinal and poisonous plants of Southern and Eastern Africa. 2nded. London: Livingstone; 1962. 1467 p.
20. Viljoen AM, Demirci B, Baser KH, van Wyk B-E. The essential oil composition of the roots and rhizomes of *Siphonochilus aethiopicus*. S. Afr. J. Bot. 2002; 68: 115–116.
21. Nwaji MN, Onyiriuka SO, Taylor DA. 6-(3=Methylbuta-1 ,3=dienyl)indole from *Monodora tenuifolia*. J. Chem. Soc. Chem. Commun. 1972; 327–327.
22. Burkill HM, Dalziel JM. The Useful Plants of West Tropical Africa. vol. 1. Kew: Royal Botanic Gardens; 1985. 976 p.
23. Enweruzoh RO, Okafor DC, Uzoukwu AE, Ukanwoke MO, Nwakaudu AA, Uyanwa CN. Flavour Extraction from *Monodora myristica* and Tetrapleura tetraptera and Production of Flavoured Popcorn from the Extract [Internet]. 2015. Available from: http://www.eajournals.org/wp-content/uploads/Flavour-Extraction-from-Monodora-myristica-And-Tetrapleura-tetraptera-and-Production-of-Flavored-Popcorn-from-the-Extract.pdf [Accessed: 2016-09-07].
24. Eze-Steven PE, Ishiwu CN, Udedi SC, Ogeneh BO. Evaluation of antioxidant potential of *Monodora myristica* (African Nutmeg) [Internet]. 2013. Available from: http://www.ijcmas.com/vol-2-11/P.E.Eze-Steven, %20et %20al.pdf [Accessed: 2016-09-07].
25. Nwozo SO, Kasumu TF, Oyinloye BE. African Nutmeg (*Monodora Myristica*) Lowers Cholesterol and Modulates Lipid Peroxidation in Experimentally Induced Hypercholesterolemic Male Wistar Rats [Internet]. 2015. Available from: https://www.ncbi.nlmnih.gov/pmc/articles/PMC4502738/ [Accessed: 2016-09-07].
26. Keay RWJ. Trees of Nigeria. A revised version of Nigerian trees (1960, 1964) by Keay RWJ, Onochie CFA, Stanfield DP. Oxford: Clarendon; 1989. 5 p
27. Kew Science. *Monodora myristica* (calabash nutmeg) Kew Science [Internet]. 2011? Avail–able from: http://www.kew.org/science-conservation/plants-fungi/monodora-myristica-calabash-nutmeg [Accessed: 2016-09-07].
28. Feyisayo AK, Oluokun OO. Evaluation of antioxidant potentials of *Monodora myristica*(Gaertn) dunel seeds. Afr. J. Food Sci. 2013; 7: 317–324.
29. Burkill HM. The Useful Plants of West Tropical Africa. vol. 1. Families A-D. Kew: Royal Botanic Gardens; 1985. 331–333p.
30. Flowers of India. Garlic Pear Tree [Internet]. 2008? Available from: www.flowersofIndia.net/catalog/slides/Garlic%20Pear%20Tree.html [Accessed: 2016-09-13].
31. Igoli NP, Clements CJ, Singla RK, Igoli JO, Nzekwe U, Gray AI. Anti-trypanosomal activity & docking studies of *Crateva adansonii* DC leaves: novel multifunctional scaffolds. Curr. Top. Med. Chem. 2014; 14: 981–990.
32. Abudullahi A, Hamzah RU, Jigam AA, Yahya A, Kabiru AY, Muhammad H, Sakpe S, Adefolalu FS, Isah MC, Kolo MZ. Inhibitory activity of xanthine oxidase by fractions *Crateva adansonii*. J. Acute Dis. 2012; 1:126–129.
33. Sivarajan VV, Balachandran I. Ayurvedic Drugs and their Plant Sources. Delhi: Oxford and IBH Publishing; 1994. pp. 234–456.
34. Shukla Y, Singh M. Cancer preventive properties of ginger: a brief review. Food Chem. Toxicol. 2007; 45: 683–690.
35. Igoli NP, Obanu ZA. The volatile components of wild ginger (*Siphonochilus aethiopicus* (Schweinf) B.L. Burtt). Afr. J. Food Sci. 2011; 5: 541–549.
36. Belitz H-D, Grosch W. Food Chemistry. Berlin: Springer-Verlag; 1999. 319p.

37. Kirk RS. Pearson's Composition and Analysis of Foods. Kirk RS, Sawyer R, editors. 9th ed. Essex: Longman; 1991. 393 p.
38. Bruice PY. Organic Chemistry. 5th ed. New Jersey: Pearson; 2007. 732p.
39. Igoli NP, Obanu ZA, Gray AI, Clements C. Bioactive diterpenes and sesquiterpenes from the rhizomes of wild ginger (*Siphonochilus aethiopicus* (Schweinf) B.L. Burtt). Afr. J. Trad. CAM. 2012; 9: 88–93.
40. Onyenekwe PC, Ogbadu GH, Deslauriers H, Gagnon M, Collin GJ. Volatile constituents of the essential oil of *Monodora myristica* (Gaertn) dunal. J. Sci. Food Agric. 1993; 61: 379–381.
41. Owokotomo IA, Ekundayo O. Comparative study of the essential oils of *Monodora myristica* from Nigeria [Internet]. Available from: http://www.eurchembull.com/index.php/ecb/article/view/48/108 Accessed: 2016-09-14.
42. Nguefack J, Letha V, Amvam Zollo PH, Mathur SB. Evaluation of five essential oils from aromatic plants of Cameroon for controlling food spoilage and mycotoxin producing fungi. Int. J. Food Microbiol. 2004; 94: 329–334.
43. Igoli JO, Gray AI, Clements CJ, Mouad HA. Antitrypanosomal activity and Cytotoxicity of some Compounds and Extracts from Nigerian Medicinal Plants. In: Phytochemical-Bioactivities and impact on Health. Croatia: Intech; 2011. pp. 375–388.
44. Nkunya MH, Makangara JJ, Jonker SA. Prenylindoles from Tanzanian *Monodora* and *Isolona* Species. Nat. Prod. Res. 2004; 18: 253–258.
45. Ogunwade IA, Ogunbinu AO, Flamini G, Cioni PL, Okeniyi SO. Essential oil profiles of some Nigerian medicinal plants. J. Essent. Oil-Bearing Plant. 2009; 12: 225–235.
46. Cantrell CL, Berhow MA, Phillips BS, Duval SM, Weisleder D, Vaughn SF. Bioactive crude plant seed extracts from the NCAUR oilseed repository. Phytomedicine. 2003; 10: 325–333.
47. Tchimene MK, Nwaehujor CO, Ezenwali M, Okoli CC, Iwu MM. Free radical scavenging activity of lupeol isolated from the methanol leaf extract of *Crateva adansonii* Oliv. (Capparidaceae). Int. J. Pharmacogn. Phytochem. Res. 2016; 8: 419–426.
48. Coopoosamy RM, Naidoo KK, Buwa L, Mayekiso B. Screening of Siphonochilus aetiopicus (Schweinf.) B. L. Burtt for antibacterial and antifungal properties. J. Med. Plants Res. 2010; 4: 1228–1231.
49. Stafford GI, Jäger AK, van Staden J. Effect of storage on the chemical composition and biological activity of several popular South African medicinal plants. J. Ethnopharmacol. 2005; 97: 107–115.
50. Dolara P, Corte B, Ghelardini C, Pugliese AM, Cerbal E, Menichetti S, Lo Nostro A. Local anaesthetic, antibacterial and antifungal properties of sesquiterpenes from myrrh. Planta Med. 2000;66: 356–358.
51. Morita H, Itokawa H. Cytotoxic and antifungal diterpenes from the seeds of *Alpinia galanga*. Planta Med. 1988; 54: 117–120.
52. Noudogbessi JP, Delort L, Chalard P, Billard H, Figueredo G, Ruiz N, Chalchat JC, Sohounhloue D, Chézetquée FC. Anti-proliferative activity of four aromatic plants of Benin. J. Nat. Prod. I. 2013; 6: 123–131.
53. Fouche G, Nieuwenhuizen N, Maharaj V, van Rooyen S, Harding N, Nthambeleni R, Jayakumar J, Kirstein F, Emedi B, Meoni P. Investigation of in vitro and in vivo anti-asthmatic properties of *Siphonochilus aethiopicus* [Internet]. 2011. Available from: <http://researchspace.csir.co.za/dspace/bitstream/10204/5647/1/Fouche_2011.pdf> [Accessed: 2016-09-14].
54. Zschocke S, Rabe T, Taylor JL, Jäger AK, van Staden J. Plant part substitution—a way to conserve endangered medicinal plants? J. Ethnopharmacol. 2000; 71: 281–292.
55. Lindsey K, Jäger AK, Raidoo DM, van Staden J. Screening of plants used by Southern African traditional healers in the treatment of dysmenorrhea for prostaglandin-synthesis inhibitors and uterine relaxing activity. J. Ethnopharmacol. 1999; 64: 9–14.

56. Pillay P, Vleggar R, Maharaj VJ, Smith P, Lategan CA, Chouteau F, Chibale K. Antipla–smodial hirsutinolides from *Vernonia staehelinoides* and their utilization towards sim-plified pharmacophore. Phytochemistry. 2007; 68: 1200–1205.
57. Iweala EEJ, Liu F-F, Cheng R-R, Li Y, Omonhinmin CA, Zhang Y-J. Anti-cancer and free radical scavenging activity of some Nigerian food plants in vitro. Int. J. Cancer Res. 2015; 11: 41–51.
58. Magori N, Nyandoro SS, Munissi JJE, Heydenreich M. Antimycobacterial and cytotoxicity evaluation of the constituents of *Monodora carolinae* [Internet]. 2013. Available from: http://www.ajol.info/index.php/tjs/article /download /105696/95712
59. Onguéné PA, Ntie-Kang F, Lifongo LL, Ndom JC, Sippl W, Mbaze LM. The potential of anti-malarial compounds derived from African medicinal plants. Part I: A pharmacological evaluation of alkaloids and terpenoids. Malaria J. 2013; 12: 449.
60. Erukainure OL, Oke OV, Owolabi FO, Kayode FO, Umanhoulen EE, Aliyu M. Chemical properties of *Monodora myristica* and its protective potentials against free radicals in vitro. Oxid. Antioxid. Med. Sci. 2012; 1:127–132.
61. Oyinloye BE, Adenowo AF, Osunsanmi FO, Ogunyinka BI, Nwozo SO, Kappo AP. Aqueous extract of *Monodora myristica* ameliorates cadmium-induced hepatotoxicity in male rats. Springerplus. 2016; 5: 641.
62. Uwakwe AA, Nwaoguikpe RN. *In-vitro* antisickling effects of *Xylopiaa ethiopica* and *Monodora myristica*. J. Med. Plant Res. 2008; 2: 119–124.
63. Igoli NP, Gray AI, Clements C, Igoli JO, Nzekwe U, Singla RK. Scientific Investigation of Anti-trypanosomal Activity of *Crateva adansonii* DC Leaves. Indo Global J. Pharm. Sci. 2012; 2: 226–229.
64. Agboke AA, Attama AA, Momoh MA. Evaluation of the antimicrobial activities of crude extract of *Cryptolepis sanguinolenta* and *Crateva adansonii* leaves and their interactions. J. Appl. Pharm. Sci. 2011; 1: 85–89.
65. Lagnika L, Anago E, Atindehou M, Adjahoutonon B, Dramane K, Sanni A. Antimicrobial activity of *Crataeva religiosa* Forst against bacteria isolated from *Thryonomysswinderianus* Temminck. Afr. J. Biotechnol. 2011; 10: 10034–10039.
66. Udeh NE, Onoja SO. Analgesic and free radical scavenging activities of hydromethanolic extract of *Crateva adansonii* stem bark. J. Intercult. Ethnopharmacol. 2015; 4: 224–227.

CHAPTER 10

A Review on Geranium wallichianum D-Don Ex-Sweet: An Endangered Medicinal Herb from Himalaya Region

Sabah Shaheen, Yamin Bibi, Mubashir Hussain, Muhammad Iqbal*, Hafiza Saira, Irum Safdar, Hira Mehboob, Qura Tul Ain, Kiran Naseem and Sadaf Laraib

Department of Botany, PMAS-Arid Agriculture University, Rawalpindi, Pakistan

ABSTRACT

Geranium wallichianum D. Don ex sweet belongs to Family Geraniaceae, distributed in Himalaya region of Pakistan. It is highly medicinal. It is native to Pakistan due to its restricted geographic ranges, specific habitats and anthropogenic activities more than 75% population has been reduced. Major threats are grazing of pasture, forest encroachment, soil-slope erosion, poverty, forest fire, illegal trade, habitat loss, invasive species and the lack of adequate training. In this review efforts are made to document the conservation status of important medicinal plant *G. wallichianum*. It is most threatened species and deserves special attention. Due to insufficient information, there is no organization strategy for conservation of this plant. With great ethnomedicinal value of this plant species draw the attention of the researchers to conserve it. Several conservation studies and efforts are making to conserve it, but comprehensive and long lasting conservation policies are necessary to ensure its conservation.

Keywords: Endemic plants; Ethno-medicinal; Endangered; *Ex-situ* & *In-situ* conservation

INTRODUCTION

Plants fulfill the basic needs of human with materials for existence, which is medicinal, economic and fodder values [1]. From a region plant species is lost, the knowledge enclosed in and also destroyed slowly until become disappear forever [2]. In Pakistan, more than 10% of the flora is endangered [3]. The data is limited on the conservation status of indigenous plant species; however, there is a lot of argument in the presented information. In Pakistan 709 plants are reported as threatened and endangered [4]. Among them about 580 are flowering plants [5]. In contrast, more recently reports represent only 21 flowering plants that are threatened in Pakistan [6]. Conservation of a plant species are initially hot problem [2]. Due to over harvesting, medicinal plant population are become severally reduced [7-9]. In Pakistan, more than 10% endangered flora is present [3]. In this review efforts are made to document the conservation status of important medicinal plant *G. wallichianum*. At local level this plant is used *G. wallichianum* , is most threatened species and need special attention. Overexploitation of plant resources combined with improper harvesting and post harvesting techniques are increases pressure on this plant.

GEOGRAPHICAL DISTRIBUTION

In Pakistan its distribution is Swat, Murree hills, Baluchistan, Hazara, Chitral, Dir, Gilgit, Shogran valley and Lowari top. While in world distribution is East Asia-Himalayas from Nepal, Afghanistan to Bhutan [10]. It's very common in mountainous areas of Pakistan like Khyber Pakhtunkhwa, Azad Jammu Kashmir and Murree (**FIGURE 1**) [11].

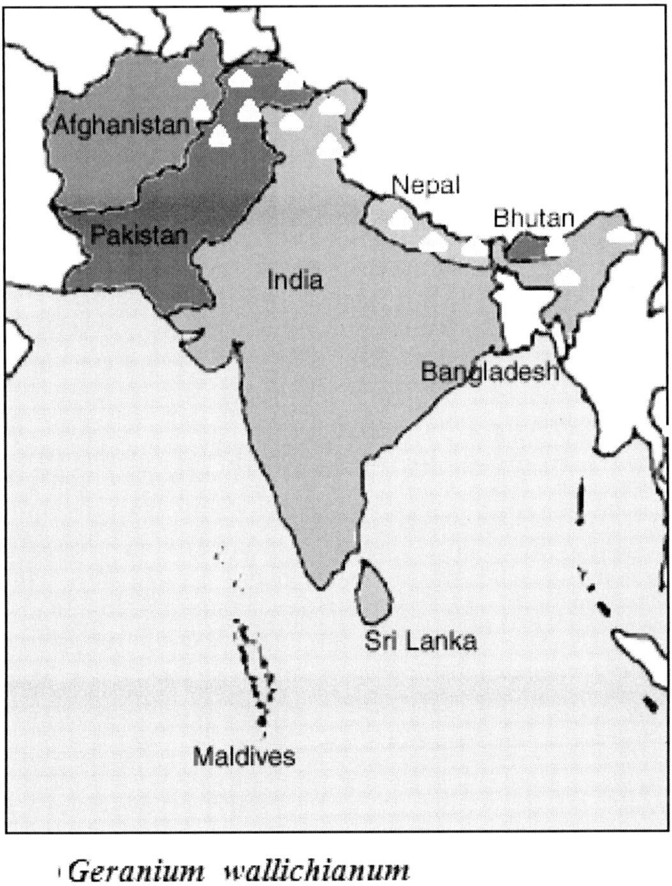

FIGURE 1: Geograpical Distribution of Geranium wallichianum.

Occurrence

Commonly found in Forests, shrubberies and open slopes, 2400-3600 m [10]. It is found in moist damp places [12]. Indian Himalaya region the altitude range of *G. wallichianum* is 2500-4000 m [13]. In Azad Kashmir, its altitude range is 6000-8000 [12].

Taxonomy and Morphology

Geranium wallichianum D. Don ex sweet usually named as "Shepherd's needles" and locally as "Ratanjot" [14]. It's commonly called Wallach cranes bill [15]. Its English name is Buxton's blue and in Tib it is known as Rus Jot [10]. It is also known as Sra zeal [16]. It is also called as mamekh [17]. It also known srazela [18]. It is also known as ratinloog and rattenjot [19].

Morphological and Taxonomic Features

Geranium wallichianum D. Don ex-sweet belongs to Family Geraniaceae. It comprises 5 genera and 750 species, it is worldwide in distribution but commonly in mild areas [15]. It comprises 4 genera and 26 native species in Pakistan [20]. There are two tribes in family i.e., Geraniaceae and Pelargonieae [21]. The length of plant is 45 cm or more. Stem ascending, adolescent and 1-4 ft. length. Leaves palmate partite, reniform, 5.5-7 cm wide, lamina pubescent-villous, 5-angled lobes rhomboid cuneate; stipules 0. 9-1.4 cm in length, pubescent to lightly villous, widelance late, sometimes apex 2-fid. Size of Petiole is 7 cm, at the base of lamina dense villous. Peduncles 18.5 cm long, 2-flowered, lance late, bracts 0.7-1.4 cm in length, appressed pubescentglandular, Size of Sepals is 10-12 mm long, glandular hairy, elliptic ovate, awn 1 mm long; base villous 1.4-2 cm long Petals [13]. It is a perennial herb, blue-colored flowers ranges from 3-8. 5 cm in diameter [22]. March to April is its flowering period [10]. It is observed that its flowering period is June to September [23]. It is reported that its flowering period is August-September [14].

Conservation Status

G. wallichianum D. Don ex sweet is endemic to Pakistan [24]. In Kalam it is rare [25]. It is endangered in AJK [26]. In swat, it is rare [27]. It is rare in poonch valley [14]. In Dir Kohistan valley it is fall in most threatened species and need special attention [28]. It is near endemic in Nanda Devi [13].

Market Value

G. wallichianum (**FIGURE 2A**) is commonly brought to the market as medicinal plant, from the mountain zone, mainly collected (2500-3100 m.) under *Picea smithiana* (**FIGURE 2B**) and *Abies pindrow* forest; a few species are collected from (sub-) alpine areas [29]. At altitude between 2,800-3,800 m, the vegetation is subalpine type represented by *G. wallichianum*. Mostly rhizome is sold at national and international market from 500-1000 Rs/kg (**FIGURE 3**) [30].

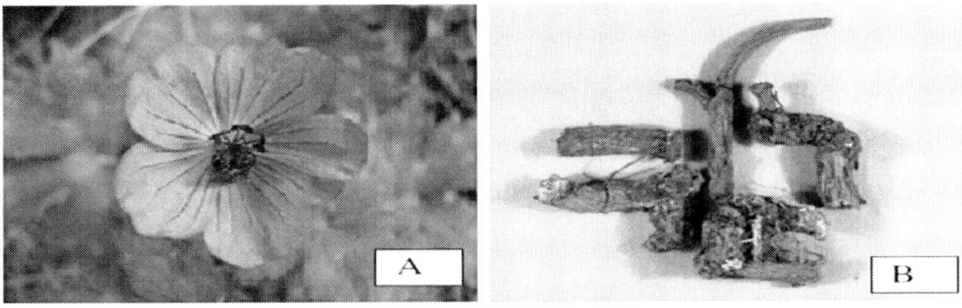

Figure 2: A: *G. wallichianum* B: *Picea smithiana* .

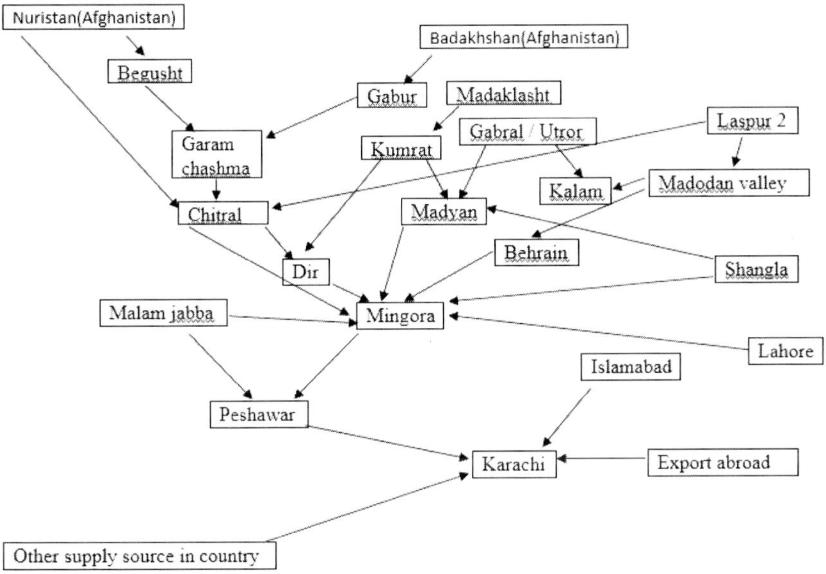

Figure 3: Supply routes from collection sites and other areas to local, national and international markets.

Ethanomedicinal Uses

For primary health care 75-80% of the world's population are used herbal medicines because of superior traditional adequacy, better compatibility with human body and minor side effects.

In Pakistan, this species is endemic. More than 75% population is reduced. Unsustainable assembly of the rhizome and over exploitation is its main threats. As medicinal plant, there must be bane on the unsustainable collection and over exploitation of this species. Owing to over harvesting and extreme use of this plant, it needs to be conserved before it is vanish [24]. This plant need to be preserved before they disappear because of excessive commercial harvest and local uses. The major factors which take part in plant diversity loss found are grazing of pasture, forest encroachment, soil-slope erosion, collection of medicinal plant, agro-system threats, poverty, forest fire and invasive species intensify the environment. Further research is needed in extensive field conservation. Climate variations are strongly influences the dispersal and richness of plant variety margins of equally in latitude and altitude [38,39]. Deforestation, illegal trade, habitat loss, invasive species, growing demand for natural resources, industrial pollution, and the lack of adequate training are major threats of this plant [40]. Through illicit means collection of this plant continues from the forest (**TABLE 1**) [41,42].

Table 1: Medicinal uses of Ethanomedicinal plants.

Ethanomedicinal Uses	Citations
Root is used in backache, sexual debility, joint pain, colic, jaundice, and kidney and spleen disorder.	[10]
In different assay the crude extracts and different fractions of rhizomes and leaves showed varied degree of antimicrobial activities and enzyme inhibitions	[11]
Root is used for astringent, ear & eye disorder and toothache	[13]
In medicine its root and fruit is used. It may be Yellow, red, and brown in colour. It shows astringent property and used in toothache and eye infection	[15]
Root is tonic and is used against jaundice and also effective to reduce blood pressure.	[16]
Pods of *Pistacia chinensis* with root decoction to treat cough and fever and urinary complaints. It also used for Fever, headache, analgesic and Respiratory disorder, Genital and sexual diseases.	[18]
For vision problem and blood purification its floral extract is used. *Daucuscarrota* with its root powder is used for jaundice, spleen problems and kidney	[19]
The root stock is used in eye troubles and also in the cure of toothache.	[20]
The roots of the plant are used in mouth ulceration, dysentery, diarrhea, passive hemorrhage, and leucorrhoea.	[21]
Geranium wallichianum D. Don ex Sweet is usually used as tonic by women especially for physical fitness and other internal body complaints.	[23]
Root is used as astringent and clotting agent.	[26]
Whole plant is used for curing stomach acidity, gastric problems stomachache, arthritis, back pain and rheumatism.	[27]
The decoctions of the plant with hydrastine are used in disorders like gonorrhea, leucorrhoea and diabetes	[31]
Aqueous ethanolic extract obtaining from *G. wallichianum* showed antibacterial activity against Staphylococcus aureus.	[32]
The rhizomes of the plant are used to treat ulceration, dysentery, diarrhea, hemorrhage and leucorrhoea.	[33]
Crushed mixture of dried root is mixed with milk and sugar and is used in back pain, gout and also used in strengthening of the body muscles and bones.	[34]
Isolated for the first time from *Geranium wallichianum* compounds are identified as ursolic acid, β-sitosterol, stigma sterol, β-sitosterolgalactoside, herniarin, and 2, 4, 6-trihydroxyethylbenzoate which is highly used	[35]
The roots are used in mouth ulceration. An efficient astringent which is given for chronic dysentery, diarrhea and leucorrhoea.	[36]
Leaves are effective for liver problem.	[37]

Conservation Strategies39

The cultivation of *G. wallichianum*, ex-situ management is ignored in the past [43]. An ex-situ conservation effort was carried out by cultivating threatened and medicinally important plant species. *G. wallichianum* survived in the new habitat. Similar ex-Situ conservation effort was carried out by Ahmad [44]. An ex-situ conservation effort was made by cultivating medicinal plant *G. wallichianum* [17]. In 2002, an ex-situ conservation effort was made and plant is shifted from Swat to Tindodog for cultivation but it is failed [27]. The most effective route to conserve plants in their homeland and confirm use of plant resources sustainable, it is necessary to educate the people, give concept about sustainability and provide the chances for well settlement. In this way, the misuse of the native flora might be reduced. Excessive use of *G. wallichianum* and due to overgrazing plants gets threatened, endemic or vulnerable. Some steps are taken to reduce the negative effects of misuse and mismanagement of this plant like revolving greasing method introduced, cultivation increases, sustainable use, public awareness through media and stop illegal trade and don't up root the plant. *G. wallichianum* is an important plant with respect to its ethnobotanical importance. It is reduced more than 75% population it is listed as endangered plant. Few of the problems that arise during various conservation activities include; poverty, lack of education and awareness, lack of marketing opportunity, heavy human and animal pressure on plant resources. Communities should be given exposure and awareness that they start working for the environment and all the plant biodiversity that needs to be protected. They can join institutions to support this purpose and play their role in conserving these massive endangered plant species. Furthermore, government should ensure provision and promotion of advanced facilities for research and study in different aspects of plant conservation and should also promote and arrange courses, conferences, lectures and training courses at different levels for those already engaged in this field as well as general public.

Several conservation studies and efforts are make to conserve it, but a more comprehensive and multiple-approach conservation policies are required to ensure its conservation. Indirect methods like callogenesis and somatic embryogenesis can be applied for production of plantlets helping in establishment of large scale plant nurseries [45-47]. Ali et al. established a protocol for synthetic seed production by using artificial coating material (sodium alginate) and complexing agent (calcium chloride) [48]. This synthetic seed technology can also be very helpful for medicinal plants such as *G. wallichianum* species. These seeds can be stored and viable for comparatively longer period if provided suitable conditions. Germination of seeds can be enhanced by treating seeds with exogenous hormone like chemicals dissolved in water [49].

Long term monitoring programs and propagation studies may prove fruitful in conserving such critically endemic species. There recommendations are suggested for further research in future are biodiversity conservation awareness and information must be provided by people, anthropogenic activities should be avoided, for plant conservation protected area should be formed and afforestation plan should be introduced. Efforts must be made to ensure public participation in conservation programs and awareness through training or utilization of mass media coupled with permanent monitoring programs. Permanent monitoring programs should be developed and conservation strategies should be planned to maintain threatened species [50-53].

CONCLUSION

G. wallichianum is an important plant with respect to its ethno botanical importance. It is reduced more than 75% population it is listed as endangered plant. Number of factors is present for declining of their population. The plant is used to cure a wide range of disorders from simple toothache to complex illnesses of blood pressure and diabetes. Although the plant is recorded as endemic, due to insufficient information, there is no organization strategy for conservation of this plant. So, this plant need to conserve by long term monitoring, reducing man made pressure on plant resources, proper documentation is best method for its conservation. Proper conservation also demands of extinction actions and education about the significance of plants as well as planned cultivation initiatives.

REFERENCES

1. Morgan WT (1981) Ethnobotany of the Turkana: Use of Plants by a Pastoral People and their Livestock in Kenya. Journal of Economic Botany 35: 96-130.
2. Cunningham AB (1996) Professional Ethics and Ethnobotanical Research. In: Alexiades MN, Sheldon JW (eds.) Selected Guidelines for Ethnobotanical Research: A Field Manual the New York. Botanical Garden Bronx, New York US, pp: 19-51.
3. Shinwari ZK, Gilani SS, Shoukat M (2002) Ethnobotanical resources and implications for curriculum. In: Shinwari ZK, Hamilton A, Khan AA (eds.) Proceedings of Workshop on Curriculum Development in Applied Ethnobotany Nathiagali Abbottabad WWF and Pakistan, pp: 34.
4. Chaudhri M, Qureshi R (1991) Pakistan Endangered Flora II: A Checklist of rare and seriously threatened taxa of Pakistan PakistanSystem 5: 1-84.
5. Nasir YJ (1991) Threatened Plants of Pakistan. In: Ali SI, A Ghaffa (eds.) Plant Life of South Asia: Proceedings of the International Symposium, Karachi, pp: 229-234.
6. Ali H, Qaiser M (2010) Contribution to the Red List of Pakistan: A case study of Astragalusgahiratensis Ali (Fabaceae-Papilionoideae) Pakistan. Journal of Botany 42: 1523-1528.
7. Sher HZD, Khan A, Khan U, Hussain F (2005) In-situ conservation of some selected medicinal plants of Upper Swat. Pakistan Acta Botanica Yunnanica 27: 27-36.
8. Ahmad I, Hussain M, Ahmad MSA, Hameed M (2008) Spatio-temporal effects on association of plant species in soone valley of Pakistan Pakistan. Journal of Botany 40: 1865-1876.
9. Ahmad I, Ahmad MSA, Hussain M, Hameed M, Ashraf MY, et al. (2009) Spatio-temporal effects on species classification of medicinal plants in Soone valley of Pakistan International. Journal of Agriculture Biology 11: 64-68.

10. Ahmad M, Khan MA, Zafar M, Arshad M (2010) Use of chemotaxonomic markers for misidentified medicinal plants used in traditional medicines. Journal of Medicinal Plants Research 4: 1244-1252.
11. Ismail M, Hussain J, Khan A, Khan AL, Ali L, et al. (2012) Antibacterial Antifungal Cytotoxic Phytotoxic Insecticidal and Enzyme Inhibitory Activities of Geranium wallichianum. Journal List Evidence Based Complement Alternative Medicine 1: 113-118.
12. Khan MAM, Khan A, Hussain M, Mujtaba G (2014) Plant diversity and conservation status of Himalayan Region Poonch Valley Azad Kashmir (Pakistan) Pakistan. Journal of Pharmaceutical Science 127: 1215-1239.
13. Vikram SR (2014) Indigenous uses of medicinal and edible plants of Nanda Devi biosphere reserve-a review based on previous studies. Global Journal of Research on medicinal Plants and indigenous medicine 3: 57-66
14. Nadkarni KM (1976) Indian Materia Medica 3rd edn. Popular Book Prakashan; Bombay India: Depootapapeshwar Prakashan Panvel, pp: 1050.
15. Siva R (2007) Status of natural dyes and dye yielding plants in India Current Science 92: 10.
16. Razzaq A, Hadi F, Rashid A, Ibrar M, Ali U (2015) Exploration of Medicinal Plants and Their Conservation Status at Higher Altitude of District Shangla Khyber Pakhtunkhwa Pakistan American-Eurasian. Journal of Agriculture and Environmental Science 15: 328-331.
17. Shah GM (2007) Plant and plant resources of siran valley. Mansehra NWFP Pakistan PhD thesis, Quaid-i-Azam University Islamabad Pakistan, pp: 68.
18. Akhtar N, Rashid A, Murad W, Bergmeier E (2013) Diversity and use of ethno-medicinal plants in the region of Swat North Pakistan. Journal of Ethnobiology and Ethnomedicine 9: 25-27.
19. Chaudhary MI, Ahmed F, Maqbool M, Hussain T (2013) Ethnomedicinal Inventory of Flora of Maradori Valley District Forward Khahuta Azad Kashmir Pakistan American. Journal of Research Communication 1: 239-261.
20. Watt G (1972) In: The Dictionary of Economic Products of India. Vol 3. Cosmo Publications, New Delhi India, pp: 448-490.
21. Shinwari ZK, Gilani SS (2003) Swat profile. In: Medicinal and other useful plants of District Swat Pakistan ZK Shinwari, AA Khan, Nakaike T, Al-Aziz (eds.) Communication Peshawar Pakistan, pp: 9-17.
22. Council of Scientific and Industrial Research (1968) The Wealth of India: A Dictionary of Indian Raw Materials and Industrial Products New Delhi India: National Instituteof Science Communication.
23. Qureshi RA, Ghufran MA, Gilani SA, Yousaf Z, Abbas G, et al. (2009) Indigenous Medicinal Plants Used By Local Women in Southern Himalayan Regions of Pakistan Pakistan. Journal of Botany 41: 19-25.
24. Ullah A, Rashid A (2014) Conservation status of threatened medicinal plants of Mankial Valley Hindukush Range Pakistan International. Journal of Biodiversity and Conservation 6: 59-70.
25. Ali H, Ahmad H, Marwat KB, Yousaf M, Gul B, et al. (2012) Trade Potential and Conservation Issues of Medicinal Plants in District Swat Pakistan Pakistan. Journal of Botany 44: 1905-1912.
26. Ahmad SK, Qureshi AR, Hameed M, Ahmad F, Nawaz T (2012) Conservation Assessment and Medicinal Importance of some Plants Resources from Sharda Neelum Valley Azad Jammu and Kashmir Pakistan International. Journal of Agriculture and Biology 2: 12-389.
27. Hamyun MS, Khan A, Kim HY, Leechae IJ (2006) Traditional knowledge and ex-situ conservation of some threatened medicinal plants of swat kohistan Pakistan International. Journal of botany 38: 205-209.
28. Jan GF, Jan G, Hamayun M, Khan K, Khan K (2014) Diversity and conservation status of vascular plants of DirKohistan valley Khyber Pakhtunkhwa Province. Journal of Biodiversity and Environmental Sciences 5: 164-172.

29. Saqib ZR, Malik N, Shinwari MI, Shinwari ZK (2011) Species Richness Ethnobotanical Species Richness and Human Settlements along a Himalayan Altitudinal Gradient: Prioritizing Plant Conservation in Palas Valley Pakistan Pakistan. Journal of Botany 43: 129-133.
30. Sher H, Yousaf S (2014) Collection and marketing of high value pharmaceutically and therapeutically important plants from the Swat District Pakistan African. Journal of Pharmacy and Pharmacology 8: 507-513.
31. Nadkarni KM (1954) Indian MateriaMidica 3rd Ed Popular Book Prakashan Bombay India Dhootapapeshwar Prakashan Panvel, pp: 1050.
32. Ahmad B, Ismail M, Iqbal Z, Chaudhry MI (2003) Biological Activities of Geranium wallichianum Asian. Journal of Plant Sciences 2: 971-973.
33. Zabta S, Ashiq AK, Toshiyuki N (2003) Medicinal and other useful plants of District Swat Pakistan WWF-Pakistan, pp: 68.
34. Gilani SAR, Qureshi A, Gilani SJ (2006) Indigenous Uses of Some Important Ethnomedicinal Herbs of Ayubia National Park Abbottabad Pakistan Ethnobotanical Leaflets 10: 285-293.
35. Ismail M, Ibrar M, Iqbal Z, Hussain J, Hussain H, et al. (2009) Chemical Constituents and Antioxidant Activity of Geranium wallichianum Records of Natural Products 3: 193-197.
36. Hazrat A, Nisar M, Shah J, Ahmad S (2011) Ethnobotanical Study of Some Elite Plants Belonging To Dir Kohistan Valley Khyber Pakhtunkhwa Pakistan Pakistan. Journal of Botany 43: 787-795.
37. Rana PK, Kumar P, Singhal VK, Rana JC (2014) Uses of Local Plant Biodiversity among the Tribal Communities of Pangi Valley of District Chambain Cold Desert Himalaya India. The Scientific World Journal, pp: 15.
38. Hickling RD, Roy B, Hill KJ, Fox R, Thomas CD (2006) The distributions of a wide range of taxonomic groups are expanding pole wards Global Change Biology 12: 450-455.
39. Tryjanowski P, Sparks HT, Profus P (2005) Uphill shift in the distribution of the white stork Ciconia Diversity and Distribution 11: 219-223.
40. Alam J, Ali SI (2009) Conservation Status of Astragalusgilgitensis Ali (Fabaceae): A Critically Endangered Species in the Gilgit District Pakistan Phyton(Horn Austria) 48: 211-225.
41. Chauhan RS (2004) Ecophysiology Agro-technology and Trade of an Endangered Medicinal herb. Nardostachys jatamansi DC .Ph D Thesis HNB Garhwal University Srinagar Garhwal.
42. Kala CP (2005) Indigenous use population density and conservation of threatened medicinal plants in protected areas of the Indian Himalaya Conservation Biology 19: 368-378.
43. Sher H, Ali H, Rehman S (2012) Identification And Conservation Of Important Plant Areas (Ipas) For The Distribution of Medicinal Aromatic And Economic Plants In The Hindukush-Himalaya Mountain Range Pakistan. Journal of Botany 44: 187-194.
44. Ahmad B, Ismail M, Iqbal Z, Chaudhry MI (2003) Biological Activities of Geranium wallichianum Asian. Journal of Plant Sciences 2: 971-973.
45. Iqbal M, Aamir A, Naima HN, Umair AK, Muhammad NAF, et al. (2016) Effect of Explants and Growth Regulators on the Expression of Callogenesis Somatic Embryogenesis and Plantlets Formation in Sugarcane (Saccharum officinarum L) International. Journal of Biosciences 9: 147-156.
46. Iqbal M, Raja NI, Asif S, Ilyas N, Hussain M, et al. (2016) In Vitro Study of Callogenesis and Regeneration Potential of Elite Wheat (Triticum aestivum L) Cultivars American. Journal of Plant Sciences 7: 2515-2526.
47. Hussain M, Raja NI, Iqbal M, Iftikhar A, Sadaf HM, et al. (2016) Plantlets regeneration via somatic embryogenesis from the nucellus tissues of Kinnow mandarin (Citrus reticulata L) American. Journal of Plant Sciences 7: 798-805.
48. Ali A, Muhammd I, Abdul M, Naima HN, Abdul R, (2013) In vitro conservation and production of vigorous and desiccate tolerant synthetic seed formation in sugarcane

(Saccharum officinarum L) Conference paper published in proceedings of 47th annual conference of Pakistan Society of Sugar Technologists Rawalpindi Pakistan, pp: 9-10.

49. Iqbal M, Asif S, Ilyas N, Raja NI, Hussain M, et al. (2016) Effect of Plant Derived Smoke on Germination and Post Germination Expression of Wheat (Triticum aestivum L) American. Journal of Plant Sciences 7: 806-813.

50. Hussain M, Yamin B, Naveed IR, Muhammad I, Sumaira A, et al. (2016) A review of therapeutic potential of Ajuga bracteosa: A critically endangered plant from Himalaya. Journal of Coastal Life and Medicine 4: 918-924.

51. Tahir N, Bibi Y, Iqbal M, Hussain M, Laraib S, et al. (2016) Overview of Dioscorea deltoidea Wall Ex Griseb: An endangered medicinal plant from Himalaya region. Journal of Biodiversity Environmental Sciences 9: 13-24.

52. Iqbal M, Bibi Y, Raja NI, Ejaz M, Hussain M, et al. (2017) Review on Therapeutic and Pharmaceutically Important Medicinal Plant Asparagus officinalis L. J Plant Biochem Physiol 5: 180.

53. Hussain M, Raja NI, Akram A, Iftikhar A, Ashfaq D, et al. (2017) A status review on the pharmacological implications of Artemisia absinthium: A critically endangered Plant Asian Pacific. Journal of Tropical Disease 7: 185-192.

CHAPTER 11

Vanillin and its Detection in Air

Saaya Minematsu[1] and Xing-Zheng Wu[2*]

[1] *Environmental Chemistry, Fukuoka Technical High School, Fukuoka, Japan*

[2] *Department of Life, Environment and Materials Science, Fukuoka Institute of Technology, Fukuoka, Japan*

ABSTRACT

Vanillin (4-hydroxy-3-methoxybenzaldehyde) is an important flavor and aroma molecule, which has been widely used in not only foods and beverages such as chocolate and dairy products, but also masking unpleasant tastes in medicines or livestock fodder. Its chemical properties, manufacturing methods, novel applications, and developments in fast detections in air are discussed in detail.

Keywords: vanillin, properties, detection, capillary electrophoresis

1. INTRODUCTION

Vanillin (4-hydroxy-3-methoxybenzaldehyde) is an important flavor and aroma molecule but is also of interest because of its biogenetic relationship to the phenylpropanoid pathway and to other molecules of physiological significance, notably salicylate [1]. Vanillin is the most important ingredient of the well-known vanilla, which is a complex blend of flavor and fragrance ingredients extracted from the seed pods of the vanilla orchid. As a flavorings agent, vanillin is used in not only foods and beverages such as chocolate and dairy products, but also masking unpleasant tastes in medicines or livestock fodder [2].

Here, its properties, manufacturing methods, and novel applications are discussed. Furthermore, its detection in air is introduced.

2. MOLECULAR STRUCTURE AND PROPERTIES OF VANILLIN

Vanillin is the common name for 3-methoxy-4-hydroxybenzaldehyde, and its molecular structure is shown in **Figure 1** .

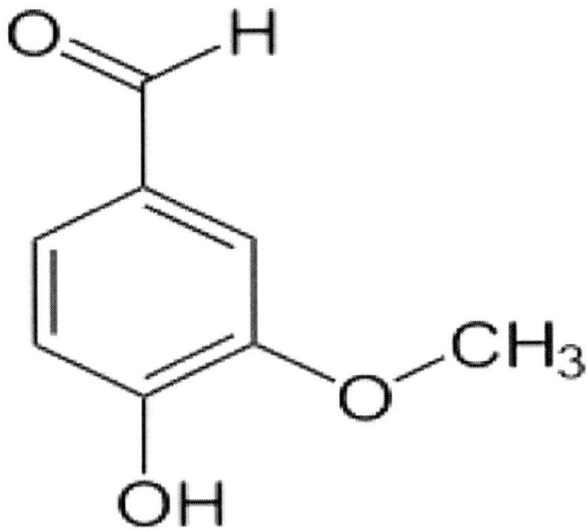

Figure 1. Molecular structure of vanillin.

2.1. Toxicity About Vanillin

Toxicity about vanillin was studied as early as 1940 [3]. Generally, vanillin does not give human skin irritation and produce no sensitization reactions. Vanillin is considered to be a secondary allergen because sensitivity was found only in patients sensitive to vanilla, isoeugenol, and coniferyl benzoate. Animal studies showed that vanillin was not carcinogenic.

2.2. Antioxidant, antifungal or antimicrobial, and antimutagenic properties of vanillin and vanillin derivatives

Vanillin and vanillin derivatives have antioxidant and antimutagenic properties. Antifungal activities of vanillin and 33 vanillin derivatives against the human fungal pathogen Cryptococcus neoformans, which was the main pathogen of cryptococcal meningitis in immunocompromised patients, have been studied [4]. Functional groups in the vanillin derivatives seemed to affect antifungal activity. The hydroxyl or alkoxy group seemed to be more effective than the halogenated or nitrated group in benzaldehyde in antifungal ability. O-vanillin and o-ethyl vanillin were with the highest antifungal activity against *C. neoformans* in the vanillin derivatives. O-Vanillin was further found to be able to cause mitochondrial dysfunction and trigger oxidative stress. These antifungal mechanisms of o-vanillin were experimentally confirmed by the significantly reduced growth of the mutants lacking the genes involved in mitochondrial functions and oxidative stress response.

Evaluation in which structural elements of the vanillin molecule are responsible for its antifungal activity was also investigated [5]. Minimum inhibitory concentrations (MICs) of vanillin, its six direct structural analogs, and several other related compounds were determined in yeast extract peptone dextrose broth against a total of 18 different food spoilage molds and yeasts. Experimental results showed that the antifungal order of isomers of hydroxybenzaldehyde and anisaldehyde was 2- > 3- > 4- and 3- > 2- > 4-, respectively. The aldehyde moiety of vanillin seems to play a key role in its antifungal activity, but side-group position on the benzene ring also influences this activity.

Antimicrobial activities and the MICs of solutions containing vanillin and vanillic acid against *Staphylococcus aureus, Staphylococcus epidermidis, Bacillus cereus, Enterobacter aerogenes, Escherichia coli,* and *Yersinia enterocolitica* were experimental studied by the agar well-diffusion method [6]. The experimental results showed that the vanillin and vanillic acid was with inhibitory activity against all of the bacteria. Moreover, the MIC of the vanillin and vanillic acid decreased with the vanillic acid concentration. It suggested that thermal treatment of vanillin-containing food may lead to products with higher antioxidant and antimicrobial properties.

2.3. Protection Human Keratinocyte Stem Cells Against Ultraviolet-B Irradiation

Ultraviolet-B (UVB) irradiation is one of major factors, which induce cellular damages in the epidermis. Protective effects and mechanisms of vanillin against UVB-induced cellular damages in keratinocyte stem cells (KSC) have been investigated recently [7]. Experimental results indicated that vanillin significantly decreased the UVB irradiation-induced cytotoxicity. Also, vanillin seemed to be able to induce production of pro-inflammatory cytokines. It was explained that vanillin could significantly reduce phosphorylation of ataxia telangiectasia mutated (ATM), tumor suppressor protein 53 (p53), serine threonine kinase checkpoint kinase 2 (Chk2), c-Jun N-terminal kinase/stress-activated protein kinase (JNK), S6 ribosomal protein (S6RP), p38/mitogen-activated protein kinase (p38), and histone 2A family member X (H2A.X) generated by the UVB. Vanillin also could inhibit UVB-induced activation of p53 luciferase reporter. The results suggested that vanillin protects KSC from UVB irradiation. Vanillin may play its role through the suppression of downstream step of MDM2 in UVB irradiation-induced p53 activation.

3. MANUFACTURING METHODS OF VANILLIN

A simple laboratory synthesis is illustrated in **Figure 2** to make a small amount of vanillin. This synthesis scheme involves electrophilic bromination of 4-hydroxybenzaldehyde, followed by copper-catalyzed methoxylation.

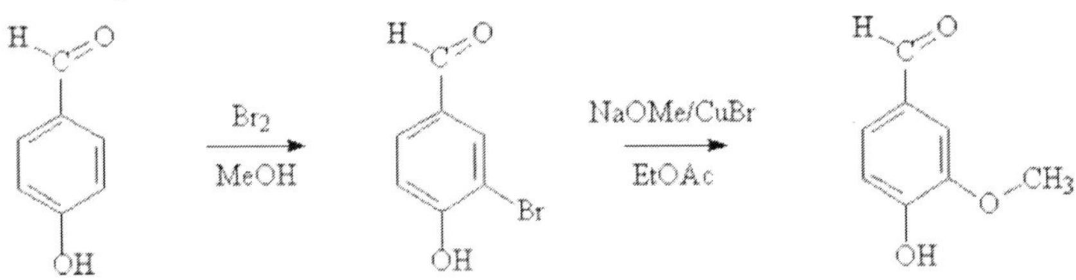

Figure 2. Laboratory synthesis scheme of vanillin.

For large-scale industrial syntheses, a classic early method starts from eugenol, which occurs naturally in cloves, nutmeg, and cinnamon. This isomerizes to isoeugenol in alkaline solution, and this in turn can be oxidized (by nitrobenzene) to vanillin (**Figure 3**).

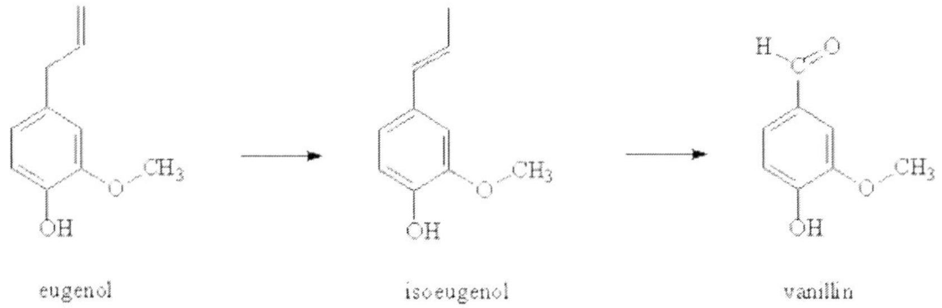

Figure 3. Industrial synthesis scheme of vanillin.

Lignin was known to be a source of vanillin as early as at the beginning of the twentieth century. Lignin is a well-known polymer, which plays strengthening role in woods and in the cell walls of plants. Since the 1920s, much of the world's vanillin was extracted from lignin waste from the cellulose industry [8].

Recently, potential of industrial Eucalyptus globulus sulfite liquor and kraft liquors was evaluated for the production of syringaldehyde and vanillin by oxidation with O_2 in alkaline medium [9]. The Eucalyptus globulus sulfite liquor and kraft liquors were collected at different stages of processing before the recovery boiler. Under controlled temperature and pressure, the oxidations were performed in a jacketed reactor by two methods. One was the direct oxidation of pulping liquors, and the other was the one of kraft lignins isolated from liquors. Products profiles were established, as well as the yields, temperature and O_2 uptake during the reaction. Results showed that sulfite liquor was the best raw material leading to the highest yield by direct oxidation. Thin kraft liquor (KL) was with the second high yield. Proportion of by-products such as syringic and vanillic acids was low.

Natural vanillin was obtained by plant tissue culture early. Molecular biology and microbial biotransformation techniques can also be used to produce natural vanillin [10]. These techniques rely on natural vanillin precursor molecules (eugenol, isoeugenol, curcumin, or ferulic acid), and their enzymatic reaction pathways are very different. Among them, microbial biotransformation method seems to be the most promising for large amount of natural vanillin production with high efficiency and high quality.

Screening of bacteria to produce vanillin and/or vanillic acid from isoeugenol was carried out [11]. *Achromobacter, Aeromonas, Agrobacerium, Alcaligenes, Arthrobacter, Bacillus, Micrococcus, Pseudomonas*, *Rhodobacter*, and *Rhodococcus* were found to be able to produce vanillin and/or vanillic acid, in addition of isoeugenol to the culture medium [11]. In particular, a soil isolate strain IE27 showed the highest vanillin-producing activity, and it was identified as *Pseudomonas putida*. Under the optimized culture conditions, *P. putida* IE27 cells produced 16.1 g/l vanillin from 150 mM isoeugenol. The molar conversion yield from isoeugenol to vanillin was as high as 71% at 20°C after a 24-h incubation. Therefore, it is expectable to produce natural vanillin with high efficiency.

Production of vanillin from vanillic acid and *O*-benzylvanillic acid was investigated by using whole cells and enzyme preparations of *Nocardia* sp. strain NRRL 5646 [12]. With growing cultures of the whole cells, 69 and 11% of vanillic acid were found to be decarboxylated to guaiacol and reduced to vanillyl alcohol, respectively. On the other hand, no decarboxylation of 4-*O*-benzylvanillic acid was found in conversion to the corresponding alcohol product. Purified *Nocardia* carboxylic acid reductase, an ATP and NADPH-dependent enzyme, was found to be able to reduce vanillic acid to vanillin quantitatively.

In addition to make use of the microbial biotransformation, enzymatic synthesis of natural vanillin was studied [13]. Flavoprotein vanillyl alcohol oxidase (VAO) could convert both creosol and vanillylamine to vanillin with high production yield. This conversion of creosol was realized via a two-step process. The first step was to convert creosol to vanillyl alcohol, and then, the second step was the oxidation of the vanillyl alcohol to vanillin. In the second step, the conversion of vanillyl alcohol to vanillin was inhibited by the competitive binding of creosol. The VAO-catalyzed conversion yield of vanillylamine to vanillin was high at alkaline solutions. Furthermore, mechanism study showed that vanillylamine was firstly converted to a vanillylimine intermediate product. The intermediate product was then hydrolyzed to vanillin nonenzymatically.

4. NOVEL APPLICATIONS OF VANILLIN

4.1. Preparation of benzoxazine resin and reactive monomeric surfactant containing oxazine ring

Vanillin is used to synthesize polybenzoxazine with the expected desirable benzoxazine properties as well as a high char yield of 55.3% [14]. The synthesized monomer provides an unused aldehyde group from vanillin. The aldehyde can be further reacted with other materials to enhance properties. As a model, the unused aldehyde is reacted with amine terminated poly (ethylene oxide) to form a surfactant, which retains 1,3-benzoxazine's reactivity. The chemical structure of the synthesized monomers, surfactant, and polymers is characterized by Fourier transform infrared spectroscopy (FT-IR) and proton nuclear magnetic resonance spectroscopy (^{1}H NMR). Thermal properties are also characterized by differential scanning calorimetry (DSC) and thermogravimetric analysis (TGA). Miniemulsions with stability up to 2 weeks are created with the newly synthesized surfactant and polystyrene. Dynamic light scattering (DLS) indicates 627 nm as the average diameter of the emulsion droplets.

4.2. Renewable Polymers Prepared from Vanillin and Its Derivatives

Methacrylated derivatives of vanillin and vanillyl alcohol are synthesized and used as two monomers, respectively. The two monomers were further polymerized by a free-radical process [15]. Rheokinetics of their polymerization were studied to determine the cure behaviors. Thermomechanical properties of the resulting polymers affected by the structure and functionalities of the monomers were discussed. In comparison with methacrylated vanillin monomer, the methacrylated vanillyl alcohol gave a higher cross-linking density of the polymers, which in turn resulted a higher storage modulus and glass transition temperature, a better thermal resistance. Methacrylated vanillyl alcohol monomer was also with low viscosity at room temperature. Therefore, methacrylated vanillyl alcohol monomer is an expectable bio-based reactive diluent for unsaturated polyester resins and vinyl esters.

5. DETECTION AND ANALYSIS OF VANILLIN AND VANILLIN-TYPE AROMATIC COMPOUNDS

Vanillin and other purely olfactory odorants such as coumarin, octanoic acid, and phenylethyl alcohol cannot be identified when presented oral cavity only (OCO), because the oral cavity trigeminal system is fully unresponsive to these odorants in vapor phase [16]. Therefore, modern analytical methods are usually required for detection and analysis of vanillin and its derivatives.

5.1. Detection of Real Natural Vanilla From Synthetic One

Because of the extremely different in price of natural real vanilla from synthetic one, detection of fake vanilla is required. Basically, gas or liquid chromatography can tell the real natural from synthetic vanilla because some impurities such as 4-hydroxybenzaldehyde present in natural vanilla essence could be detected. Another way is to investigate amount of radioactive carbon-14. Natural plant-sourced vanillin contains a certain level of carbon-14, whose half-life is 5730 years. On the other hand, vanillin derived from crude oil has no radiocarbon since it has decayed away over the millions of years; the oil was trapped underground. Ratio of the natural isotopes, carbon-13 to carbon-12, also can be used for identify the real natural vanilla from the synthetic ones, because the vanilla orchid uses a different biosynthetic pathway to other plants. Orchid-derived vanillin has a greater ratio of carbon-13 to carbon-12 than synthetic vanillin [17].

A solid phase micro-extraction (SPME)-GC-MS method seems to be able to distinguish the natural real vanilla extracts from the synthetic one too [18]. The fiber material in SPME, sampling time, desorption time, and other experimental conditions were optimized. Under the optimized conditions, a relative standard deviation (RSD) of 2.5–6.4% indicated good reproducibility of the method. Because GC profile of the natural extracts was different from synthetic ones, it is easily to determine whether the sample is natural or synthetic. The method is also applicable to identify the type of vanilla extract/flavoring used in flavor foods.

5.2. Analysis of Vanillin

The presence of vanillin in orange, tangerine, lemon, lime, and grapefruit juices could be easily identified and confirmed using GC-MS [19]. Vanillin concentrations in the orange, tangerine, lemon, lime, and grapefruit juices were determined to be 0.20, 0.35, 0.41, 0.35, and 0.60 ppm, respectively.

A headspace-solid phase micro-extraction (HS-SPME) GC-MS method was also proposed to determine of vanillin in vanilla products [20]. Detection limits were reported to be 1.33–13.2 ppb. Furthermore, LC-ESI-MS determinations of the vanillin were carried out at the same time, and the results were compared. Totally, 24 commercially available vanilla products were analyzed with the two methods. Vanillin was detected in all of the 24 products. Also, 18 other flavor related compounds were detected in the samples.

As illustrated in **Figure 4**, the sampling system allowed real-time and continuously sampling of the aroma volatiles from model liquid foods [21]. The sample samples could be kept at a certain temperature (e.g., at 37°C), and aroma volatiles released into the headspace. A carrier gas was flowed into the headspace and further into a quadrupole MS via a jet separator. The MS analysis could monitor and identify the volatile molecular weight. Also, this sampling system can examine the dynamic flavor releasing process of liquid samples.

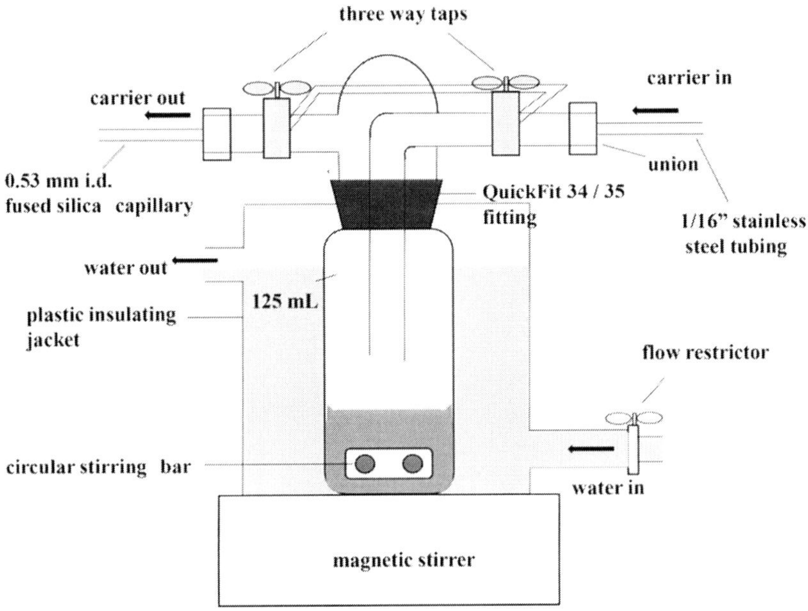

Figure 4. Illustration of a sampling system for the measurement of dynamic flavor release.

Recently, odor imaging sensing with multi-probe film is reported [22]. Odor substances are determined by fluorescent quenching with imaging film and a cooled CCD camera. The system could detect gaseous odor flow and visualize shape, spread, and concentration distribution of odor. A multi-film and FRET probes consisted of a certain combination of fluorescence dyes such as tryptophan and vanillin have been used for high sensitive and selective detection of odor. This approach is expectable in near future for more sensitivity and selectivity.

There are a lots of other analytical methods for the analysis of vanillin, based on spectrophotometric [23, 24], FIA [25], ion selective electrodes [26], flourimetric [27], thin layer chromatography [28], GC-MS [20, 29], HPLC [30, 31], and capillary electrophoresis (CE) [32–35]. Because CE is not only fast (usually shorter than 10 min), but also with nano-liter amount of injection volume for samples, it is particularly noticeable. Recently, direct sampling method for CE determination of vanillin in indoor air has been developed [36]. Here, the CE determination of vanillin is discussed in detail.

5.3. Ce Detection of Vanillin in Indoor Air

Generally, a fused silica capillary is used in CE. When running buffer solution is filled into the capillary, silanol groups (**Figure 5A**) in the surface of capillary dissociate, and inner surface of the capillary is charged negatively (**Figure 5B**). Cations in the running buffer solution are pulled toward the inner surface. As a result, an electric double layer is formed (**Figure 5C**). When an electric voltage is applied across the two ends of the capillary, an electroosmotic flow (EOF) toward cathode arises.

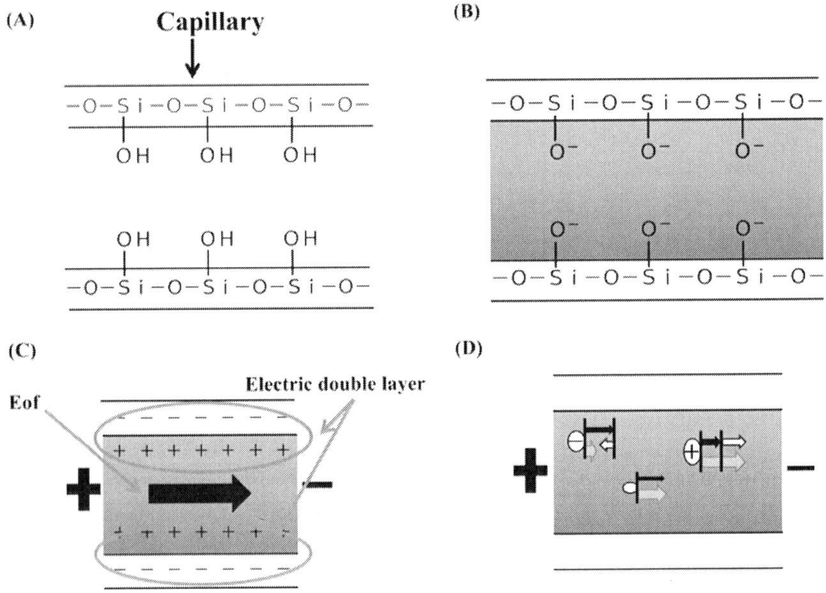

Figure 5. Illustration of molecular structure of the capillary inner surface and principle of CE separation. (A) molecular structure of inner surface of capillary; (B) dissociation of silanol group; (C) formation of electric double layer; (D) electromigration of analytes in CE.

$$EOF = (\varepsilon Z/(4\pi\rho))E \quad (1)$$

$$Vep = (q/(6\pi\rho r))E \quad (2)$$

where ε, Z, ρ, and E are the dielectric constant, zeta potential, solution viscosity, and electric field, respectively.

On the other hand, electrophoretic velocity (Vep) of an ion toward the electrode of opposite charge is:

where q and r are the electric charge and radius of the ion, respectively. Usually, EOF is larger than Vep for most of ions, and detector is usually close to the cathode end of the capillary. Electromigration velocity of a cation will be (EOF + Vep) while that of an anion (EOF–Vep). Neutral molecule will move with the electroosmotic flow (EOF). Then, cations, anions, and neutral molecules are separated (**Figure 5D**). Because different ions are with different q/r, their Vep is different. Thus, ions can be separated from each other by CE.

CE instrument is relatively simple. Basically, it consists of an electric power supply, a capillary, and a detector. **Figure 6** illustrates a typical laboratory-built CE apparatus [35]. It was consisted of a 30 kV high-voltage power supply and an UV-absorbance detector. Wavelength of the UV-absorbance detector could be set at 254 nm for detection of aromatic compounds. A capillary (inner diameter of 50–100 μm, out diameter of 364 μm) could be used. Its total length and the effective length (length from the anode end to the detector) could be about 30–70 and 20–60 cm, respectively. The capillary was usually cleaned thoroughly by subsequently flushing 1 mol/L NaOH, distilled-deionized water, and finally running buffer. Sample injection could be performed with either an electrophoretic method (e.g., injection voltage 1 kV, injecting time 30 s) or a hydrodynamic flow method with a height difference (e.g., 1 cm) between the two ends of the capillary.

Buffer solutions such as phosphate buffers or boric buffers with certain pH and concentration can be used in CE. Samples are usually dissolved or diluted with the buffers. For example, vanillin stock solution was prepared by dissolving a certain amount of vanillin into distilled-deionized water directly [35]. Its concentration was 10^{-3} mol/L. This stock solution was diluted to required concentrations with the running buffer when used. Vanilla perfume was also diluted with the buffer solutions to concentrations of 1 and 10%. For vanillin spiked vanilla perfume sample, a certain amount of the vanillin standard solution was added into the diluted vanilla perfume sample.

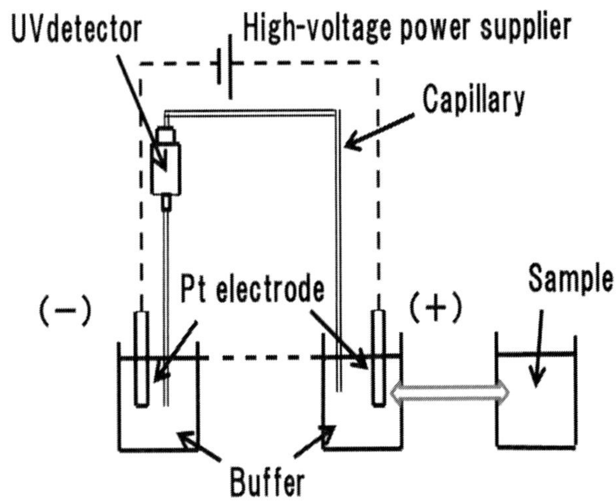

Figure 6. Illustration of a laboratory-built CE instrument.

5.3.1. CE of vanillin standard solution [35]

Figure 7 showed the CE results of vanillin at running buffers with different pH. It is well known that the higher the pH of the running buffer, the faster the EOF. Therefore, EOF was the fastest in running buffer of pH 11.5, while slowest in running buffer of pH 7.2. On the other hand, vanillin is a weak acid, and its acid dissociation constant Ka is about $10^{-9.25}$. At pH 11.5, almost all of vanillin molecules behavior as anions, while at pH 7.2 most of them as neutral molecules. At pH 9.3, about half vanillin molecules dissociated to anions. Therefore, Vep of vanillin was the largest at pH 11.5, slowest at pH 7.2. As a result of V = EOF-Vep, the vanillin was detected at about 500, 450, and 600 s at pH of 11.5, 9.3, and 7.2, respectively.

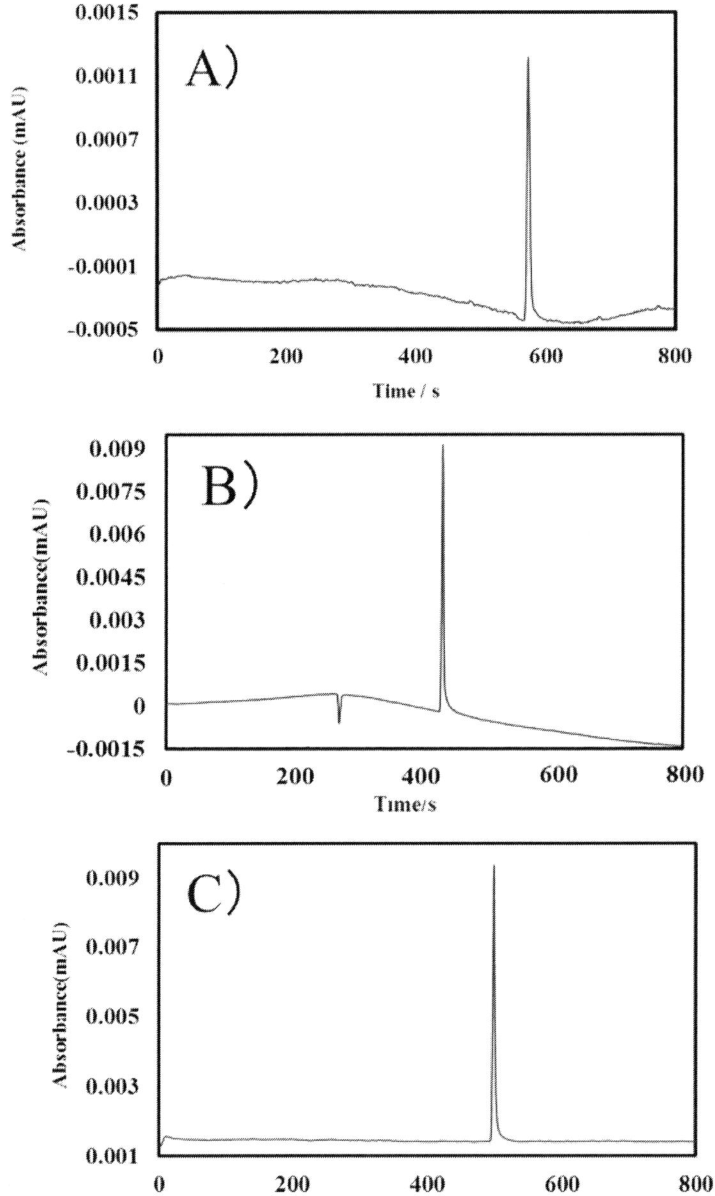

Figure 7. Electropherograms of vanillin standard solution (10^{-4} mol/L) in running buffers of pH 7.2 (A), 9.3 (B), and 11.5 (C).

Figure 8 showed the calibration curve of vanillin at the running buffer of pH 9.3. It can be seen that the peak area is proportional to the vanillin concentration in the range of 10^{-6}–10^{-2} mol/L. The detection limit was about 10^{-6} mol/L.

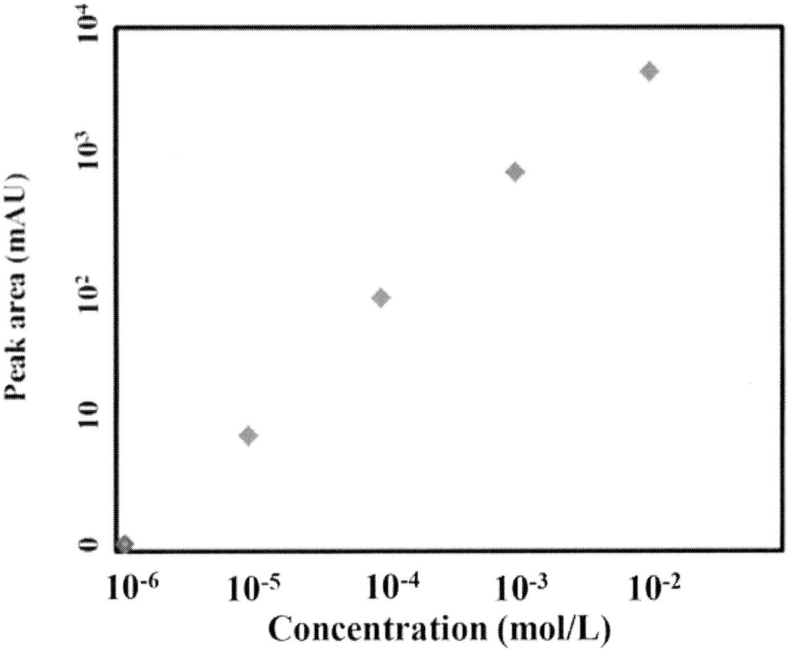

Figure 8. Vanillin calibration curves in running buffers of pH 9.3.

5.3.2. CE Of Vanilla Perfume [35]

Figure 9 showed electropherogram of 10% vanilla perfume sample solution (A) and 10% vanilla perfume sample solution spiked with 10^{-3} mol/L vanillin standard solution (B). In **Figure 9A**, three peaks were detected. In order to confirm which peak was vanillin, the vanilla perfume sample was spiked with vanillin standard solution. **Figure 9B** showed that only the third peak became large in the spiked sample. Therefore, the third peak was confirmed to be vanillin in the vanilla perfume. Vanillin concentration in the 10% vanilla perfume sample was determined to be about 3×10^{-3} mol/L by a standard addition method. The first and second peaks have not been identified.

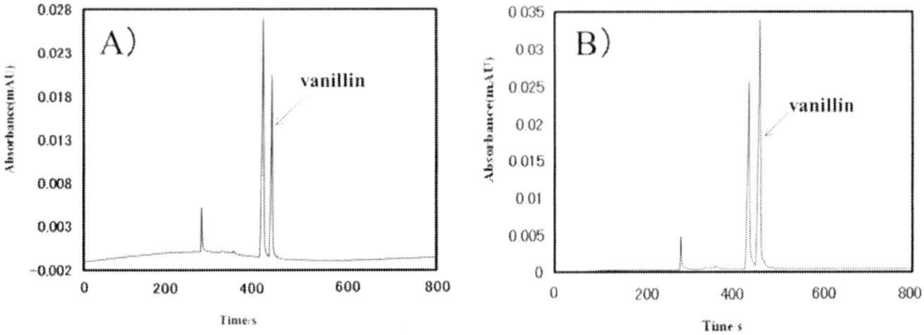

Figure 9. Electropherograms of 10% vanilla perfume sample (A) and the 10% vanilla perfume spiked with 10^{-3} (mol/L) vanillin (B) in the pH 9.3 phosphate buffer.

5.3.3. Detection of vanillin in indoor air by offline combination of ce with absorption and desorption of vanillin with active carbon [35]

In order to detect vanillin in indoor air with CE, about 0.02 g active carbon was spread on a glass slide for adsorption of vanillin in air. The glass slide was placed in a room of about 80 m^2, where a vanilla perfume of 5 ml was placed too for a certain period of time. Then, the active carbon on the glass slide was collected into a vial, and 0.5 ml ethanol or mixture of ethanol/pH 11.5 running buffer with a mixing ratio of 1:1 was added for desorption of vanillin adsorpted on the active carbon. The vial was centrifuged for 5 min at a centrifugation speed of 3000 rpm. The supernatant was directly injected into the capillary, and CE was carried out. **Figure 11** showed CE results.

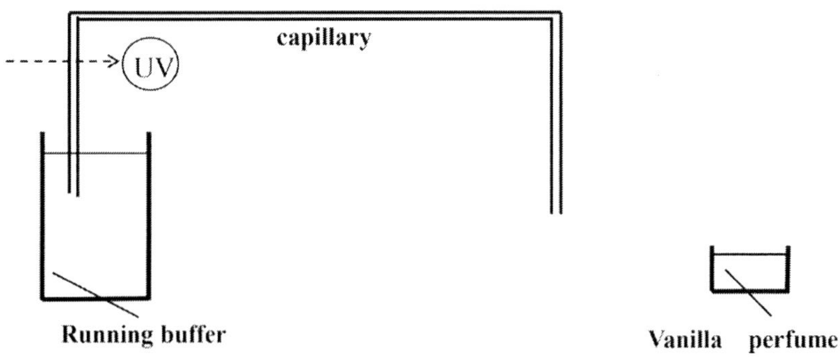

Figure 10. Electropherograms of supernatant of the ethanol/pH 11.5 running buffer after desorption from the active carbons. The active carbons adsorption time was 2 (top) and 4 (bottom) days, respectively.

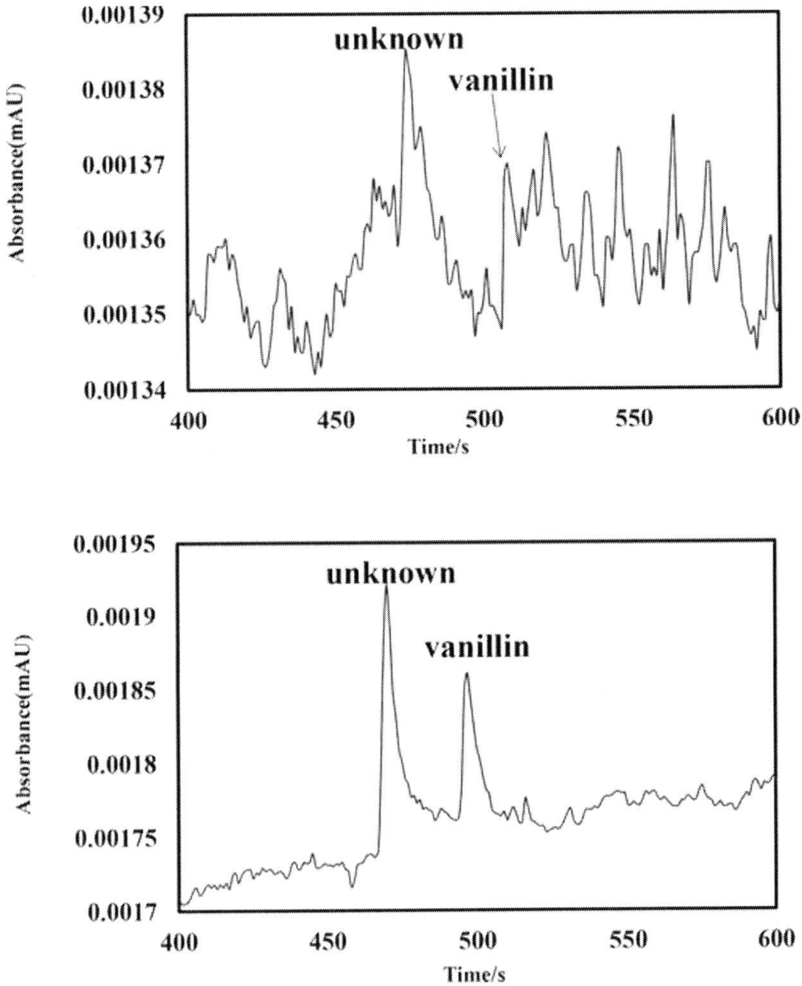

Figure 11. Illustration of the direct sampling of CE in air for vanillin in indoor air.

When the active carbon placed in the room for 2 days, the peak of vanillin was not detectable (top figure in **Figure 10**). When the active carbon placed in the room for 4 days, the vanillin peak was clearly detected (bottom figure in **Figure 10**). This meant that the CE method could be used for detection of vanillin in air by combination with the active carbon adsorption.

5.3.4. Fast detection for trace vanillin and vanillin-type aromatic compounds in air by direct sampling [36]

As stated above, CE could detect vanillin in air by offline combining with the adsorption/desorption method. However, it is usually time- and labor-consuming. Recently, a direct sampling of vanillin in the air of CE was demonstrated [36].

As shown in **Figure 11** , the inlet end of a fused silica capillary filled with a pH 7.2 phosphate buffer was directly placed in the air, while the outlet end was immersed into a buffer vial at the low electric potential side. Then, gaseous or volatile components such as vanillin and its derivatives would absorb at the air/buffer interface of the capillary inlet end. That meant a direct sampling of the vanillin in air at capillary inlet end for CE. After a certain period of sampling time, the inlet end was immersed into another buffer vial at the high electric potential side; CE was carried out by applying a high electric voltage of 20 kV. Evaporated vanillin in indoor air was detected fast.

It was found that the CE peak area increased with the direct sampling time. This was easily understood because the longer the sampling time, the more the vanillin absorbed at the running /air/buffer interface, the larger the peak area. Also, the peak area in the direct sampling-CE was considered to be proportional to the vanillin concentrations in air.

Figure 12 showed results of a conventional CE of 10% vanilla perfume sample (A) and direct sampling-CE in indoor air with a sampling time of 5 min (B), respectively. A conventional CE usually gave two peaks for vanilla perfume sample. The detection time of the two peaks was about 420–600 s. In particular, the second peak was identified to be vanillin [35].

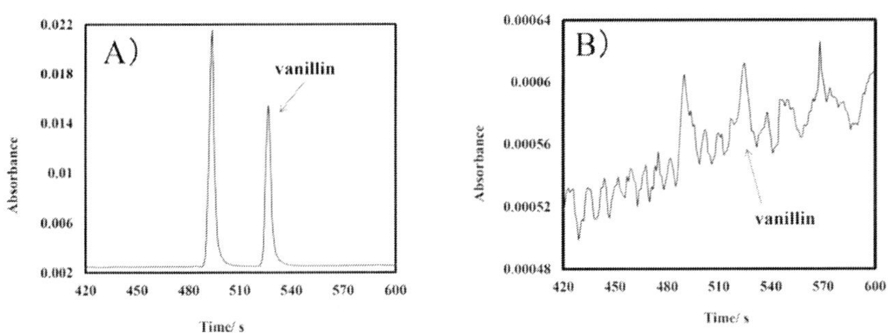

Figure 12. Results of a conventional CE of 10% vanilla perfume solution (A) and direct sampling-CE in indoor air with a sampling time of 5 min (B), respectively.

As shown in **Figure 12B** , there were also two peaks detected in the direct sampling-CE in air, and they were from vanilla perfume. Moreover, they were detected even with a sampling time as short as 5 min. This sampling time was extremely short in comparison with the offline active carbon adsorption/desorption-CE determinations [35]. In the offline active carbon adsorption/desorption-CE, it took more than 2 days to detect vanillin in indoor air [35]. Therefore, the direct sampling-CE is much faster and simpler than the offline active carbon adsorption/desorption-CE. The direct sampling-CE is promising and expectable in fast gas analysis.

REFERENCES

1. Walton N.J., Mayer M.J., Narbad A.: Phytochemistry. 2003; 63: 505–515. \
2. Vanillin: PubChem Open Chemistry Database [Internet] 2004. Available from: https://pubchem.ncbi.nlm.nih.gov/compound/vanillin
3. Deichmann W., Kitzmiller K.V.: J Am Pharm Assoc (Scientific ed.). 1940; 29(10): 425–428.
4. Kim J.H., et al.: PLoS One. 2014; 9(2): e89122.
5. Fitzgerald D.J., Gasson M.J., Narbad A.: J Agric Food Chem. 2005; 53 : 1769–1775.
6. Mourtzinos I., Kalogeropoulos N., Karathanos V.T., Konteles S.: Food Chem. 2009; 114 : 791–797.
7. Lee J., Lee K.-W., Song J.-Y., Cho J.Y., et al.: Food Chem Toxicol. 2014; 63 : 30–37.
8. Fahlbusch K.-G., et al: Flavors and Fragrances. Ullmann's Encyclopedia of Industrial Chemistry, 7th ed. (1999–2015) New York, NY: John Wiley & Sons. Online Posting Date: January 15, 2003.
9. Pinto P.C.R., Costa C.E., Rodrigues A.E.: Ind Eng Chem Res. 2013; 52 : 4421–4428.
10. Burdock G.A. (ed.): Fenaroli's Handbook of Flavor Ingredients. 6th ed. CRC Press, Taylor & Francis Group, Boca Raton, FL 2010. 2001 p.
11. Yamada M., Okada Y., Yoshida T., Nagasawa T.: Appl Microbiol Biotechnol. 2007; 73: 1025–1030.
12. Li T., Rosazza J.P.N.: Appl Environ Microbiol. 2000; 66: 684–687.
13. Robert H. H. van den Heuvel , Marco W. Fraaije , Colja Laane , and Willem J. H. van Berkel: J Agric Food Chem. 2001;49 : 2954–2958.
14. Van A., Chiou K., Ishida H.: Polymer. 2014; 55: 1443–1451.
15. Zhang C., Kessler M.R., Madbouly S.A.: Macromol Chem Phys. 2015; 216: 1816–1822.
16. Stephensin D., Halpern B.P.: Chem Senses. 2009; 34: 121–126.
17. Vanillin. Toxicology Data Nerwork [Internet] 1983. Available from http://www.rsc.org/chemistryworld/podcast/CIIEcompounds/transcripts/vanillin.asp
18. Sostaric T., Boyce M.C., Spickett E.E.: J Agric Food Chem. 2000; 12: 5802–5807.
19. Goodner K.L., Jella P, Rouseff R.L.: J Agric Food Chem. 2000; 48: 2882–2886.
20. de Jager L.S., Perfetti G.A., Diachenko G.W.: Food Chem. 2008; 107: 1701–1709.
21. Elmore J.S., Langley K.R.: J Agric Food Chem. 1996; 44: 3560–3563.
22. Frusawa Y., Yokoyama R., Liu C., Hayashi K.: Inst Electr Eng Japan. 2013; 133: 199–205.
23. Rind F.M.A., Mughal U.R., Memon A.H. et al.: Asian J Chem. 2009; 21: 2849–2856.
24. Longares-Patron A., Canizares-Macias M.P.: Talanta. 2006; 69: 882–887.
25. Linares P., Luque de Castro M.D., Valcarcel M.: Microchem. J. 1987; 35: 120–124.
26. Chan W.H., Lee W.M., Foo C.L., Tang W.K.: Analyst. 1987; 112: 845–848.

27. Katayama M., Mukai Y., Taniguchi H.: Anal Sci. 1987; 3: 369–372.
28. Belay M.T., Poole C.F.: Chromatographia. 1993; 37: 365–373.
29. Gassenmeier K.: Lebensm Wiss Technol. 2003; 36 : 99–103.
30. Sinha A.N., Verma S.C., Sharma U.K.: J Separation Sci. 2007; 30 : 15–20.
31. Guenther L.: Food Chem. 2009; 114: 1130–1134.
32. Masataka O., Hisako O., Misako H., et al.: J Chromatogr A. 2007; 1138: 262–267.
33. Toshiro W., Akira Y., Shiro N., et al.: J Chromatogr A. 1998; 793: 409–413.
34. Jozef L.B. : J Chem Educ. 2005; 82: 604–606.
35. Minematsu S., Xuan G.S., Wu X.-Z. : J Environ Sci. 2013; 25: 8–14. Minematsu S., Wu X.-Z.: Anal. Sci. 2013; 29: 373–375.

CHAPTER 12

Antioxidant Properties of Medicinal Plants from Peru

Adam Berłowski, Katarzyna Zawada, Iwona Wawer, Katarzyna Paradowska*

Faculty of Pharmacy, Medical University of Warsaw, Warsaw, Poland.

ABSTRACT

There is a wide diversity of plants and seasonal crops in Peru, due to the presence of many climatic zones. Numerous plants are used to cure or prevent diseases. These plants are promising candidates for functional foods products. The most frequent form in which they are used is an aqueous infusion or decoction. In this study, we compared the antioxidant properties of ten Peruvian plants infusions and investigated their relation to the phenolic content. The studied plants were: Uncaria tomentosa (cat's claw), Lepidium meyenii (maca), Berberis vulgaris L. (barberry, agracejo), Phyllantus niruri (chanca piedra), Annona muricata L. (graviola, soursop), Gentianella alborosea (hercampure), Geranium dielsianum (pasuchaca), Tabebuia ochracea (tahuari), Notholaena nivea ("cuti cuti") and Tiquilia paronychioides ("flor de arena"). Infusions of all studied plants have shown antioxidant activity, though there was a large diversity between the results. The antioxidant properties, determined with DPPH and ABTS scavenging assays as well as FRAP test, were strongly correlated with total phenolic content, while there was no correlation with the carotenoid content.

Keywords: Antioxidant; Total Polyphenols; Medicinal Plants

1. INTRODUCTION

There is a wide diversity of plants and seasonal crops in Peru, due to the presence of many climatic zones, including the unique areas such as Amazonian rainforest or Andean mountains. Numerous plants are used in medicine, to cure or prevent diseases. Widely known are Uncaria tomentosa (cat's claw), Phyllantus niruri (chanca piedra) or Lepidium meyenii (maca) which were already shown to have therapeutic or prophylactic potential. These plants are promising candidates for functional foods products.

U. tomentosa, a vine growing in the Amazon region, has been used medicinally by native tribes for at least 2000 years in treating inflammation, arthritis, bone pain, asthma, deep wounds, and cancer [1]. It is probably the best known medicinal plant of South America. Extracts from the bark were extensively studied [2] and found to exhibit immunostimulating, antiinflammatory and antioxidant properties [3].

L. meyenii, a tuber from the central Andes consumed by native Peruvians has also multipharmacological functions such as fertility improvement and the protection of cells against oxidative stress [4,5]. It has been demonstrated that different types of L. meyenii (differentiated by size and colour) have distinct biological properties [6,7]. L. meyenii infusion may act as an antioxidant by radical scavenging or by maintenance of intracellular ATP production in conditions of oxidative stress [8] as well as an immunomodulator [9].

Barberry (Berberis vulgaris L.) is known in Peru under the name "agracejo". Its fruits, leaves, root and stem are used for ailments of kidneys, liver, urinary and gastrointestinal tract, for the relieve of respiratory tract discomforts, and as a stimulant for the circulatory system [10, 11]. Berberis species were active antioxidants in many tests evaluating antioxidant activity, such as DPPH (2,2- diphenyl-1-picrylhydrazyl) radical scavenging, β-carotene bleaching, prevention of lipid peroxidation and oxyhaemoglobin bleaching, or protection from DNA damage [12,13].

Chanca piedra (P. niruri), a small herb indigenous to the tropical areas is widely used as medicinal plant. In Ayurveda and Traditional Chinese Medicine it was used for treating dysentery, influenza, diabetes, kidney stones, diuretics and tumours, as well as hepatotoxicity and hyperglycaemia [14]. The biologically active compounds of P. niruri are flavonoids, alkaloids, lignans, tannins, coumarins, terpenes, saponins and phenylpropanoids [15,16]. Research on this plant revealed that its aqueous extract inhibited HIV-1 reverse transcriptase [17], as well as reduced urinary calcium in patients with hypercalcuria [18]. It showed antioxidant properties in in vitro tests [19], namely free radical-scavenging, inhibition of reactive oxygen species production and peroxidation of lipids, as well as hepatoprotective properties against the paracetamol-induced injury in mice model [20].

Annona muricata L. (graviola, soursop) is a tree widely distributed in most of tropical countries. Its leaves have been traditionally used to treat headaches, hypertension, cough, asthma and as sedative [21,22]. In a model of skin papilloma in mice, the A. muricata leaves extract was able to suppress tumor initiation and tumor promotion even at lower dosage [23]. It showed antioxidant [24,25], antibacterial [26], antifungal [27] and anti-inflammatory [28] properties.

Gentianella alborosea (hercampure) is used in folk medicine for obesity treatment, in liver ailments and as colagogue, coleretic and digestive [29]. The active compounds are flavonoids, alkaloids, saponins and glycosides [30]. The extract exhibited moderate antioxidant activity and apoptotic properties on HeLa cell line [1].

Plants from the Geranium genus, to which Geranium dielsianum (pasuchaca) belongs, have been shown to have anti-influenza virus activity (G. sanguineum L.) [31], as well as antioxidant and radical scavenging capacities (G. macrorrhizum) [32]. Those properties have been attributed to their polyphenolic constituents. In traditional medicine G. dielsianum is used as blood purifier and hypoglycemic herb.

Antioxidant properties were also observed for Tabebuia ochracea (tahuari) [33]. It exhibited also antibacterial activity against Staphylococcus aureus.

Notholaena nivea (cuti cuti) is used in South America mainly as a herbal tea with the hypoglycaemic effect [34]. The lipophilic extract from aerial parts of this plant exhibited antioxidant properties [35].

T. paronychioides (flor de arena) is used in traditional medicine for treating inflammation of the ovaries [36] and the antioxidant mechanism is often at least partly responsible for antiinflammatory action. To our best knowledge, there is no data on its antioxidant properties.

The common feature of all mentioned above plants is their antioxidant activity. It is usually ascribed to the presence polyphenols, since a high content of phenolic compounds in a plant is usually connected with high antioxidant properties [37]. However, popular and the most frequent form in which those plants are used is an aqueous infusion or decoction. Its preparation can induce the degradation of polyphenols. Therefore, in this study, we compared the antioxidant properties of ten Peruvian plants infusions and investigated their relations to the phenolic content.

The antioxidant properties were studied with DPPH and ABTS (2,2'-azino-bis(3-ethylbenzothiazoline-6-sulphonic acid)) radical scavenging and ferric reducing antioxidant power (FRAP) tests. For DPPH test we have chosen

EPR spectroscopy technique, as it gives more reliable results than spectrophotometry [38]. We determined also the carotenoid content of those plants, to check whether this group of antioxidants can be also partly responsible for antioxidant properties.

2. EXPERIMENTAL

2.1. Plant Material

Ten samples of dried Peruvian plants: Uncaria tomentosa (cat's claw), Lepidium meyenii (maca), Berberis vulgaris L. (barberry, agracejo), Phyllantus niruri (chanca piedra), Annona muricata L. (graviola, soursop), Gentianella alborosea (hercampure), Tabebuia ochracea (tahuari), were obtained from Uncaria Institute (Warsaw, Poland).

2.2. Infusion Preparation

The 2.5 g of dry powdered plant material was weighted and 250 ml of distilled boiling water was added. Then infusions were left for 10 hours at darkness.

2.3. DPPH Scavenging (EPR Test)

100 μl of an infusion was mixed with 1 ml of 1.3 mM DPPH methanolic solution. After vortexing the samples were kept for 30 minutes at darkness and then EPR spectra were recorded. The samples with distilled water (100 μl) in place of an infusion were prepared as intensity standards. The intensity was taken as the double integral of the spectra. Results were expressed as Trolox equivalents (TEAC, milimoles per 100 ml) with the use of previously prepared standard curve. All experiments were performed in triplicate.

ESR measurements were performed on a Miniscope MS200 spectrometer (Magnettech GmbH). Parameters were as follows: central field 334 mT, sweep range 8 mT, sweep time 30 s, microwave power 10 mW, modulation amplitude 0.1 mT.

2.4. ABTS Scavenging

ABTS assay was performed according to Re et al [39] with small modifications. Briefly, 1500 μl of ABTS cation radical solution, prepared by mixing 7 mM ABTS reagent and 2.45 mM potassium persulfate in equal volumes, equlibrating the mixture for 16 hours and diluting with ethanol to absorbance value of 0.70, was added to 15 µl of plant infusion or standard (trolox) solution. The absorbance reading was taken at 734 nm in sixth minute after adding the radical solution. All experiments were performed in triplicate and results were expressed as milimoles of trolox for 100 ml (TEAC) of an aqueous infusion.

2.5. FRAP Assay

FRAP assay was done according to Benzie and Strain procedure [40]. Briefly, 50 µl of plant infusion or 50 µl of freshly prepared $FeSO_4$ standard solution was mixed with 1500 µl of working FRAP reagent, and absorbance reading at 593 nm was taken after 4 minutes of thermostatting at 37°C. The working FRAP reagent was prepared daily by mixing $FeCl_3$ and TPTZ (2,4,6-Tripyridyl-sTriazine) solutions with acetate buffer (pH 3.6). The results were taken as a mean of three replicates and expressed as milimoles of reduced Fe^{3+} per 100 ml of plant infusion.

2.6. Total Phenolic Content Total phenolic content (TP) was determined by modified Folin-Ciocalteu colorimetric method [41]. Briefly, to 20 µl of an infusion 1580 µl of Millipore water and 100 µl of Folin-Ciocalteu reagent was added. After 5 minutes at room temperature, 300 µl of 20% sodium carbonate was added, and the reaction mixture was thermostated for 20 minutes at 37°C and the absorbance at 765 nm was taken using Evolution 60S spectrophotometer (Thermo Scientific). Results were expressed as gallic acid equivalents (GAE [mg/100 ml]) with the use of the standard curve, prepared in parallel with measurements. All experiments were performed in triplicate.

2.7. Total Carotenoid Content

Total carotenoid content (TC) determination was carried out as beforehand reported [42] with small modifications. The extraction was realized by adding 30 mL n-hexaneacetone mixture (6:4) to 2.5 g of the powdered sample. After shaking for 10 minutes it was filtered through a paper filter and the absorbance at 450 nm was measured immediately with Evolution 60S spectrophotometer (Thermo Scientific), with up to 10 fold dilution where appropriate. Results were expressed as µg of carotenoids per 1 g of plant material, calculated with average carotenoid absorbance coefficient $A_{1\%}$ of 2500 [43]. All experiments were performed in duplicate.

2.8. Statistical Analysis

Pearson correlation analysis was performed using a Statistica (Statistical Statsoft, Tulsa, OK) software; P-values < 0.05 were considered significant.

3. RESULTS AND DISCUSSION

Among studied plants, the best DPPH-scavenging properties were observed for U. tomentosa and G. dielsianum (1.327 ± 0.034 and 1.234 ± 0.031 mmol trolox/100 ml, respectively), followed by N. nivea, T. paronychioides and A. muricata L. (**Figure 1**(a)). The DPPH-TEAC value obtained for those samples was over three times higher than the value for the weakest DPPH scavengers, i.e. L. meyenii and T. ochracea (0.434 ± 0.005 and 0.413 ± 0.016, respectively). In case of the ABTS radical, G. dielsianum and U. tomentosa still showed the highest scavenging activity (0.645 ± 0.027 and 0.513 ± 0.061 mmol trolox/100 ml, respectively), although all results obtained from this test were lower than those obtained from the DPPH test (**Figure 1**(b)). However, the best result in the ABTS test (for G. dielsianum) was over six times better than those for the weakest scavengers, i.e. L. meyenii and T. ochracea (0.067 ± 0.010 and 0.079 ± 0.006 mmol trolox/100 ml, respectively). The result for U. tomentosa was also quite high, but for other plants it was distinctly lower. The difference in results from those two tests can be tentatively ascribed to the different structure of those radicals and especially to different charge, since the DPPH molecule has no charge and the ABTS radical is a cation. This can result in different reaction mechanism both for active compounds in plant infusions and for standard antioxidant, i.e. trolox.

For the FRAP test results, the highest value was obtained for T. paronychioides infusion (0.607 ± 0.022 mmol Fe/100 ml), followed by G. dielsianum (0.562 ± 0.017 mmol Fe/100 ml) and U. tomentosa (0.507 ± 0.016 mmol Fe/100 ml) (**Figure 1**(c)). The lowest result was given by L. meyenii (0.011 ± 0.001 mmol Fe/100 ml), similar to the radical scavenging tests. In this test the variability among studied samples was distinctly bigger than for previous tests, with the value obtained for T. paronychioides over fifty times higher than the value for L. meyenii.

The compounds responsible for DPPH and ABTS scavenging, as well as for iron-reducing activity, are mainly polyphenols, as can be seen from coefficients of correlations (r) between results of those tests and polyphenol content (**Table 1**), ranging from 0.803 to 0.930. It is worth stressing that correlations between FRAP and ABTS tests results and total polyphenols was significant with $p < 0.005$, and between DPPH test results and polyphenol content with $p = 0.005$.

The polyphenol amount per 100 ml of an infusion varies from 4.6 GAE/100 ml for L. meyenii to 75.7 GAE for T. paronychioides (**Figure 2**(a)). The high polyphenol content and good antioxidant properties of U. tomentosa bark infusion is consistent with results obtained by Gon- çalves et al. [1], Pilarski et al. [44] and Ranilla et al. [17]. These works also have shown that potent radical scavenging activity strongly correlated with the presence of proanthocyanidins and phenolic acids. The main phenolic acid was either caffeic acid [1,32] or chlorogenic acid [32].

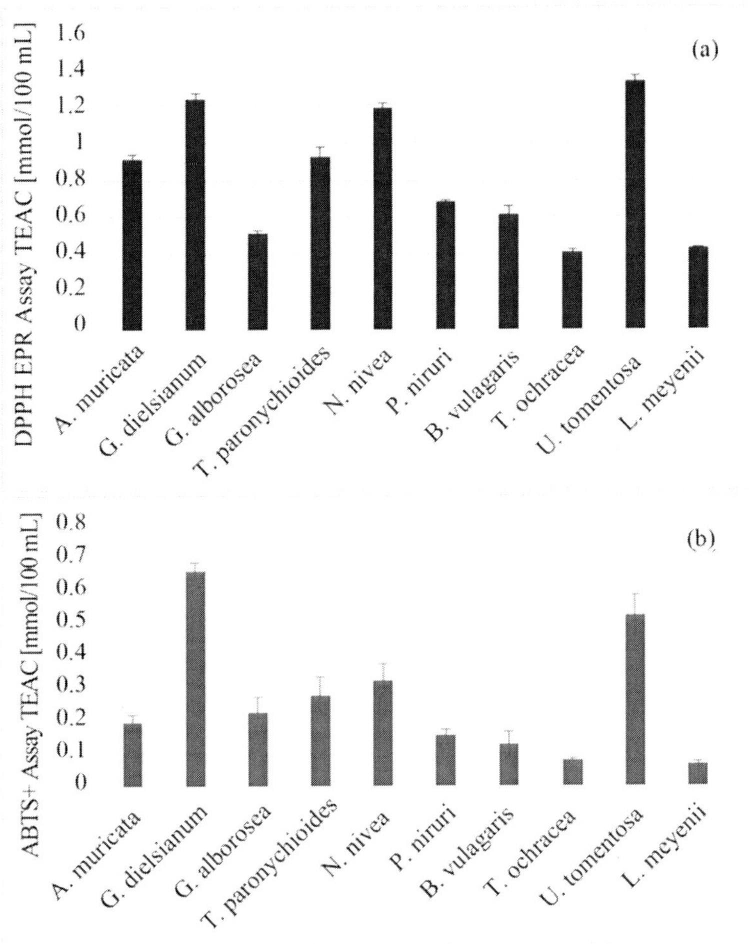

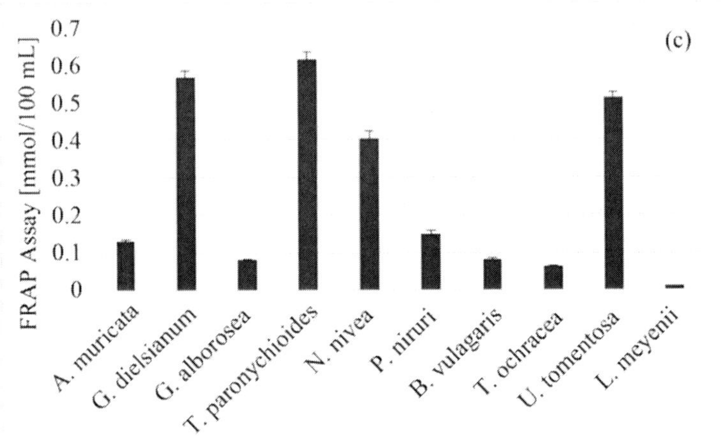

Figure 1. Antioxidant properties of Peruvian plants infusions. (a) DPPH-EPR; (b) ABTS; (c) FRAP. Results are presented as means of three experiments results with standard deviation.

In our study the total phenolic content obtained for P. niruri (14.9 GAE/100 ml) was lower than that for U. tomentosa (58.9 GAE/100 ml), as opposed to the results of Ranilla [17]. However, this difference can be due to different methods of extraction.

The highest total carotenoid content was observed for A. muricata L., the lowest for T. ochracea (**Figure 2**(b)). However, it should be noted that for all studied plants the carotenoid content in plant material was low (0.001 - 0.310 μg/1 g). Whatsmore, carotenoids are lipophilic compounds and therefore they would be extracted only in very small part into aqueous infusion. It can be the reason behind the lack of correlation between carotenoids content of plant material and antioxidant properties of aqueous infusions (**Table 1**).

Overall, the studied plant infusions could be divided into three groups: good antioxidant sources (U. tomentosa, P. niruri, T. paronychioides and N. nivea), which gave high results in all antioxidant tests (DPPH, ABTS and FRAP), weak antioxidant sources, with low values obtained from all those tests (L. meyenii, T. ochracea, B. vulgaris L., P. niruri) and selective antioxidant sources, which had high or at least moderate values only in one or two tests (A. muricata L., G. alborosea). Good scavengers have also a high content of total polyphenols, while the phenolic content of plants infusions with low TEAC values is also low. The exception is hercampuri (G. alborosea), which is a weak scavenger of ABTS radical and has low FRAP value despite its relatively high polyphenol content.

Table 1. Pearson's correlation coefficients. An asterisk indicates significant correlations.

	ABTS	DPPH	TP	TC
FRAP	0.824*	0.851*	0.930*	−0.289
ABTS		0.868*	0.819*	−0.297
DPPH			0.803*	0.012
TP				−0.297

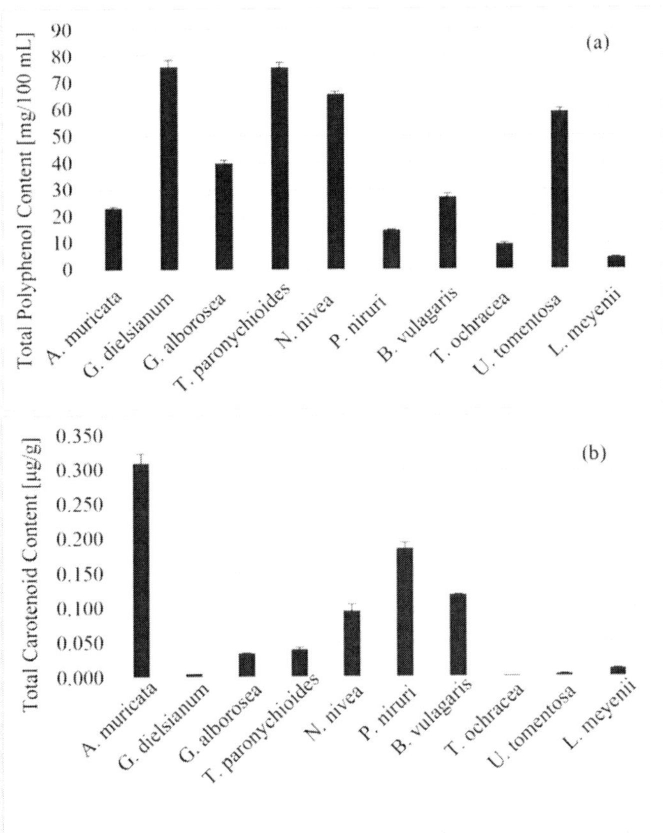

Figure 2. Total phenolic (a) and total carotenoid; (b) content of Peruvian plants infusions and in plant material, respectively. Results are presented as means of three or two respectively experiment results with standard deviation.

4. CONCLUSIONS

Infusions of all ten studied plants have shown antioxidant activity, though there was a large diversity between the results. Therefore, they can be used as a valuable antioxidant component of human diet not only in areas where they are endemic to, but in the whole world.

Adam Berłowski, Katarzyna Zawada, Iwona Wawer, Katarzyna Paradowska The antioxidant properties were correlated with polyphenols content; there was no correlation with the carotenoids content. It supports the hypothesis that the polyphenol group is the dominating group of antioxidants in those plants, at least in their most popular serving form, i.e. aqueous infusion.

REFERENCES

1. K. Keplinger, G. Laus, M. Wurm, M. P. Dierich and H. Teppner, "Uncaria tomentosa (Willd.) DC.-Ethnomedicinal Use and New Pharmacological, Toxicological and Botanical Results," Journal of Ethnopharmacology, Vol. 64, No. 1, 1999, pp. 23-34.
2. M. E. Heitzman, C. C. Neto, E. Winiarz, A. J. Vaisberg and G. B. Hammond, "Ethnobotany, Phytochemistry and Pharmacology of Uncaria (Rubiaceae)," Phytochemistry, Vol. 66, No. 1, 2005, pp. 5-29.
3. C. Gonçalves, T. Dinis and M. T. Batista, "Antioxidant Properties of Proanthocyanidins of Uncaria tomentosa Bark Decoction: A Mechanism for Anti-Inflammatory Activity," Phytochemistry, Vol. 66, No. 1, 2005, pp. 89-98.
4. Y. Wang, Y. Wang, B. McNeil and L. M. Harvey, "Maca: An Andean Crop with Multi-Pharmacological Functions," Food Research International, Vol. 40, No. 7, 2007 pp. 783-792.
5. G. F. Gonzales, C. Gonzales and C. Gonzales-Castañeda. "Lepidium meyenii (Maca): A Plant from the Highlands of Peru—From Tradition to Science," Forsch Komplementmed, Vol. 16, No. 6, 2009, pp. 373-380.
6. C. Gonzales, J. Rubio, M. Gasco, Y. Nieto, S. Yucra and G. F. Gonzales, "Effect of Short Term and Long Term Treatments with Three Ecotypes of Lepidium meyenii (maca) on Spermatogenesis in Rats," Journal of Ethnopharmacology, Vol. 103, No. 3, 2006, pp. 448-454.
7. J. Rubio, M. Caldas, S. Dávila, M. Gasco and G. F. Gonzales, "Effect of Three Different Cultivars of Lepidium meyenii (maca) on Learning and Depression in Ovariect-Omized Mice," BMC Complementary & Alternative Medicine, Vol. 6, No. 1, 2006, pp. 23-30.
8. M. Sandoval, N. N. Okuhama, F. M. Angeles, V. M. Melchor, L. A. Condezo, J. Lao and M. J. S. Miller, "Antioxidant Activity of the Cruciferous Vegetable Maca (Lepidium meyenii)," Food Chemistry, Vol. 79, No. 2, 2002, pp. 207-213.
9. K. J. Lee, K. Dabrowski, J. Rinchard, C. Gomez, L. Guz and C. Vilchez, "Supplementation of Maca (Lepidium meyenii) Tuber Meal in Diets Improves Growth Rate and Survival of Rainbow Trout Oncorhynchus Mykiss (Walbaum) Alevins and Juveniles," Aquaculture Research, Vol. 35, No. 3, 2004, pp. 215-223.
10. M. Blumenthal, W. R. Busse, A. Goldbert, J. Gruenwald, T. Hall, S. Klein, C. W. Riggins and R. S. Rister, "The Complete German Commission E Monographs: Therapeutic Guide to Herbal Medicine," American Botanical Council, Austin, 1998.

11. J. Brunetton, "Pharmacognosy, Phytochemistry, Medicinal Plants," Lavoisier Publishing, Paris, 1999.
12. P. Hanachi, S. H. Kua, R. Asmah, G. Motalleb and O. Fauziah, "Cytotoxic Effect of Berberis vulgaris Fruit Extract on the Proliferation of Human Liver Cancer Line (HepG2) and Its Antioxidant Properties," International Journal of Cancer Research, Vol. 2, No. 1, 2006, pp. 1-9.
13. S. Schaffer and M. Heinrich, "Understanding Local Mediterranean Diets: A Multidisciplinary Pharmacological and Ethnobotanical Approach," Pharmacological Research, Vol. 52, No. 4, 2005, pp. 353-366.
14. R. N. Chopra, S. L. Nayar and I. C. Chopra, "Glossary of Indian Medicinal Plants," Catholic Press, Ranchi, CSIR, New Delhi, 1986.
15. M. L. Dhar, M. M. Dhar, B. N. Dhawan, B. N. Mehrotra and C. Ray, "Screening of Indian Plants for Biological Activity: Part 1," Indian Journal of Experimental Biology, Vol. 6, No. 4, 1968, pp. 232-247.
16. G. Bagalkotkar, S. R. Sagineedu, M. S. Saad and J. Stanslas, "Phytochemicals from Phyllantus niruri L. and Their Pharmacological Properties: A Review," Journal of Pharmacy and Pharmacology, Vol. 58, No. 12, 2006, pp. 1559- 1570. s
17. A. D. Naik and A. R. Juvekar, "Effect of Alkaloidal Extract of Phyllantus niruri on HIV Replication," Indian Journal of Medical Science, Vol. 57, No. 9, 2003, pp. 387-393.
18. J. L. Nishiura, A. H. Campos, M. A. Boim, I. P. Heilberg and N. Schor, "Phyllantus niruri Normalizes Elevated Urinary Calcium Levels in Calcium Stone Forming (CSF) Patients," Urological Research, Vol. 32, No. 5, 2004, pp. 362-366.
19. L. G. Ranilla, Y.-I. Kwon, E. Apostolidis and K. Shetty, "Phenolic Compounds, Antioxidant Activity and in Vitro Inhibitory Potential against Key Enzymes Relevant for Hyperglycemia and Hypertension of Commonly Used Medicinal Plants, Herbs and Spices in Latin America," Bioresource Technology, Vol. 101, No. 12, 2010, pp. 4676- 4689.
20. S. M. Sabir and J. B. T. Rocha, "Water-Extractable Phytochemicals from Phyllanthus niruri Exhibit Distinct in Vitro Antioxidant and in Vivo Hepatoprotective Activity against Paracetamol-Induced Liver Damage in Mice" Food Chemistry, Vol. 111, No. 4, 2008, pp. 845-851.
21. L. Taylor, "Technical Data Report for Graviola, Annona Muricata," Sage Press, Austin, 2002.
22. C. A. Lans, "Ethnomedicines Used in Trinidad and Tobago for Urinary Problems and Diabetes Mellitus," Journal of Ethnobiology and Ethnomedicine, Vol. 2, No. 52, 2006, pp. 45-55.
23. H. Sulaiman, A. H. Roslida, O. Fezah, K. L. Tan, Y. S. Tor and C. I. Tan, "Chemopreventive Potential of Annona Muricata L. Leaves on Chemically-Induced Skin Papillomagenesis in Mice," Asian Pacific Journal of Cancer Prevention, Vol. 13, No. 6, 2012, pp. 2533-2539.
24. J. G. de Melo, T. A. de Sousa Araújo, V. T. N. de Almeida e Castro, D. L. de Vasconcelos Cabral, M. do Desterro Rodrigues, S. C. do Nascimento, E. L. Cavalcanti de Amorim, U. P. de Albuquerque, "Antiproliferative Activity, Antioxidant Capacity and Tannin Content in Plants of Semi-Arid Northeastern Brazil," Molecules, Vol. 15, No. 12, 2010, pp. 8534-8542.
25. R. Baskar, V. Rajeswari and T. S. Kumar, "In Vitro Antioxidant Studies in Leaves of Annona Species," Indian Journal of Experimental Biology, Vol. 45, No. 5, 2007, pp. 480-485.
26. K. Sundarrao, I. Burrows and M. Kuduk, "Preliminary Screening of Anti-Bacterial and Anti-Tumor Activities of Papuan New Guinean Active Medicinal Plants," Pharmaceutical Biology, Vol. 31, No. 1, 1993, pp. 3-6.
27. M. Heinrich, M. Kuhnt and C. W. Wright, "Parasitological and Microbiological Evaluation of Mixe Indian Medicinal Plants (Mexico)," Journal of Ethnopharmacology, Vol. 36, No. 1, 1992, pp. 81-85.
28. O. V. de Sousa, G. D. V. Vieira, J. J. R. G. de Pinho et al., "Antinociceptive and Anti-Inflammatory Activities of the Ethanol Extract of annona muricata L. Leaves in

Animal Models," International Journal of Molecular Sciences, Vol. 11, No. 5, 2010, pp. 2067-2078.
29. K. Żurowska, "Amazonian and Andean Phytotherapy," Tower Press, Gdańsk, 2001.
30. N. Acero, F. Llinares, A. Galán de Mera, B. Oltra and D. Muñoz-Mingarro, "Apoptotic and Free Radical Scavenging Properties of the Methanolic Extract of Gentianella alborosea," Fitoterapia, Vol. 77, No. 8, 2006, pp. 475- 477.
31. J. Serkedjieva and A. J. Hay, "In Vitro Anti-Influenza Virus Activity of a Plant Preparation from Geranium sanguineum L.," Antiviral Research, Vol. 37, No. 2, 1998, pp. 221-230.
32. M. Sokmena, M. Angelova, E. Krumova, S. Pashova, S. Ivancheva, A. Sokmen and J. Serkedjieva, "In Vitro Antioxidant Activity of Polyphenol Extracts with Antiviral Properties from Geranium sanguineum L.," Life Sciences, Vol. 76, No. 25, 2005, pp. 2981-2993.
33. A. F. Ospina, J. P. C. Guerrero, Y. C. O. Buendía, I. B. P. Bolívar and F. D. Castillo, "Antiinflammatory, Antioxidant and Antibacterial Activity of Two Species of Tabebuia Genus," Revista Cubana de Plantas Medicinales, Vol. 18, No. 1, 2013, pp. 34-46.
34. R. W. Bussmann and A. Glenn, "Traditional Knowledge for Modern Ailments—Plants Used for the Treatment of Diabetes and Cancer in Northern Peru," Journal of Medical Plants Research, Vol. 5, No. 31, 2011, pp. 6916- 6930.
35. G. Cioffi, P. Montoro, O. Lock de Ugaz, A. Vassallo, L. Severino, C. Pizza and N. de Tommasi, "Antioxidant Bibenzyl Derivatives from Notholaena Nivea Desv," Molecules, Vol. 16, No. 3, 2011, pp. 2527-2541.
36. R. W. Bussmann and A. Glenn, "Medicinal Plants Used in Northern Peru for Reproductive Problems and Female Health," Journal of Ethnobiology and Ethnomedicine, Vol. 6, No. 1, 2010, pp. 30-41.
37. M. P. Kähkönen, A. I. Hopia, H. J. Vuorela, J. P. Rauha, K. Pihlaja, T. S. Kujala and M. Heinonen, "Antioxidant Activity of Plant Extracts Containing Phenolic, Compounds," Journal of Agricultural and Food Chemistry, Vol. 47, No. 10, 1999, pp. 3954-3962.
38. D. Sanna, G. Delogu, M. Mulas, M. Schirra and A. Fadda, "Determination of Free Radical Scavenging Activity of Plant Extracts through DPPH Assay: An EPR and UVVis Study," Food Analytical Methods, Vol. 5, No. 4, 2012, pp. 759-766.
39. R. Re, N. Pellegrini, A. Proteggente, A. Pannala, M. Yang and C. Rice-Evans, "Antioxidant Activity Applying an Improved ABTS Radical Cation Decolorization Assay," Free Radical Biology and Medicine, Vol. 26, No. 9-10, 1999, pp. 1231-1237.
40. I. F. Benzie and J. J. Strain, "The Ferric Reducing Ability of Plasma (FRAP) as a Measure of 'Antioxidant Power': The FRAP Assay," Analytical Biochemistry, Vol. 239, No. 1, 1996, pp. 70-76.
41. V. Singleton, R. Orthofer and R. M. Lamuela-Raventos, "Analysis of Total Phenols and Other Oxidation Substrates and Antioxidants by Means of Folin-Ciocalteu Reagent," Methods in Enzymology, Vol. 299, No. 39, 1999, pp. 152-175.
42. I. Ferreira, E. Aires, J. C. M. Barreira and L. Estevinho, "Antiooxidant Activity of Portuguese Honey Samples: Different Contributions of the Entire Honey and Phenolic Extract," Food Chemistry, Vol. 114, No. 4, 2009, pp. 1438- 1443.
43. Current Protocols in Food Analytical Chemistry, "Detection and Measurement of Carotenoids by UV/VIS Spectrophotometry," 2001.
44. R. Pilarski, H. Zielinski, D. Ciesiolka and K. Gulewicz, "Antioxidant Activity of Ethanolic and Aqueous Extracts of Uncaria Tomentosa (Willd.) DC," Journal of Ethnopharmacology, Vol. 104, No. 1-2, 2006, pp. 18-23.

CHAPTER 13

Potent and Synergistic Extract Combinations from *Terminalia Catappa*, *Terminalia Mantaly* and *Monodora tenuifolia* Against Pathogenic Yeasts

Thierry Kammalac Ngouana [1,2,3], Cedric Derick Jiatsa Mbouna [1], Rufin Marie Toghueo Kuipou [1], Marthe Aimée Tchuente Tchuenmogne [4], Elisabeth Menkem Zeuko'o [1], Vincent Ngouana [1], Michèle Mallié [3], Sebastien Bertout [3] and Fabrice Fekam Boyom [1,*]

[1]*Antimicrobial Agents Unit, Laboratory for Phytobiochemistry and Medicinal Plants Studies, Department of Biochemistry, Faculty of Science, University of Yaoundé I, P.O. Box 812, Messa-Yaoundé, Cameroon*

[2]*Clinical Biology Laboratory, Yaoundé Central Hospital, P.O. Box 87, Messa-Yaoundé, Cameroon*

[3]*IRD UMI 233 TransVIHMI—UM INSERM U1175 TransVIHMI Laboratoire de Parasitologie et Mycologie Médicale, UFR Pharmacie, Université de Montpellier 1, 15, Avenue Charles Flahault—BP 14491 34093 Montpellier, France*

[4]*Laboratory of Natural Products and Organic Synthesis, Department of Organic Chemistry, Faculty of Science, University of Yaoundé 1, P.O. Box 812, Messa-Yaoundé, Cameroon*

ABSTRACT

Mycoses caused by *Candida* and *Cryptococcus* species, associated with the advent of antifungal drug resistance have emerged as major health problems. Improved control measures and innovative therapies are needed. This paper describes results from the screening of bio-guided fractionated extracts alone and combinations of *Terminalia catappa, Terminalia mantaly* and *Monodora tenuifolia* harvested in Cameroon. Crude ethanolic, hydro-ethanolic and aqueous extracts and bio-guided fractions were screened for antifungal activity against isolates of *C. albicans, C. glabrata, C. parapsilosis* and *Cr. neoformans* and the reference strain *C. albicans* NR-29450. Minimal inhibitory concentrations (MIC) were determined using a broth micro dilution method according to the Clinical & Laboratory Standards Institute (CLSI). Time kill kinetics of extracts alone and in combination were also evaluated. Extracts from *T. mantaly* stem bark were the most active with the best MIC values ranging from 0.04 mg/mL to 0.16 mg/mL. Synergistic interactions were observed with combinations of sub-fractions from *M. tenuifolia*, *T. mantaly* and *T. catappa.* Combination of sub-fractions from *M. tenuifolia* and *T. mantaly* (C36/C12) showed synergistic interaction and fungicidal effect against four out of five tested yeasts. These results support further investigation of medicinal plant extracts alone and in combination as starting points for the development of alternative antifungal therapy.

Keywords: *Terminalia catappa*; *Terminalia mantaly*; *Monodora tenuifolia*; antifungal activity; bio-guided fractionation; combinations

1. INTRODUCTION

The knowledge of the epidemiology and pathogenesis of yeasts affecting humans has increased in the last three decades [1,2,3,4]. Many conditions have promoted the increase of opportunistic fungal infections in humans, especially candidiasis and cryptococcosis [1,3,4,5,6,7,8,9,10,11,12].

Candidiasis is the most common fungal infection in both immunocompromised and immunocompetent persons [10,13,14,15, 16,17]. Invasive candidiasis is responsible for the high mortality among intensive care patients with a frequency of 40%–75% [3,9,15,18]. *C. albicans* is the most frequent yeast involved in these infections, although other species are also frequently isolated such as *C. parapsilosis*, *C. glabrata*, *C. tropicalis*, *C. krusei*, *C. lusitaniae*, and *C.*

dubliniensis. Furthermore, some other pathogenic yeasts are emerging, including *C. famata*, *C. guilliermondii*, *C. kefyr* and more recently *C. africana* [9,19]. Moreover, Cryptococcal meningitis caused by *Cr. neoformans* is the most dangerous systemic mycosis associated with AIDS [20,21,22].

Fungal infections are treated using different families of drugs such as polyens, azoles, pyrimidine analogues, echinocandins and allylamines [2,9,23]. However, these drugs have some limitations such as high toxicity and low bioavailability (polyens) and reduced spectrum of action (echinocandins) [9,23]. Resistance has already been described for all these antifungal drugs and is more often associated with therapeutic failure [23,24,25]. These limitations in treatments associated with the increased knowledge of the epidemiology and pathogenicity of yeasts underline the urgent need for new antifungal agents with improved potency and innovative modes of action that could be further developed as antifungal treatments.

Natural products have proven efficacy in the treatment of a wide range of ailments. Ethnobotanically selected plants such as *T. mantaly* (Combretaceae), *T. catappa* (Combretaceae) and *M. tenuifolia* (Annonaceae) are frequently used in traditional medicine for the treatment of various infectious diseases [26,27]. This paper describes the antifungal activity of their extracts, fractions and sub-fractions against *Candida* spp and *Cr. neoformans* isolates.

2. EXPERIMENTAL SECTION

2.1. Collection of Plant Materials

Organs from *T. catappa*, *T. mantaly* (Combretaceae) and *M. tenuifolia* (Annonaceae) were collected in the Yaoundé area and identified at the Cameroon National Herbarium where voucher specimens are kept under registration numbers 51244/HNC (*Terminalia catappa* Linné), 64212/HNC (*Terminalia mantaly* H. Perrier) and 30549/HNC (*Monodora tenuifolia* Benth). Leaves, twigs, fruits, fruits pericarp, pulp, seeds, seeds pericarp, branches and stem bark were collected and separately dried at room temperature and ground using a blender.

2.2. Yeasts Isolates and Reference Strain

Yeast isolates were provided by the laboratory of clinical biology, Yaoundé Central Hospital (Yaoundé, Cameroon), and consisted of clinical isolates of *C. albicans*, *C. glabrata*, *C. parapsilosis* and *Cr. neoformans*. *C. albicans* NR-29450 was obtained from BeiResources (Manassas, VA, USA) and used as positive control. These yeasts were maintained at room temperature and cultured at 37 °C for 24 h on Sabouraud Dextrose Agar (Oxoid) slants prior to use.

2.3. Plant Extraction and Preliminary Screening of Antifungal Activity

The crude extracts were prepared by maceration of individual plant material in each solvent; ethanol (95%), distilled water and ethanol-water mixture (70:30, *v/v*), for 48 h. Organic solvents were subsequently evaporated from filtrates using a rotary evaporator (Büchi 011, Flawil Switzerland). Water filtrates were dried using a desiccator at 60 °C (Memmert UN30, Schwabach, Germany). Each plant powder was extracted separately three times using the same amount of solvent, and the dried extracts were pooled and weighed. The yields of extraction were calculated in percentage relative to the weight of the starting plant material.

For the preliminary antifungal screening, the crude extracts were dissolved in 10% dimethylsulfoxide (DMSO) and tested using the agar dilution method, according to the CLSI M44-A2 protocol [28,29] with some modifications. Briefly, sterile Mueller Hinton agar medium (Oxoid) supplemented with 2% glucose, was maintained at 45 °C and the extract added to obtain a final concentration of 40 mg/mL. The mixture was then transferred in a Petri dish and let to solidify at room temperature. Plates were inoculated with 2.0×10^4 yeast cells/mL in sterile normal saline solution and incubated at 37 °C for 48 h. Growth control consisted of microorganisms cultured on the Mueller Hinton agar medium without plant extracts. Fluconazole (Sigma-Aldrich, Heidenheim, Germany) was used as positive control at 10 µg/mL. Experiments were run simultaneously and each test performed in duplicate. At incubation time, yeast colonies were enumerated and inhibition percentages (IP) were calculated using the formula:

IP (%) = (colonies in growth control − colonies in test dish) × 100/colonies in growth control

Crude extracts that showed an inhibition percentage of ≥75% against at least one of the tested yeasts were selected for the determination of minimal inhibitory concentration (MIC), defined as the lowest concentration at which there was no visible growth of the yeasts.

MIC values of the selected extracts were determined according to the CLSI M27-A3 protocol [30] with little modifications. Briefly, a serial two-fold dilution of each crude extract using Mueller Hinton broth medium (Oxoid) supplemented with 2% glucose was done in 96-wells microtiter plates starting with 40 mg/mL up to 0.04 mg/mL. One hundred microliters of fungal inocula at the final concentrations of 0.5–2.5 × 10^3 CFU/mL were then added into each well of the plate to achieve a final volume of 200 μL. Plates were incubated at 37 °C for 48 h. Fluconazole was tested as positive control at the highest concentration of 64 μg/mL. MICs were determined based on turbidity in plate wells through macroscopic observation. Crude extracts that showed MIC ≤ 1.25 mg/mL against at least half of the tested yeasts were selected for progression to the bio-guided fractionation.

2.4. Bio-Guided Fractionation of Selected Crude Extracts

Extract fractionation consisted in partitioning 10 g of each selected extract in 300 mL mixture of methylene chloride/water (*v/v*). After exhaustion of the methylene chloride phase, fractions were concentrated and dried. Change in MICs of the dried fractions was determined as described above and promising fractions were selected for progression. Five grams of each selected fraction was further partitioned into 150 mL mixture of hexane/methanol (*v/v*) and the afforded sub-fractions tested for activity.

2.5. Antifungal Activity of Combined Sub-Fractions

Seven sub-fractions from the three plants were combined to give 11 interspecies combinations, with no combination between extracts from the same plant. Checkerboard tests were used to determine fractional inhibitory concentration indexes (FICIs) of combinations of sub-fractions against each test microorganism. The checkerboard broth microdilution method based on CLSI recommendations [31] consisted of diluting sub-fractions in the two directions of a 96-wells microplate. Mixed concentrations in wells ranged from 1/8 × MIC to 8 × MIC and 1/128 × MIC to 8 × MIC for the two sub-fractions, respectively. The fungal inoculum was added to give an ultimate concentration of 0.5–2.5 × 10^3 CFU/mL in a final volume of 100 μL. Plates were therefore incubated at 37 °C for 48 h and each test performed in duplicate.

Changes in the Fractional Inhibitory Concentration Indexes (FICIs) were calculated using the following formula, and the type of interaction was determined according to previously described criteria [31,32].

$FICI_{(X1/X2)}$ = MIC_{X1} in combination/MIC_{X1} alone + MIC_{X2} in combination/MIC_{X2} alone [33].

Based on these criteria, a combination was considered to be synergistic when the FICI was ≤ 0.5, additive when it was > 0.5 to ≤ 1, indifferent when it was > 1.0 to ≤ 4.0. Antagonism was obtained with FICI > 4.0.

2.6. Time Kill Kinetic Assay of the Most Active Combinations

The time kill kinetic assay of the selected combinations was assessed using the method described by Klepser *et al.* [34] with some modifications. The assays were performed in duplicate in 1.5 mL conical tubes at final concentrations of 4xMIC for each sub-fraction in the Mueller Hinton broth. For each combination, three tubes were used, two containing sub-fractions alone and one containing the combination of sub-fractions. The final volume in each tube was 1 mL, containing a fungal inoculum of $1–2.10^5$ cells/mL. After incubation under orbital shaking (IKA-Vibrax-VXR, Radnor, PA, USA) (32× *g*) at 37 °C, and at time intervals of 0, 2, 4, 6, 8, 12, 16 and 24 h, a 50 µL aliquot was collected from each culture to inoculate 950 µL of sterile broth medium and incubated at 37 °C for 24 h to determine the number of cells/mL using a Malassez haemocytometer (Thermo Fisher Scientific, Darmstadt, Germany). Time kill kinetic curves were plotted as log10 (number of viable cells) *versus* time. The following criteria were used to interpret the interactions: Synergy was obtained when the fungicidal effect led to ≥2log10 decrease in cells/mL for the combination compared to the most active sub-fraction; additivity was defined as <2log10 decrease in cells/mL for the combination compared to the most active sub-fraction; indifference as <2log10 increase in cells/mL for the combination compared to the least active sub-fraction; and antagonism as ≥2log10 increase in cells/mL for the combination compared to the least active sub-fraction [9,34].

2.7. Phytochemical Screening of Selected Crude Extracts and Sub-Fractions

A qualitative phytochemical analysis of promising extracts and sub-fractions was carried out to identify groups of secondary metabolites known as having antifungal activity, including flavonoids, tannins, anthraquinones, alkaloids, steroids, glucosides, saponins, triterpenes, anthocyanins and phenols and using the previously described protocols [35,36,37,38].

3. RESULTS

3.1. Plant Extraction and Preliminary Screening of Antifungal Activity

Thirty-four crude extracts were obtained with yields ranging from 1.7%–36.7% relative to the weights of the starting plant materials. Results from the preliminary screening of antifungal activity enabled the selection of 23 extracts based on their inhibitory percentages against clinical isolates and the reference *Candida albicans* NR-29450 (≥75% against at least one of the tested yeasts). Change in MIC values in liquid medium ranging from 0.04 mg/mL to 40.0 mg/mL was observed as described in Table 1. Based on our selection criteria (MIC ≤ 1.25 mg/mL against at least half of tested yeasts), 15 crude extracts were selected and progressed to the bio-guided fractionation, including extracts A1 to A15.

Table 1. Minimal inhibitory concentrations (MIC) of crude extracts, fractions and sub-fractions.

Codes	Crude Extracts	[a] MIC (mg/mL)				
		Reference Strain	Isolates			
		C. albicans NR-29450	*C. albicans*	*C. glabrata*	*C. parapsilosis*	*Cr. neoformans*
A1	Te C L H_2O	0.31	2.50	0.63	2.50	0.63
A2	Te C L H_2O/EthOH	0.08	0.31	0.08	0.31	0.16
A3	Te C L EthOH	0.31	0.16	0.08	0.31	0.31
A4	Te C Sb H_2O	0.16	0.63	0.31	0.31	0.63
A5	Te C Sb H_2O/EthOH	0.08	2.50	0.63	0.31	0.63
A6	Te C Sb EthOH	0.16	0.16	0.16	0.16	0.16
A7	Te M L H_2O	0.31	2.50	0.31	0.31	0.16
A8	Te M L H_2O/EthOH	0.04	0.04	0.04	0.08	0.08
A9	Te M L EthOH	0.04	0.08	0.08	0.16	0.16
A10	Te M Sb H_2O	0.08	0.08	0.08	0.08	0.08
A11	Te M Sb H_2O/EthOH	0.08	0.31	0.16	0.16	0.08
A12	Te M Sb EthOH	0.08	0.16	0.16	0.08	0.04
A13	Mo T L H_2O	10.00	1.25	0.63	10.00	1.25
A14	Mo T L H_2O/EthOH	5.00	0.63	0.63	1.25	1.25
A15	Mo T L EthOH	2.50	5.00	1.25	1.25	0.08
A16	Mo T Tw H_2O	20.00	10.00	5.00	5.00	2.50
A17	Mo T Br H_2O	10.00	5.00	5.00	2.50	5.00
A18	Mo T Pu H_2O	2.50	10.00	5.00	5.00	10.00
A19	Mo T Pu H_2O/EthOH	5.00	5.00	5.00	10.00	2.50
A20	Mo T Pu EthOH	5.00	2.50	5.00	10.00	2.50
A21	Mo T Se H_2O/EthOH	40.00	10.00	10.00	40.00	40.00
A22	Mo T FrPe EthOH	10.00	5.00	5.00	10.00	2.50
A23	Mo T PSe EthOH	20.00	10.00	20.00	40.00	20.00

B1	Te C L H_2O	Fr H_2O	1.25	1.25	5	0.63	0.16
B2		Fr CH_2Cl_2	0.63	1.25	5	0.16	0.16
B3	Te C L H_2O/EthOH	Fr H_2O	0.31	2.5	0.63	0.31	0.31
B4		Fr CH_2Cl_2	0.31	5	0.63	0.31	0.31
B5	Te C L EthOH	Fr H_2O	0.63	>5.000	0.63	0.31	0.16
B6		Fr CH_2Cl_2	0.63	5	0.63	0.63	0.31
B7	Te C Sb H_2O	Fr H_2O	1.25	2.5	2.5	0.63	0.31
B8		Fr CH_2Cl_2	1.25	5	5	2.5	0.63
B9	Te C Sb H_2O/EthOH	Fr H_2O	0.31	5	2.5	2.5	0.63
B10		Fr CH_2Cl_2	0.16	0.63	1.25	1.25	0.63
B11	Te C Sb EthOH	Fr H_2O	0.16	0.63	2.5	1.25	0.63
B12		Fr CH_2Cl_2	0.31	>5.000	5	5	1.25
B13	Te M L H_2O	Fr H_2O	2.5	1.25	0.31	0.31	0.16
B14		Fr CH_2Cl_2	2.5	0.63	0.16	0.08	0.08
B15	Te M L H_2O/EthOH	Fr H_2O	2.5	1.25	0.31	0.16	0.04
B16		Fr CH_2Cl_2	2.5	0.16	0.31	0.16	0.08
B17	Te M L EthOH	Fr H_2O	2.5	0.63	0.63	0.31	0.31
B18		Fr CH_2Cl_2	0.63	0.31	0.08	0.04	0.04
B19	Te M Sb H_2O	Fr H_2O	1.25	0.63	0.16	0.08	0.08
B20		Fr CH_2Cl_2	0.63	0.31	0.16	0.16	0.08
B21	Te M Sb H_2O/EthOH	Fr H_2O	0.31	1.25	0.31	0.31	0.16
B22		Fr CH_2Cl_2	0.63	1.25	0.63	0.31	0.31
B23	Te M Sb EthOH	Fr H_2O	0.63	1.25	0.31	0.16	0.16
B24		Fr CH_2Cl_2	1.25	1.25	0.31	0.16	0.16
B25	Mo T L H_2O	Fr H_2O	5	5	0.08	>5.000	>5.000
B26		Fr CH_2Cl_2	2.5	2.5	0.63	2.5	2.5
B27	Mo T L H_2O/EthOH	Fr H_2O	2.5	5	5	5	2.5
B28		Fr CH_2Cl_2	2.5	2.5	1.25	0.63	0.31
B29	Mo T L EthOH	Fr H_2O	2.5	>5.000	2.5	5	5
B30		Fr CH_2Cl_2	2.5	2.5	0.63	0.63	0.63

Codes	Sub-fractions	[a] MIC (mg/mL)					
		Reference Strain		Isolates			
		C. albicans NR-29450		*C. albicans*	*C. glabrata*	*C. parapsilosis*	*Cr. neoformans*
C2	Te C L H_2O/EthOH Fr H_2O	sFr CH_3OH	0.31	0.62	0.31	0.31	0.16
C4	Te C L H_2O/EthOH Fr CH_2Cl_2	sFr CH_3OH	0.31	0.62	0.31	0.62	0.31
C5	Te C L EthOH Fr CH_2Cl_2	sFr C_6H_{12}	0.62	1.25	0.62	1.25	0.62
C6		sFr CH_3OH	1.25	1.25	1.25	>1.25	1.25
C8	Te C Sb H_2O/EthOH Fr CH_2Cl_2	sFr CH_3OH	0.16	1.25	0.31	0.62	0.08
C10	Te M L H_2O Fr H_2O	sFr CH_3OH	0.16	0.62	0.31	0.31	0.16
C12	Te M L H_2O Fr CH_2Cl_2	sFr CH_3OH	0.31	0.62	0.08	0.08	0.08
C14	Te M L H_2O/EthOH Fr H_2O	sFr CH_3OH	0.31	0.62	0.16	0.16	0.16
C15	Te M L H_2O/EthOH FrCH_2Cl_2	sFr C_6H_{12}	0.31	0.31	0.31	0.31	0.16
C16		sFr CH_3OH	0.16	0.16	0.16	0.62	0.16
C18	Te M L EthOH Fr H_2O	sFr CH_3OH	0.31	0.16	0.31	0.31	0.62
C20	Te M L EthOH Fr CH_2Cl_2	sFr CH_3OH	0.16	0.16	0.31	0.31	0.31
C22	Te M Sb H_2O Fr H_2O	sFr CH_3OH	0.16	0.08	0.08	0.16	0.04
C24	Te M Sb H_2O Fr CH_2Cl_2	sFr CH_3OH	0.16	0.08	0.08	0.16	0.04
C26	Te M Sb H_2O/EthOH Fr H_2O	sFr CH_3OH	0.04	0.08	0.08	0.16	0.08
C28	Te M Sb H_2O/EthOH FrCH_2Cl_2	sFr CH_3OH	0.31	1.25	1.25	1.25	0.62
C30	Te M Sb EthOH Fr H_2O	sFr CH_3OH	0.31	0.31	0.31	0.31	0.31
C32	Te M Sb EthOH Fr CH_2Cl_2	sFr CH_3OH	0.16	0.08	0.08	0.16	0.08
C33	Mo T L H_2O/EthOH Fr CH_2Cl_2	sFr C_6H_{12}	>1.25	>1.25	>1.25	>1.25	>1.25
C34		sFr CH_3OH	1.25	1.25	1.25	1.25	>1.25
C35	Mo T L EthOH Fr CH_2Cl_2	sFr C_6H_{12}	1.25	>1.25	>1.25	>1.25	1.25
C36		sFr CH_3OH	1.25	0.62	0.62	>1.25	1.25
	Fluconazole (µg/mL)		0.50	1.00	8.00	8.00	2.00

[a] Minimal Inhibitory Concentration (lowest concentration at which there is no visible growth of yeast); Mo T: *Monodora tenuifolia*; Te C: *Terminalia catappa*; Te M: *Terminalia mantaly*; L: leaves; Sb: stem bark; Tw: twigs; Br: branches; Pu: pulp; Se: seeds; PSe: seeds pericarp; FrPe: fruit pericarp; H_2O: distilled water; H_2O/EthOH: hydroethanol; EthOH: ethanol; Fr H_2O: aqueous fraction; Fr CH_2Cl_2: methylene chloride fraction; sFr C_6H_{12}: Hexanic sub-fraction; sFr CH_3OH: Methanolic sub-fraction. A1–A23: crude extracts; B1-B30: fractions; C2–C36: sub-fractions.

3.2. Activity of Bio-Guided Fractions

The first partition of the 15 selected extracts generated 30 fractions out of which 18 were progressed based on their MIC values (median MIC ≤ 0.62 mg/mL), including B3–B6, B10, B13–B24, and B28. Subsequent to the partition of the 18 fractions, seven sub-fractions (C2, C12, C22, C24, C26, C32 and C36) were afforded with median MIC ≤ 0.31 mg/mL. The most active sub-fraction (C26) showed a median MIC of 0.08 mg/mL. MICs of fractions and sub-fractions are summarized in Table 2.

Globally, there was a significant increase in activity from crude extracts to sub-fractions, with median MICs = 0.31 mg/mL for crude extracts; median MICs = 0.31 mg/mL for fractions; and median MICs = 0.08 mg/mL for sub-fractions. Extracts from *T. mantaly* showed more potency, while those from *M. tenuifolia* were the least active against the tested yeasts. There was no correlation between observed activities and solvents used for extraction.

Table 2. Effect of combined sub-fractions on the tested yeasts.

Combinations of sub-fractions	*C. albicans* NR-29450		*C. albicans*		*C. glabrata*		*C. parapsilosis*		*Cr. neoformans*	
	FICI	Int	FICI	Int	FICI	Int	FICI	Int	FICI	Int
C2/C12	1.06	I	1.06	I	1.06	I	1.00	A	0.62	A
C2/C22	0.75	A	0.50	S	0.75	A	0.62	A	0.31	S
C2/C24	0.56	A	0.75	A	0.75	A	0.53	A	0.31	S
C2/C26	0.62	A	0.75	A	0.75	A	1.12	I	0.62	A
C2/C32	0.56	A	1.12	I	0.75	A	1.00	A	0.37	S
C36/C2	0.50	A	0.62	A	0.50	S	0.50	S	0.50	S
C36/C12	0.37	S	0.62	A	0.56	A	0.37	S	0.25	S
C36/C22	0.53	A	0.62	A	0.75	A	0.75	A	0.37	S
C36/C24	1.00	S	1.12	I	0.50	S	0.75	A	0.50	S
C36/C26	0.75	A	0.75	A	0.62	A	0.56	A	0.25	S
C36/C32	0.62	S	0.50	S	0.75	A	0.56	A	0.62	A

FICI: Fractional Inhibitory Concentration Index; Int: interaction; I: indifference; A: additivity; S: synergy; C2: Te CL H_2O/EthOH Fr H_2O sFr CH_3OH; C12: Te ML H_2O Fr CH_2Cl_2 sFr CH_3OH; C22: Te MSb H_2O Fr H_2O sFr CH_3OH; C24: Te MSb H_2O Fr CH_2Cl_2 sFr CH_3OH; C26: Te MSbH_2O/EthOH Fr H_2O sFr CH_3OH; C32: Te MSb EthOH Fr CH_2Cl_2 sFr CH_3OH; C36: Mo TL EthOH Fr CH_2Cl_2 sFr CH_3OH.

3.3. Activity of Combined Sub-Fractions

Eleven combinations were prepared from seven selected sub-fractions (C2, C12, C22, C24, C26, C32 and C36) and tested for their bioactivity. The seven sub-fractions used consisted of five from *T. mantaly*, one from *T. catappa* and one from *M. tenuifolia.* All the calculated fractional inhibitory concentrations (FICIs) showed values between 0.25 and 1.12, presenting synergy, additivity or indifference (Table 2). No case of antagonism was observed. Two combinations (C36/C2 and C36/C12) exerted above 50% synergistic interaction associated with more than 75% of MIC reduction. These combinations were prepared from a *M. tenuifolia* sub-fraction (C36) associated with each sub-fraction of *T.*

mantaly (C12) and *T. catappa* (C2), respectively. They were therefore submitted to time kill kinetic assays.

3.4. Time Kill Kinetics of the Promising Sub-Fractions and Their Combinations

The results obtained from the time kill assays indicated globally that all sub-fractions tested alone were fungistatic against all the microorganisms. On the contrary, the combinations exhibited fungicidal activity in most cases against the tested yeasts (Figure 1). Particularly, *C. albicans* and *C. glabrata* isolates were the most susceptible to the combinations C36/C2 and C36/C12, with all cells killed within 6 and 8 h of exposure, respectively. Comparatively, *C. albicans* NR-29450 was less sensitive to the combinations than the isolate. Overall, a fungicidal effect was obtained against the other microorganisms beyond 16 h. Only C36/C12 showed fungicidal action against *C. parapsilosis,* and the two combinations were fungistatic on *Cr. neoformans*.

4. DISCUSSION

The preparations of medicines by traditional healers are mainly carried out by boiling different organs of plants in water or macerating in alcoholic white wine. This information oriented the choice for extraction solvents. The antifungal activities obtained for *T. catappa*, *T. mantaly* and *M. tenuifolia* confirmed the traditional use of these plants in the treatment of fungal infections and also corroborated what had been previously found [39,40,41]. Aqueous and ethanolic extracts of stem bark of *T. catappa* and *T. mantaly* growing in the Cote d'Ivoire were shown to have anticandidal activity on clinical isolates of *C. albicans* with IC_{50} ranging from 0.02 to 0.55 mg/mL [39,40]. Furthermore, a comparative study showed that extracts from *T. mantaly* were more active than those from *T. catappa* against *Aspergillus fumigatus* [40]. To the best of our knowledge, this is the first scientific report on the antifungal potential of *M. tenuifolia* extracts. Also, the present work is pioneering on the aspect of optimizing activities by coupling bio-guided fractionation and extracts combinations. Based on modern trends in combining drugs for activity optimization [34,42,43], combinations of *T. mantaly* and *T. catappa* with *M. tenuifolia* sub-fractions have considerably increased the anti-yeast activity. This optimized action of the combinations is more likely due to components of the sub-fractions acting synergistically against the pathogens [33,44,45,46]. Moreover, the

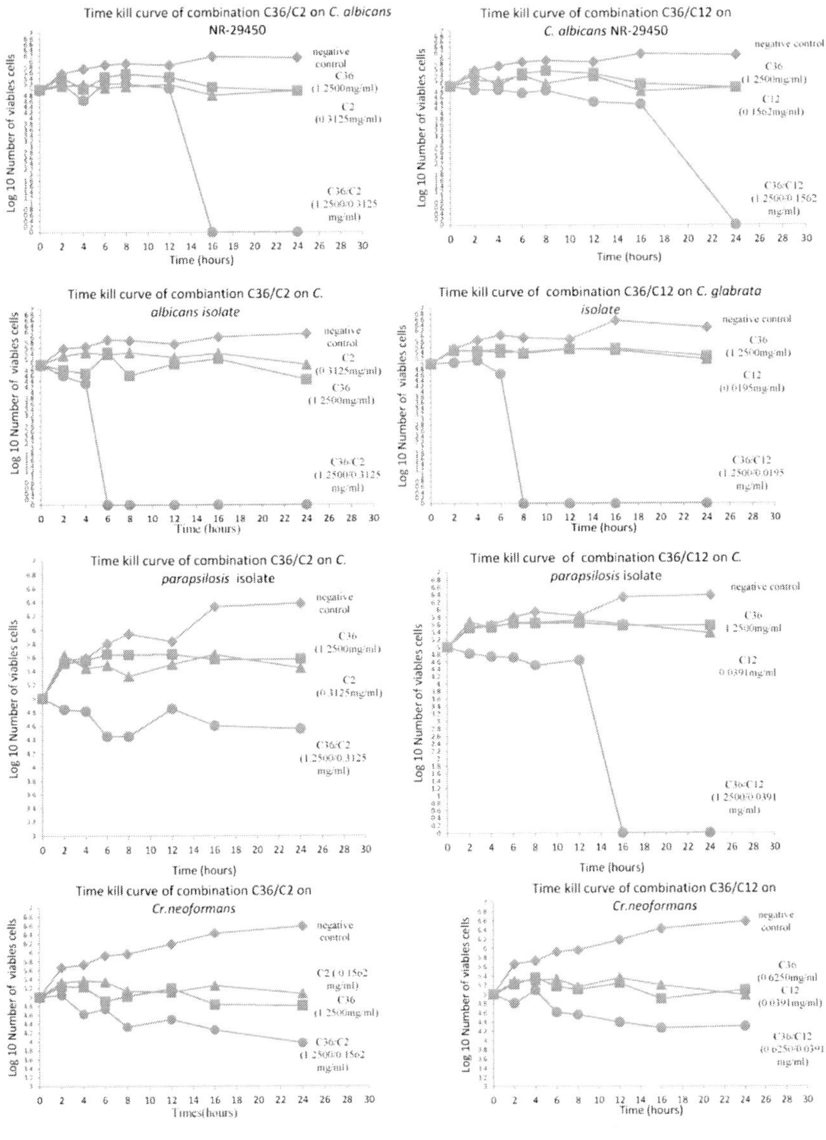

Figure 1. Time kill kinetic curves of combinations C36/C2 and C36/C12 against the tested yeasts. Sub-fractions were assessed alone and in combination for bioactivity at different time intervals.

simultaneous actions of metabolites on different targets enhance their bioactivity and might also reduce the advent of resistance by the fungi.

3.5. Phytochemical Screening of Crude Extracts and Selected Sub-Fractions

The qualitative phytochemical screening of crude extracts and sub-fractions of interest showed the presence of alkaloids, flavonoids, tannins, saponins and steroids. However only two types of secondary metabolites (glucosides and steroids) were identified in all three most active sub-fractions (C2, C12, and C36), suggesting their implication in the exerted anti-yeast activity.

Results from the time kill kinetics assays showed that combinations increased the antifungal activity and also changed the pharmacokinetics of extracts from fungistatic to fungicidal. This finding is particularly significant, given that among the drawbacks of azole drugs, they are rather fungistatic. This profile probably contributed to the development of resistance of clinical isolates from immunocompromised patients to azoles, given that residual yeast cells are not cleared, enabling a positive selection of drug-resistant mutants [47]. The qualitative phytochemical screening of the promising sub-fractions (C2, C12 and C36) led to the identification of steroids and glucosides as common classes of secondary metabolites. Steroids have been reported to have antibacterial and antifungal properties. The correlation between membrane lipids and sensitivity for steroidal compounds indicates the mechanism through which steroids specifically associate with membrane lipids and exert their action by causing leakages from liposomes [48]. Apart from this, steroidal glucosides were shown to possess potent antifungal activity [49]. Needed antifungal drugs should be ideally fungicidal. To achieve this, promising combinations identified within the framework of this study should be rationally studied in detail and potentially developed into novel drugs to control yeast infections.

5. CONCLUSIONS

The results achieved from this work indicate that *T. mantaly*, *T. catappa* and *M. tenuifolia* are potential sources of new antifungal drugs. The results also show improved potency in correlation with fractionation of extracts, as well as the combination of sub-fractions. The more promising combinations exerted synergistic and fungicidal actions against the tested yeasts. Overall, these findings support the continued investigation of extracts from the studied plants using the bio-guided fractionation and combination strategies to develop innovative therapies against mycoses.

ACKNOWLEDGMENT

The following reagent was obtained through BEI Resources, NIAID, NIH: *C. albicans* strain ATCC P37037. The authors thank the Yaoundé Central Hospital for providing the microorganisms. They also thank the Cameroon National Herbarium for the identification of plants.

AUTHORS CONTRIBUTIONS

F.F.B. and T.K.N. conceived and designed the work. T.K.N. participated in all manipulations, analyzed and interpreted the results and drafted the manuscript. C.M.J. and R.M.K.T. participated in plant collection, and antifungal and phytochemical tests. M.A.T.T. helped with bio-guided fractionation. E.M.Z., V.N., M.M. and S.B. read and critically revised the manuscript. All authors read and approved the final manuscript.

CONFLICTS INTEREST

The authors declare no competing interests

REFERENCES

1. Sganga, G. Fungal infections in immunocompromised patients: Diagnosis, Therapy and Prophylaxis of Fungal Diseases. Mycoses 2011, 54, 1–3.
2. David, W.D.; William, W. Hope Therapy for fungal diseases: Opportunities and priorities. Trends Microbiol. 2010, 18, 195–204.
3. Pemán, J.; Cantón, E.; Gobernado, M.; Spanish ECMM Working Group on Candidaemia. Epidemiology and antifungal susceptibility of Candida species isolated from blood: Results of a 2-year multicentre study in Spain. Eur. J. Clin. Microbiol. Infect. Dis. 2005, 24, 23–30.
4. Pfaller, M.A.; Diekema, D.J. Rare and emerging opportunistic fungal pathogens: Concern for resistance beyond Candida albicans and Aspergillus fumigatus. J. Clin. Microbiol. 2004, 42, 4419–4431.
5. Kumar, B.V.; Padshetty, N.S.; Bai, K.Y.; Rao, M.S. Prevalence of Candida in the Oral Cavity of Diabetic Subjects. Iran J. Microbiol. 2011, 53, 599–602.

6. Abu-Elteen, K.H.; Hamad, M.A.; Salah, S.A. Prevalence of Oral Candida Infections in Diabetic Patients. Bahr. Med. Bull. 2006, 28, 1–8.
7. Mlinariæ-Missoni, E.; Kaleniæ, S.; Vukelic, M.; de Syo, D.; Belicza, M.; Vaziæ-Babiæ, V. Candida infections of diabetic foot ulcers. Diabetol. Croat. 2005, 34, 29–35.
8. Park, B.J.; Wannemuehler, K.A.; Marston, B.J.; Govender, N.; Pappas, P.G.; Chiller, T.M. Estimation of the current global burden of cryptococcal meningitis among persons living with HIV/AIDS. AIDS 2009, 23, 525–530.
9. Lewis, R.E. Current Concepts in Antifungal Pharmacology. Mayo Clin. Proc. 2011, 86, 805–817.
10. De Almeid, O.P.; Scully, C. Fungal infections of the mouth. Braz. J. Oral Sci. 2002, 1, 19–26.
11. Laurent, M.; Gogly, B.; Tahmasebi, F.; Paillaud, E. Les candidoses oropharyngées des personnes âgées. Geriatr. Psychol. Neuropsychiatr. Vieil. 2011, 9, 21–28.
12. Magryś, A.; Kozioł-Montewka, M.; Starosławska, E.; Gabczyńska, B. The Prognostic and Diagnostic Markers of Invasive Candidiasis in Patients during Chemotherapy. Pol. J. Microbiol. 2005, 54, 207–213.
13. Fidel, P.L. History and update on host defence against vaginal candidiasis. Am. J. Reprod. Immunol. 2007, 57, 2–12.
14. Nelesh, G. HIV-associated opportunistic fungal infections: A guide to using the clinical microbiology laboratory. South. Afr. J. HIV Med. 2007, 1, 18–23.
15. Álvaro-Meca, A.; Jensen, J.; Micheloud, D.; Asunción, D.; Gurbindo, D.; Resino, S. Rate of candidiasis among HIV infected children in Spain in the era of highly active antiretroviral therapy (1997–2008). J. Infect. Dis 2013, 13, 1–15.
16. Brissaud, O.; Guichoux, J.; Harambat, J.; Tandonnet, O.; Zaoutis, T. Invasive fungal disease in PICU: Epidemiology and risk factors. Ann. Intensive Care 2012, 2, 6.
17. Pappas, P.G.; Kauffman, C.A.; Andes, D.; Benjamin, D.K.; Calandra, T.F.; Edwards, J.E.; Filler, S.G.; Fisher, J.F.; Kullberg, B.-J.; Ostrosky-Zeichner, L.; et al. Clinical Practice Guidelines for the Management of Candidiasis: 2009 Update by the Infectious Diseases Society of America. Clin. Infect. Dis. 2009, 48, 503–535.
18. Lass-Flörl, C.; Mayr, A.; Perkhofer, S.; Hinterberger, G.; Hausdorfer, J.; Speth, C.; Fille, M. Activities of Antifungal Agents against Yeasts and Filamentous Fungi: Assessment according to the Methodology of the European Committee on Antimicrobial Susceptibility Testing. Antimicrob. Agents Chemother. 2008, 52, 3637–3641.
19. Ruan, E.; Kock, J.L.F.; Pohl, C.H. Candida albicans or Candida dubliniensis? Mycoses 2009, 54, 1–16.
20. Busari, O.; Adeyemi, A.; Agboola, S. Cryptococcal Meningitis in AIDS. Internet J. Infect. Dis. 2009, 7, 1. Available online: http://print.ispub.com/api/0/ispub-article/8148 (accessed on 18 August 2015).
21. Maslin, J.; Morand, J.J.; Menard, G.; Camparo, P. Les Cryptococcoses. Med. Trop. 2002, 62, 480–484.
22. Millogo, A.; Ki-Zerbo, G.A.; Andonaba, J.B.; Lankoandé, D.; Sawadogo, A.; Yaméogo, I.; Sawadogo, A.B. La cryptococcose neuroméningée au cours de l'infection par le VIH au Centre hospitalier de Bobo-Dioulasso (Burkina Faso). Bull. Soc. Pathol. Exot. 2004, 97, 119–121.
23. Petrikkos, G.; Skiada, A. Recent advances in antifungal chemotherapy. Int. J. Antimicrob. Agents 2007, 30, 108–117.
24. Kanafani, Z.A.; Perfect, J.R. Antimicrobial resistance: Resistance to antifungal agents: Mechanisms and clinical impact. Clin. Infect. Dis. 2008, 46, 120–128.
25. Tscherner, M.; Schwarzmüller, T.; Kuchler, K. Pathogenesis and Antifungal Drug Resistance of the Human Fungal Pathogen Candida glabrata. Pharmaceuticals 2011, 4, 169–186.
26. Neelavathi, P.; Venkatalakshmi, P.; Brindha, P. Antibacterial activities of aqueous and ethanolic extracts of Terminalia catappa leaves and bark against some pathogenic bacteria. Int. J. Pharm. Pharm. Sci. 2013, 5, 114–120.
27. Neuwinger, H.D. African Traditional Medicine, A Dictionary of Plant Use and Applications; Medpharm Scientific: Stuttgart, Germany, 2000.

28. Therese, K.L.; Bagyalakshmi, R.; Madhavan, H.N.; Deepa, P. In vitro susceptibility testing by agar dilution method to determine the minimum inhibitory concentrations of amphotericin B, fluconazole and ketoconazole against ocular fungal isolates. Ind. J. Med. Microbiol. 2006, 24, 273–279.
29. NCCLS. Reference Method for Antifungal Disk Diffusion Susceptibility Testing of Yeasts; Approved guideline; NCCLS document M44-A2; National Committee for Clinical Laboratory Standards: Wayne, PA, USA, 2004.
30. Clinical and Laboratory Standards Institute (CLSI). Reference Method for Broth Dilution Antifungal Susceptibility Testing of Yeasts, 3rd ed.; CLSI: Wayne, PA, USA, 2008.
31. Iten, F.; Saller, R.; Abel, G.; Reichling, J. Additive antimicrobial effects of the active components of the essential oil of Thymus vulgaris-chemotype carvacrol. Planta Med. 2009, 75, 1231–1236.
32. Berenbaum, M.C. A method for testing for synergy with any number of agents. J. Infect. Dis. 1978, 137, 122–130.
33. Van Vuuren, S.; Viljoen, A. Plant-Based Antimicrobial Studies Methods and Approaches to Study the Interaction between Natural Products. Planta Med. 2011, 77, 1168–1182.
34. Klepser, M.E.; Ernst, E.J.; Lewis, R.E.; Ernst, M.E.; Pfaller, M.A. Influence of test conditions on antifungal time-kill curve results: Proposal for standardized methods. Antimicrob. Agents Chemother. 1998, 42, 1207–1212.
35. Aromedee, C.; Wichitchote, P.; Jantakun, N. Spectrophotometric Determination of Total Lactones in Andrographis paniculata Nees. Songklanakarin J. Sci Technol. 2005, 27, 1227–1231.
36. Odebeyi, O.O.; Sofowara, F.H. Antimicrobial alkaloids from a Nigerian Chewing Stick (Fagara zanthoxyloides). Planta Med. 1978, 36, 204–207.
37. Sofowara, A. Medicinal Plants and Traditional Medicine in Africa, 2nd ed.; Spectrum Books Limited: Ibadan, Nigeria, 1993.
38. Trease, G.E.; Evans, W.C. Pharmacognosy, 13th ed.; Bailliere Tindall: London, UK, 1989.
39. Ackah, J.A.B.; Kra, A.K.M.; Zirihi, G.N.; Guede-Guina, F. Évaluation et essais d'optimisations de l'activité anticandidosique de Terminalia catappa Linné (tekam3), un extrait de Combretaceae de la pharmacopée ivoirienne. Bull. Soc. R. Sci. Liège 2008, 77, 120–136.
40. Yapi, G.Y.; Adou, K.M.; Ackah, J.A.B.; Djaman, A.J. Évaluation de l'activité antifongique et essai de purification des principes actifs des extraits de Terminalia mantaly (H Perrier), une eaCombretaceae, sur la croissance in vitro de Candida albicans. Bull. Soc. R. Sci. Liège 2011, 80, 953–964.
41. Zirihi, G.N.; N'guessan, K.; Kassy, N.J.; Coulibaly, K.; Djaman, A.J. Evaluation and comparison of antifungal activities of Terminalia catappa and Terminalia mantaly (Combretaceae) on the in vitro growth of Aspergillus fumigatus. J. Med. Plant Res. 2012, 6, 2299–2308.
42. Lewis, R.E.; Diekema, D.J.; Messer, S.A.; Pfaller, M.A.; Klepser, M.E. Comparison of E-test, chequerboard dilution and time-kill studies for the detection of synergy or antagonism between antifungal agents tested against Candida species. J. Antimicrob. Chemother. 2002, 49, 345–351.
43. Lin, Y.T.; Labbe, R.G.; Shetty, K. Inhibition of Vibrio parahaemolyticus in seafood systems using oregano and cranberry phytochemical synergies and lactic acid. Innovat. Food Sci. Emerg. Technol. 2005, 6, 453–458.
44. Nwodo, U.U.; Ngene, A.A.; Iroegbu, C.U. Effects of Fractionation on Antibacterial Activity of Crude Extracts of Tamarindus indica. Afr. J. Biotechnol. 2010, 9, 7108–7113.
45. Ncube, B.; Finnie, J.F.; van Staden, J. In vitro antimicrobial synergism within plant extracts combinations from three South African medicinal bulbs. J. Ethnopharmacol. 2012, 139, 81–89.

46. Onyewu, C.; Blankenship, J.R.; del Poeta, M.; Heitman, J. Ergosterol biosynthesis inhibitors become fungicidal when combined with calcineurin inhibitors against Candida albicans, Candida glabrata, and Candida krusei. Antimicrob. Agents Chemother. 2003, 47, 956–964.
47. Shimada, T. Salivary proteins as a defence against dietary tannins. J. Chem. Ecol. 2006, 32, 1149–1163.
48. Raquel, F.E. Bacterial lipid composition and the antimicrobial efficacy of cationic steroid compounds. Biochim. Biophys. Acta 2007, 1768, 2500–2509.
49. Ahsan, S.; Ejaz, A.; Abdul, M. Recurvosides A and B, Antifungal Novel Steroidal Glucosides from Haloxylon recurvum. Z. Naturforsch. B. 2006, 61(9), 1148–1152.

CHAPTER 14

Inhibiting Microbial Toxins Using Plant-Derived Compounds and Plant Extracts

Abhinav Upadhyay, Shankumar Mooyottu, Hsinbai Yin, Meera Surendran Nair, Varunkumar Bhattaram and Kumar Venkitanarayanan *

Department of Animal Science, University of Connecticut, Storrs, CT 06269, USA

ABSTRACT

Many pathogenic bacteria and fungi produce potentially lethal toxins that cause cytotoxicity or impaired cellular function either at the site of colonization or other locations in the body through receptor-mediated interactions. Various factors, including biotic and abiotic environments, competing microbes, and chemical cues affect toxin expression in these pathogens. Recent work suggests that several natural compounds can modulate toxin production in pathogenic microbes. However, studies explaining the mechanistic basis for their effect are scanty. This review discusses the potential of various plant-derived compounds for reducing toxin production in foodborne and other microbes. In addition, studies highlighting their anti-toxigenic mechanism(s) are discussed.

Keywords: plant-derived compounds; plant extracts; bacterial toxins; mycotoxins; toxin inhibition; anti-virulence strategies

1. INTRODUCTION

Microbial infections are a leading cause of morbidity and mortality in humans worldwide. Every year, microbial infections cause more than 100 million illnesses worldwide [1]. Antibiotics have been traditionally used for treating microbial illnesses. However, their excessive and inappropriate use has led to the development of antibiotic resistance in microbes, thereby making the treatment of infections a formidable challenge [2,3]. Moreover, antibiotics such as fluoroquinolones are contraindicated in certain clinical conditions such as hemolytic uremic syndrome since their use leads to induction of Shiga toxin-encoding bacteriophages, increased expression of Shiga toxin genes and toxin release [4,5]. Therefore, research efforts to search for alternative therapies to control microbial infections in humans have intensified.

Microbial illnesses are orchestrated by an array of virulence factors that contribute to the pathophysiology and survival of the pathogen in the host [6]. These factors include cell surface proteins, receptors, and adhesins that facilitate host colonization; capsular polysaccharides that confer protection from host immune system, and microbial toxins that cause host tissue damage [7,8]. Reducing the expression of these virulence factors could control infections in humans. With increasing understanding of microbial virulence factors, pathogenesis, and cell-to-cell communication, interference with anti-virulence strategies is a compelling approach as it circumvents the development of resistance by applying less selective pressure on microbes as compared to traditional approaches, which are aimed at killing bacteria [9]. Since microbial toxins play a prime role in the virulence and pathogenesis of microbes in the host, they are major targets for developing such therapeutic interventions.

Plant-derived compounds represent an untapped source of safe, effective, and environmentally friendly antimicrobials. The properties of various plant-derived antimicrobials that target cellular viability of microbes have been adequately discussed previously [10,11,12,13], but very few reviews have highlighted the potential of plant-derived antimicrobials as effective anti-virulence strategy agents for reducing toxin production in microbes. In this review, we have focused on the effects of plant-derived antimicrobials on microbial toxins, both bacterial and fungal, critical for pathogenesis in the host. In addition, studies highlighting the anti-toxigenic mechanism(s) of plant compounds are presented.

2. MICROBIAL TOXINS: DIVERSITY, STRUCTURE AND GENERAL MECHANISMS OF ACTION

Microbial pathogens such as bacteria and fungi mediate host pathogenesis either through their invasive properties or toxigenic nature or both. A number of pathogenic microorganisms produce toxigenic substances that alter normal cellular metabolism of the host leading to various disease processes [14,15]. Toxins (endotoxins and exotoxins) produced by bacteria or mycotoxins synthesized by pathogenic fungi act as major virulence factors of a variety of microbial pathogens [16,17].

Why microbes produce toxins is still an interesting question. The diversity among the microbial toxins suggests that complex evolutionary pathways are involved in the emergence of toxigenic microorganisms. Moreover, a majority of the bacterial toxins exert no known physiological role in bacterial cells other than damaging host cell [6,16]. A popular school of thought is that bacterial toxin genes evolved from primitive eukaryotic functional genes that were incorporated to bacterial genome, and hence have similar functions or substrates. Another hypothesis is that bacteria evolved to produce toxigenic proteins that can target specific eukaryotic cells in order to acquire selective advantage over their host for niche or survival [16,18]. In addition, the evolutionary process through horizontal transfer of virulence/toxin genes enabled different bacterial groups to modify their genomes resulting in emergence of pathogenic strains [19,20]. Bacterial toxins are broadly classified as exotoxins and endotoxins based on their site of action/release.

2.1. Exotoxins

Biochemically, bacterial exotoxins are often proteins, mostly having an **A–B** structural and functional organization [14]. These toxins are released from the bacterial cell which enable them to interact with the site of infection or tissues distant from the focus of infection, subsequently injuring the host by a variety of mechanisms [14].

Generally, the A-subunit of the exotoxins has catalytic activity similar to enzymes, and hence is responsible for the toxigenic property of the toxin. The specific catalytic activity of the A-subunit defines the specific tissue or organ target, based upon which the bacterial toxins are classified as enterotoxins, neurotoxins, and leukotoxins. While several bacterial toxins, including tetanus and botulinum neurotoxin, are strictly tissue specific in their activity, many of the *Clostridial* toxins are broad-spectrum with less tissue specificity and potent cytotoxicity in the host cell [21,22,23]. The A-subunit of toxins such as hemolysins and leukocidins has broad spectrum phospholipase activity responsible for its non-specific membrane targeted activities [24,25]. B-subunit is responsible for toxin binding and transfer of enzymatically active A-subunit to the interior of the host cell. Both subunits are usually linked by a disulfide bond or non-covalent interactions [14].

One of the major groups of bacterial exotoxins is pore forming toxins. These toxins are produced by pathogens such as *Listeria monocytogenes* (Listeriolysin O), *Escherichia coli* (hemolysins), *Staphylococcus* (alpha toxin), and other bacteria producing cholesterol-dependent cytolysins (CDC) [14,26]. Pore formation in the plasma membranes of host cells such as erythrocytes or enterocytes causes influx and/or efflux of small molecules resulting in osmotic imbalance leading to cellular swelling, disruption and cell death, which often result in secondary inflammatory reaction in the tissues [15,26]. Another class of bacterial exotoxins is the heat stable enterotoxins (ST) that are produced by pathogenic strains of microbes such as *E. coli*, *Klebsiella*, and *Vibrio*. These toxins bind to the guanylate cyclase C (GC-C) receptor on the surface of the enterocytes, and activates protein kinases to initiate a downstream signaling cascade resulting in increased chloride ions in the extracellular space which cause fluid accumulation and diarrhea [14,27]. Toxins produced by pathogenic *Shigella* and *E. coli* (Stx 1 & Stx 2) exert their toxigenic activity by inhibiting protein synthesis [15]. *N*-glycosidase activity of the A-subunit Stx cleave 28S ribosomal rRNA and arrests protein synthesis [28]. Similarly, diphtheria toxin inhibits protein synthesis by ADP-ribosylation of elongation factor-2 [29].

A variety of bacterial toxins also act through the second messenger pathways in host cells, thereby eliciting injury in the host cell. For example, *Clostridial* toxins such as *C. difficile* toxins, and cytotoxic necrotizing factors from *E. coli* modify cellular Rho family GTPase leading to structural and functional disruption of the host cells and tissues [21,30]. Certain *Clostridial* toxins such as botulinum toxins and tetanus toxins exert their effects by inhibiting the release of neurotransmitters [21]. In addition, a group of bacterial exotoxins known as superantigens produced by bacteria such as *Staphylococcus aureus* bind major histocompatibility complex II (MHC II) of the antigen presenting cells and cause immune activation leading to toxic shock [31].

2.2. Endotoxins

Another major group of bacterial toxins are endotoxins. Bacterial endotoxins are lipopolysaccharides (LPS) present in the outer membrane of Gram-negative bacteria such as *E. coli*, *Salmonella*, *Shigella*, and *Pseudomonas*. These toxins are liberated following the lysis of the bacterial cell [32]. LPS consists of a glycolipid domain known as Lipid A, an oligosaccharide core and outer O-antigen. Lipid A domain is recognized by Toll-like receptor 4 (TLR-4) of host cells which in turn triggers downstream signaling pathways culminating in the synthesis and release of proinflammatory cytokines such as TNF-α and IFN-β. These molecules cause widespread vascular and inflammatory changes in the host cell, leading to endotoxin shock [32]. Endotoxin shock is a serious medical condition resulting from sepsis caused by bacterial infections [33].

Most bacterial toxin synthesis genes and other genes responsible for the regulation of toxin gene expression are encoded/located in distinct regions of the bacterial DNA called pathogenicity loci or pathogenicity islands [6,19,34]. Some bacterial toxins are also encoded by extrachromosomal plasmids, bacteriophages and other mobile genetic elements. As mentioned before, toxin genes in the pathogenicity loci of one bacterium can be transferred to other bacteria by horizontal gene transfer, thereby sharing/transmitting virulence factors among strains [6,19].

3. PLANT-DERIVED ANTIMICROBIALS

Since ancient times, plants have played a key role in the development of human civilization. Besides being used as food preservatives, flavor enhancers and dietary supplements, plant extracts have been used in traditional medicine for treating various metabolic and infectious diseases. Fossil records date the medicinal use of plants at least to the middle Paleolithic age around 60,000 years ago [35,36]. Today, plant extracts are extensively used for preparing drugs (e.g., digoxin, morphine, vincristine), bioactive compounds (metformin, nabilone), pharmacological tools (yohinbine, lysergic acid diethylamide) and herbal remedies (cranberry extracts, garlic) for various medicinal applications [36]. However, this usage accounts for only 6% of the total estimated number (approximately 250,000) of higher plant species on earth [37]. Thus, plants represent a vast, untapped natural resource of useful chemicals.

The antimicrobial activity of several plant-derived compounds has been previously reported [10,38,39,40], and a wide array of active components has been identified [41]. A majority of plant-derived antimicrobials are secondary metabolites, and are produced in response to chemical stimulus from pathogenic microbes or environmental cues

[42]. Being secondary metabolites, these compounds do not directly modulate plant physiology [43], however, they play a critical role in enhancing plant health and defense [44]. The generation of secondary metabolites with medicinal properties is restricted to a limited set of species and developmental period within a phylogenetic group as compared to primary metabolites, which are more commonly produced [45].

Although the exact pathways by which plant compounds exert their antimicrobial effects are not clearly defined; multiple, dose-dependent mechanisms have been reported. At bactericidal concentrations, plant compounds cause membrane damage, loss of energy production, enzyme dysfunction and leakage of cellular contents that lead to impaired cellular physiology and cell death [46]. In addition, recent research has revealed that at sub-lethal or sub-inhibitory concentrations, plant compounds affect virulence in Gram-positive [47,48,49] and Gram-negative bacteria [50] and fungal pathogens [51] by modulating gene transcription [52,53], protein expression [53] and quorum sensing [54,55]. Thus, plant compounds represent a viable strategy to develop anti-virulence agents that target toxin production.

4. STUDIES HIGHLIGHTING THE ANTI-TOXIN PROPERTIES OF PLANT-DERIVED ANTIMICROBIALS

Plant compounds have been found to be effective in reducing toxin production in both bacterial and fungal pathogens. The major studies which have investigated the antitoxin potential of plant compounds are discussed below, and are summarized in Table 1.

4.1. Effect of Plant Compounds on Gram-Positive Bacterial Toxins

4.1.1. *Clostridium difficile*

Clostridium difficile is a nosocomial pathogen that causes a severe, toxin-mediated colitis in humans, especially in patients undergoing prolonged antibiotic therapy [56,57]. In the United States, approximately 300,000 cases of *C. difficile*-associated disease are reported annually costing more than $3 billion in healthcare costs [58]. The hypervirulent, highly toxigenic *C. difficile* strain known as ribotype 027 is responsible for causing a majority of *C. difficile* infections reported in the United States [59]. Long-term antibiotic therapy results in the disruption of the normal gut microflora, leading to the germination of *C. difficile* spores and colonization of vegetative cells in colon to cause severe toxin-mediated colitis [59]. *Clostridium difficile* toxins, TcdA and TcdB, with the help of TcdE, a holin-like protein needed for toxin excretion, causes disruption of the cytoskeleton and colonic epithelial tight junction subsequently resulting in severe inflammation, epithelial necrosis, pseudomembrane formation and diarrhea [60,61]. The sigma factor TcdR positively regulates toxin production, whereas TcdC is an antagonist of TcdR, and thus helps to control toxin production [30,62,63]. In addition, CodY, a global pleiotropic repressor, reduces toxin production in *C. difficile* [64,65]. Not many studies have investigated the effect of plant compounds on toxin production in *C. difficile*. In a recent study, carvacrol (obtained from oregano oil) and *trans*-cinnamaldehyde (major component of cinnamon) were found to reduce *C. difficile* toxin production and toxin-mediated cell cytotoxicity without affecting the growth of normal gut flora *in vitro* [49]. Follow up gene expression studies revealed that these compounds down-regulated toxin production genes. In addition, the compounds were found to modulate transcription of toxin repressors favorably. Furthermore, mutant analysis revealed that the plant compounds reduce toxin production potentially through global repressor CodY [49]. Since toxin is the only major virulence factor in *C. difficile*, the antitoxigenic properties of carvacrol, *trans*-cinnamaldehyde or similar plant compounds on *C. difficile* deserve further scientific attention.

4.1.2. *Bacillus sp.*

Bacillus sp. is a significant public health concern due to their ability to form resistant endospores and toxins. Over the last 30 years, these organisms have undergone massive taxonomic improvements and now over 56 genera and over 500 species of *Bacillus* are identified [118,119]. Among the disease-causing *Bacillus*, *B. cereus* is well known as a cause of foodborne infections and intoxications. Although *B. licheniformis*, *B. subtilis* and *B. pumilus* have also been isolated from food associated illnesses, their role in infections still remains uncertain [119]. A few strains of *B. cereus* are reported to cause foodborne illness known as "fried rice syndrome" in humans. *B. cereus* foodborne intoxications are commonly contracted from fried rice stored at room temperature for several hours [119,120]. Foodborne illness caused by *B. cereus* normally manifests as diarrheal and/or emetic type. The diarrheal type is caused by a labile enterotoxin and the emetic type is caused by heat stable emetic toxin [118,119].

4. Studies Highlighting the Anti-Toxin Properties of Plant-Derived Antimicrobials

Table 1. Studies illustrating the efficacy of plant-derived compounds and plant extracts in reducing toxin production in microbes with potential mechanism(s) of action.

Microbe	Toxin(s)	Plant Compounds with Anti-Toxin Activity	Potential Mechanism of Action/Target Site	Reference(s)
		Gram positive bacteria		
Clostridium sp.	TcdA, TcdB	Carvacrol, *trans*-cinnamaldehyde,	Down-regulation of toxin production genes, modulation of transcriptional repressor	[49]
	Botulinum neurotoxin	Toosendanin		[66,67]
Bacilius sp.	Labile enterotoxin	Carvacrol,	Modification of bacterial membranes	[68–71]
	Anthrax lethal toxin	Celastrol, toosendanin	Inhibition of toxin entry to cell cytoplasm	[72,73]
Staphylococcus aureus	Hemolysin Enterotoxin A, B Toxic shock syndrome toxin	Essential oils from clove, cinnamon, oregano, *Zataria multiflora*, eugenol, 4-hydroxytyrosol	Reduced expression of toxin production genes, *sea, seb, tst, hla*	[53,74–77]
Listeria monocytogenes	Listerolysin O (LLO)	*trans*-Cinnamaldehyde, carvacrol, thymol, eugenol, oil of bay, clove, nutmeg, thyme	Down-regulation of *hly* and *prfA* genes coding for toxin production and transcriptional regulator	[47,78,79]
		Gram negative bacteria		
Vibrio cholerae	Cholera toxin	Tea catechins, Dihydroisosteviol	Modulation of transmembrane regulators	[80,81]
		RG-tannin, apple phenols	Inhibition of ADP-ribosyltransferase activity	[82,83]
		Red chilli, sweet fennel, white pepper Red bayberry, thymol, carvacrol, eugenol	Modulation of toxin production genes *ctxA, tcpA, toxT*	[84–87]
Toxin producing *E. coli*	ETEC toxin	Extracts from *Galla Chinensis* and *Berberis aristata*, leanolic acid, ursolic acid, and betulinic acid, essential oil from *Cymbopogon martini*, *C. winterianus* and *Psidium guajava*	Inhibiting intestinal secretion of ETEC enterotoxins Blocking the binding of heat labile enterotoxin to GM1 Reducing toxin binding and toxin mediated cellular pathology	[88–97]
	Verotoxin	Extracts from *Limonium californicum (Boiss.) A. Heller, Cupressus lustianica Miller, Salvia urica Epling* and *Jusiaea peruviana L.*, eugenol, catechin, epigallocatechin, cinnamon bark oil, cinnamaldehyde, *Curtisia dentata* extract *carvacrol, thymol, beta-resorcylic acid*, grape seed and pomace extracts	Decrease in toxin production Reducing the transcription of *stx1* and *stx2* genes Reducing the expression of globotriaosylceramide (Gb3) receptor by mimicking toxin receptors	[98–109]
		Fungal microbes		
Aspergillus ochraceus Penicillium verrucosum	Ochratoxin A	Oregano, mint, basil, sage, and coriander	Inhibiting fungal growth	[110]
Penicillium expansum	Patulin	Garlic, thyme, lavender oils, *Azadirachta indica* extracts	Inhibiting fungal growth, mycelium formation and sporulation	[111,112]
Fusarium graminearum	Zearalenone	Cinnamon, clove, oregano, palmarosa, and lemongrass oils	An aromatic nucleus and phenolic OH group of plant compounds disrupting fungal cell membrane	[113]
	Deoxynivalenol	Clove		[113]
Fusarium proliferatum	Fumonisin B	Extracts from maize, garlic, and pea	Inhibiting primary and secondary metabolism of the fungal pathogens, decreasing biomass formation	[114]
Aspergilius flavus Aspergillus parasiticus	Aflatoxins	Clove, cinnamon, *Zataria multiflora*, carvacrol, *trans*-cinnamaldehyde	Down-regulation of aflatoxin production genes including *aflc, ver1, nor1* and *norA* genes	[51,115–117]

Ultee and coworkers studied the effect of carvacrol on *B. cereus* growth, cellular physiology and toxin production *in vitro* [68,69,70] as well as its ability to reduce toxin production in food matrices, including rice and soup [70,71]. These researchers observed that carvacrol at a concentration below its MIC was able to reduce *B. cereus* toxin production *in vitro* [70]. Follow up studies in food matrices showed that the effect of carvacrol in reducing toxin production persisted in soup and rice [69,70]. In addition to its effect on toxin production, carvacrol was found to modulate the fluidity and lipid composition in the membranes of the pathogen, which could be responsible for reduced survival in food matrices [69].

In addition to inhibiting *B. cereus* toxin production, plant compounds have also shown efficacy against *B. anthracis* toxins. For example, Chapelsky and coworkers [72] reported that celastrol, a quinone methide triterpene derived from *Tripterygium wilfordii* Hook F. (Chinese thunder of God vine), inhibited anthrax lethal toxin-induced cytolysis of murine macrophages RAW264.7. Similarly, Slater and coworkers [73] identified novel compounds that protect from anthrax lethal toxin-induced cell death. These researchers observed that Toosendanin (triterpenoid from *Melia azedarach*) was effective in inhibiting the entry of anthrax lethal toxin into the cell cytoplasm without affecting cellular physiology. Toosendanin was also shown to possess *in vivo* efficacy against botulinum neurotoxin [66,67].

4.1.3. *Staphylococcus aureus*

Staphylococcus aureus produces a wide range of exotoxins that facilitate its pathogenesis in the host. These include hemolysins (alpha, beta, gamma, and delta), nucleases, proteases and collagenase that mediate host tissue digestion for bacterial nourishment and growth. In addition, *S. aureus* produces enterotoxins, exfoliative toxins and toxin shock syndrome toxin-1 that exert stress on host immune system and also on vital organs [121]. Controlling the production of many of these toxins either in food products or in the host would significantly contribute to reducing *S. aureus* virulence and pathogenesis.

As observed with *Clostridium* and *Bacillus sp.*, plant compounds at their sub-inhibitory concentrations were found to modulate toxin production in *S. aureus*. Essential oils from clove and cinnamon reduced *S. aureus* alpha-hemolysin, enterotoxin A and enterotoxin B production *in vitro* [74]. Qiu and coworkers [53] reported similar findings and observed that eugenol (active component in clove oil) significantly reduced the production of alpha hemolysin, enterotoxin A and B, and toxic shock syndrome toxin *in vitro*. In addition, eugenol was found to reduce the expression of toxin producing genes, including *sea, seb, tst, and hla* coding for various stages of toxin production in *S. aureus.* Similar antitoxin effect has been observed with low concentrations of other plant compounds, including *Origanum vulgare* or oregano [75], *Zataria multiflora* Boiss, a spice plant found in mid-west Asia [76] and 4-hydroxytyrosol [77].

4.1.4. *Listeria monocytogenes*

The foodborne pathogen *Listeria monocytogenes* produces a non-enzymatic, cytolytic, thiol-activated, pore-forming toxin called Listeriolysin O. The toxin is most active at an acidic pH of 5.5 [122] that is usually found inside a phagolysosome. The toxin ruptures the phagolysosome and helps in the escape and survival of the pathogen in the cytoplasm. Listeriolysin O is encoded by the *hly* gene and is activated by the master transcriptional regulatory protein PrfA. Several studies have investigated the efficacy of plant-derived antimicrobials on Listeriolysin production in *L. monocytogenes.* Filgueiras and Vanetti [78] investigated the effect of sub-lethal concentrations of eugenol on Listeriolysin production in *L. monocytogenes.* Using sheep red blood cells based titration assay, these researchers observed a dose-dependent reduction in Listeriolysin production with eugenol. Similar results were obtained with sub-lethal concentrations of oil of *bay, clove, cinnamon, nutmeg* and *thyme* [79]. Follow up gene expression analysis revealed that plant compounds such as *trans*-cinnamaldehyde, carvacrol, and thymol (obtained from oregano oil) at sub-inhibitory or sub-lethal concentrations modulated the expression of *hly* and *prfA* genes that code for the toxin and master regulator, respectively, without affecting growth of the pathogen [47].

4.2. Effects of Plant-Derived Antimicrobials on Toxins Produced by Gram-Negative Microbes

As compared to Gram-positive bacteria, relatively a larger number of studies has investigated the efficacy of plant compounds in reducing toxin production in Gram-negatives, especially *Vibrio* spp. and *E. coli* O157:H7.

4.2.1. *Vibrio cholerae*

Vibrio cholerae is a water-borne pathogen that is the causative agent of human cholera. It is a significant problem in countries where the sanitary conditions are poor. Manifestation of cholera in humans is through watery diarrhea accompanied by abdominal cramps and severe dehydration. The virulence in cholera is primarily mediated by an

AB_5protein toxin called cholera toxin [123,124]. After ingestion, the pathogen binds to intestinal epithelial cells using toxin co-regulated pilus (tcpA). Once attached, the pathogen secretes cholera toxin that is internalized by intestinal epithelial cells, followed by processing in the endoplasmic reticulum. Dissociation of A and B subunit leads to ADP-ribosylation of the alpha subunits of the heteromeric G_s proteins, resulting in the activation of the adenylate cyclase pathway [125,126,127]. This causes an increase in chloride ion secretion and blocking of sodium influx resulting in an increased solute concentration in the lumen leading to severe diarrhea.

In the past, plant extracts and/or their active components have been used to reduce toxin production in *Vibrio* spp. Toda and coworkers [80] investigated the effect of tea catechins on toxin production by *V. cholerae.* These researchers observed that the plant compounds significantly reduced fluid accumulation induced by cholera toxin in adult mice and intestinal loops of rabbits. Similar results were observed by Pariwat and coworkers [81] who found that dihydroisosteviol reduced cholera toxin mediated intestinal fluid secretion *in vitro*. Apical chloride current measurement indicated that dihydroisosteviol reversibly targeted the cystic fibrosis transmembrane regulator. Follow up *in vivo* studies using a mouse closed-loop model showed that intraluminal injection of dihydroisosteviol significantly reduced intestinal fluid secretion without altering cell viability or fluid absorption. Other plant compounds such as RG-tannin and apple phenols have also shown efficacy in reducing toxin induced fluid accumulation in mouse ileal loop assay [82,83]. Mechanistic studies revealed that the RG-tannin and apple phenols inhibited ADP-ribosyltransferase activity, which is critical for cholera toxin action.

In recent years, several researchers have investigated the effect of plant-derived compounds on the transcriptional profile of *V. cholerae* to elucidate their mechanism in inhibiting cholera toxin production. Yamasaki and coworkers [84] investigated the effect of spices such as red chilli, sweet fennel, and white pepper on cholera toxin production. All the spices significantly modulated cholera toxin production. Capsaicin was found to be the major component among the tested spices and significantly reduced the expression of critical virulence genes including *ctxA*, *tcpA*, and *toxT* that code for toxin production in *V. cholerae.* Interestingly, Capsaicin also enhanced the expression of *hns* gene [85] that negatively regulates toxin production genes, indicating that the effect of plant compounds in reducing toxin production in *V. cholerae* could be due to their effect on the whole transcriptome of the pathogen. A similar efficacy in modulating the expression of genes critical for toxin production has been observed with other plant compounds such as red bayberry [86], thymol, carvacrol, and eugenol [87].

4.2.2. Toxin Producing *Escherichia coli*

Escherichia coli are a large and diverse group of bacteria that are found in environment, foods and intestine of humans and animals. A majority of *E. coli* live as harmless commensals, however, some are pathogenic due to the presence of virulence factors acquired over evolution. The major pathogenic groups include enterotoxigenic (ETEC), enteropathogenic (EPEC), enterohemorrhagic (EHEC), enteroinvasive (EIEC), diffusely adherent (DAEC) and enteroaggregative *E. coli* (EAEC) [128]. These cause potentially fatal illnesses in humans characterized by gastroenteritis, septicemia, and meningitis [129]. *E. coli* O157:H7 (EHEC) is the most important serotype among the different pathogenic groups, and is responsible for approximately 73,000 foodborne illnesses annually in the US [130]. The pathogenesis of EHEC and other virulent *E. coli* except for enteroinvasive *E. coli* is chiefly mediated by the production of toxins [129,131]. In the case of EHEC, verotoxins are the major virulence factors that cause intestinal and renal epithelial damage leading to hemorrhagic colitis and hemolytic uremic syndrome, respectively [132,133]. Enteroaggregative *E. coli* adheres to intestinal epithelium and releases cytotoxins and enterotoxins in the extracellular lumen causing crypt orifice dilation, rounding and extrusion of enterocytes. Additionally, there are reports of cytoskeletal rearrangements in the host intestinal cells [134,135]. In ETEC, the production of heat liable and/or heat stable enterotoxins cause traveler's diarrhea especially in elderly and children. Enteropathogenic *E. coli* causes similar pathology by producing heat labile and heat stable enterotoxins. However, the toxins of EPEC are distinct from those produced by ETEC. In addition, EPEC associated diarrhea involves the enteroadherence and production of toxin like Shiga toxin [136].

There are several reports on the efficacy of plant extracts and plant derived antimicrobials in inhibiting *E. coli* toxin production and reducing toxin mediated cellular damage on host cells. Natural extracts from *Galla chinensis* and *Berberis aristata* have been used in traditional Chinese medicine for the treatment of diarrhea [88,89]. Chen and coworkers [90] investigated the effect of *Galla chinensis* and gallic acid, one of the major components of *Galla chinensis* extract, on ETEC toxin binding to GM1 ganglioside receptor of humans. These researchers observed that *Galla chinensis* extract and gallic acid were effective in blocking the binding of heat labile enterotoxin to GM1 receptor *in vitro*. Follow up mouse gut assay revealed that *Galla chinensis* extract was effective in reducing toxin-mediated diarrhea indicating its efficacy in reducing toxin binding and toxin mediated cellular pathology. In a similar study, Chen and coworkers [91] observed that zingerone (vanillylacetone), one of the active components of ginger, was effective in preventing binding of heat labile toxin to GM1 and prevented accumulation of diarrheal fluid in the closed ileal loops of mice. In a recent study, Dubreuil and coworkers [92] observed a significant reduction (70%) in ETEC toxin-mediated diarrhea with plant alkaloid berberine (benzyltetrahydroxyquinoline) in a rabbit ileal loop model. These researchers also found an inhibition in the secretory response of the heat-stable toxin in infant mouse [92]. A similar anti-toxic activity with other plant derived compounds namely, oleanolic acid, ursolic acid, and betulinic acid [93], λ-carragenan [94], essential oil from *Cymbopogon martini* and *C. winterianus* [95], *Psidium guajava* [96], thiols and disulfide compounds [97] has been reported.

In addition to inhibiting toxin binding with receptor, plant compounds have also been found to modulate toxin production in pathogenic *E. coli.* As mentioned earlier, the pathogenesis of EHEC is primarily associated with the production of verotoxins. Sakagami and coworkers [98], investigated the effect of extracts from four plant species namely, *Limonium californicum* (Boiss.) A. Heller, *Cupressus lustianica* Miller, *Salvia urica* Epling and *Jusiaea peruviana* L., on verotoxin production *in vitro*. These researchers observed that plant extracts at concentrations below MIC reduced verotoxin production *in vitro.* Similarly, Takemasa and coworkers [99] found a concentration dependent reduction in extracellular and intracellular verotoxins by eugenol. Likewise, a decrease in toxin production has been reported following treatment with plant extracts such as catechin [100], epigallocatechin [101], cinnamon bark oil, cinnamaldehyde, and eugenol [102]. In addition, plant extracts, including green tea extracts, tea catechins, and green tea polyphenol were found to modulate other virulence attributes in EHEC such as biofilm formation, swarming motility and quorum sensing [103]. Since the concentrations of plant compounds used in above studies were below MIC, the effects observed were not due to reduction in cell number, but probably due to modulation of gene transcription in cells. In a recent study, Doughari and colleagues [104] found that *Curtisia dentata* extract reduced the transcription of *stx1* and *stx2* genes in EHEC. A similar reduction in the expression of *stx1* and *stx2* genes was observed with sub-inhibitory concentrations of other plant compounds such as *trans*-cinnamaldehyde, eugenol, carvacrol, thymol, and beta-resorcylic acid [105]. These researchers also reported that the aforementioned plant compounds reduced the expression of globotriaosylceramide (Gb3) receptor on human lymphocytes in addition to their effect on toxin genes. A similar host protective effect was observed with grape seed and pomace extracts using a novel cell-based assay *in vitro* [106].

In addition to reducing toxin production and toxin-receptor binding, receptor-mimicking compounds are being used as a novel strategy for combating toxins and other pathological agents. A handful of studies have investigated the potential of plant compounds as phantom receptors to ameliorate the effect of toxins. Kulkarni and coworkers [107] investigated the efficacy glycan encapsulated gold nanoparticles to selectively inhibit shiga toxins 1 and 2, and observed that the designed nanoparticles were effective in reducing toxin mediated pathology by mimicking the glycans in lipid rafts of the glycolipids receptors. Similar results were obtained with shiga toxin targeting synthetic ligands [108] and trisaccharide carrying carbosilane dendrimers [109] that mimics *stx* receptor. These studies indicate that plant compounds act not only by decreasing toxin-receptor binding but also through modulation of critical toxin genes and host receptor expression, thereby leading to reduced toxin-mediated pathology.

4.3. Efficacy of Plant-Derived Antimicrobials in Reducing Fungal Toxins

The kingdom Fungi consists of eukaryotic microorganisms such as yeasts, molds and the more familiar mushrooms. Fungi are capable of producing a large number of toxins with diverse modes of action that can cause disease and death in humans and animals. The major fungal toxins include mycotoxins (produced by molds) and toxins produced by mushrooms. Mycotoxins exert their toxic effects when humans or animals inadvertently consume, or come in contact with a substrate that contains the toxin-producing fungus or the mycotoxin. Mushroom poisoning on the other hand usually occurs after deliberately ingesting the basidiocarp [137]. Since not many studies have investigated the effect of plant compounds on mushroom poisons, this review focused primarily on the effect of plant compounds on mycotoxins production in this section.

Unlike bacterial toxins, mycotoxins are secondary metabolites produced by molds, and are not essential for growth of the producing fungus. Mycotoxins are low molecular weight, differentiation products with diverse chemical structures biosynthesized by complex cellular pathways [138,139]. Mycotoxins contaminate food crops or animal feed either before or after harvest and their concentration increases during postharvest storage. Mycotoxins reach consumers through the consumption of contaminated plant materials or via contaminated animal products such as meat, milk and egg. The normal industrial processing cannot remove mycotoxins from food products since they are heat-resistant within the range of conventional food-processing temperatures (80–121 °C). This is attributed primarily to their stable chemical structure [140,141]. The economic losses associated with mycotoxin contamination of crops are estimated to range between $500 million and $1.5 billion a year in the United States [142,143].

Mycotoxins commonly found in human food products and animal feed include *ochratoxin A, patulin, zearalenone, trichothecenes, fumonisins,* and *aflatoxins.* These mycotoxins are hazardous to animal and human health due to their carcinogenic, mutagenic, and teratogenic properties [144,145]. Current methods employed to control mycotoxin contamination use synthetic fungicides. Because of the non-biodegradable nature of these synthetic chemicals, they accumulate in soil, plants, and water, eventually leading to food contamination. This poses significant public health concerns. In addition, acute toxicity and hormonal imbalance in animals and humans have been observed [141,146]. In order to reduce the utilization of synthetic chemical fungicides in foods, several alternative treatments have been studied. Among the alternative approaches, the use of plant-derived antimicrobials has gained significant attention due to their natural, eco-friendly, and bio-degradable characteristics [141,145,147]. Research highlighting the potential use of plant compounds in reducing mycotoxins is described below.

Ochratoxin A produced by *Aspergillus ochraceus* and *Penicillium verrucosum* contaminates foods and feedstuffs of cereal origins [110,148], and causes renal toxicity, nephropathy, immunosuppression and even cancer in several animal species [149,150]. Basilico and coworkers [110] tested the effect of plant-derived oils namely oregano, mint, basil, sage, and coriander on *A. ochraceus* growth and *ochratoxin A* production *in vitro.* Results revealed that oregano and mint at 1000 ppm completely inhibited the fungal growth and ochratoxin A production up to 21 days, whereas basil was only effective up to seven days.

Penicillium expansum or blue mold is a serious postharvest pathogen that causes decay on apple, pear, peach, and grape fruits. This fungus produces toxin patulin in spoiled or processed fruits [111]. *Patulin* has been reported to be genotoxic, embryotoxic, and immunosuppressive in animals and humans [151,152,153,154]. Ikeura and coworkers [111] investigated the effect of various plant-extracts on *P. expansum* growth *in vitro.* Extracts from garlic, thyme, and lavender oils were found to effectively inhibit the formation of *P. expansum* spores. Follow up studies on apples revealed that the aforementioned plant extracts reduced fungal growth on the fruit. Among the tested plant extracts, garlic extract was most effective in inhibiting *P. expansum* growth on apples. In another study, Mossini and coworkers [112] investigated the effect of *Azadirachta indica* extracts on *P. expansum* growth and patulin production *in vitro*, and found the extracts were effective to inhibiting growth of *P. expansum* as well as *patulin* production.

Several *Fusarium* species produce toxins of significant concern to livestock and poultry producers. For example, zearalenone also known as RAL is primarily produced by *Fusarium graminearum.* It contaminates cereal crops such as wheat, barley, rice sorghum and maize. Zearalenone binds to estrogen receptors and results in hormonal changes that lead to abortion, infertility and other breeding problems in animals, especially swine [155,156,157]. Deoxynivalenol, a type B trichothecene, is produced by *Fusarium* that contaminates cereal grains. Deoxynivalenol binds to the ribosome and activates critical cellular kinases involved in signal transduction related to proliferation, differentiation, and apoptosis in human cells, and eventually disrupts normal cell function by inhibiting protein synthesis [158], thereby causing immunosuppression and kidney damage [159]. *Fumonisins B* is another important class of mycotoxins synthesized by *F. proliferatum* [160]. When ingested, they can lead to liver cancer and pulmonary edema in animals and esophageal cancer in humans [161]. They are commonly found in maize [162] and other plant tissues, including asparagus [163], pineapple [164] and garlic [165].

Mari and coworkers [113] investigated the effect of *cinnamon, clove, oregano, palmarosa,* and *lemongrass* oils on zearalenone and deoxynivalenol production by *F. graminearum* in maize. *Lemongrass,* and *palmarosa* significantly reduced zearalenone production, whereas clove was found to be the most effective treatment that inhibited the production of both zearalenone and deoxynivalenol. In another study, Stepien and coworkers [114] observed that extracts from various plant sources such as maize, garlic, and pea modulated the growth and fumonisin synthesis in *F. proliferatum* strains *in vitro.* Pea extract was found to be the most efficient inhibitor of *fumonisin* synthesis in the mold.

Aflatoxins produced by *Aspergillus flavus* and *Aspergillus parasiticus* are one of the most potent mycotoxins because of their high toxicity, carcinogenicity as well as their biotransformation ability in the animal and human body. They primarily contaminate commonly used feed ingredients such as peanut, corns, and cottonseeds resulting in severe negative health effects in both animals and humans [166]. In animals, consumption of aflatoxin contaminated feed causes poor feed utilization, decreased body weight gain, reduced egg production and increased mortality [167,168,169]. Moreover, *aflatoxin* residues can be found in animal products such as milk, meat and egg [170], and are metabolized and transformed to *aflatoxin M1* and *M2*. *Aflatoxin B1* is listed as group 1 human carcinogen by the International Agency for Research on Cancer, and mainly causes liver cancer [171,172]. *Aflatoxins* are also known to interact with hepatitis B virus leading to human hepatocellular carcinoma [173]. According to the current U.S FDA regulations, the maximum permissible limit of *aflatoxins* in human and young animal food is 20 ppb, whereas for adult livestock it is 100 ppb [174].

Significant research is being conducted to investigate the potential of natural antimicrobials in inhibiting the growth and aflatoxin production by *Aspergillus* species in food products and animal feed. Clove and cinnamon oils at 100 ppm were found to inhibit *A. flavus* growth and *aflatoxin* production *in vitro* [115]. In the same study, the researchers demonstrated that cinnamon oil at 1000 ppm exhibited maximum inhibitory action and reduced *aflatoxin* production by 78% on maize. Similarly, *Zataria multiflora* was found to exert significant inhibitory effect on both *A. flavus* growth and *aflatoxin* production in cheese [116]. Follow up mass spectrometric analysis revealed that the major component (71%) of *Zataria multiflora* oil was carvacrol. The antitoxin effect of carvacrol on *A. parasiticus* was investigated by Razzaghi-Abyaneh and coworkers [117]. These researchers reported that carvacrol effectively suppressed the growth as well as the synthesis of *aflatoxin B1* and *G1* by *A. parasiticus in vitro*. In another study, carvacrol and trans-cinnamaldehyde were found to significantly reduce growth and aflatoxin production in *A. flavus* and *A. parasiticus in vitro*. In addition, these compounds also reduced the toxin production in poultry feed [51]. The two plant compounds also down-regulated the expression of *aflatoxins* synthesis genes (*aflC*, *nor1*, *norA*, and *ver1*) in the molds. Follow up *in vivo* studies showed that in-feed supplementation of carvacrol and *trans*-cinnamaldehyde was effective in reducing aflatoxicosis in broiler chickens that were fed with mold-contaminated feed. In addition, histological analysis of liver from chickens fed with plant compounds revealed significantly lesser hepatocellular degeneration, inflammation and necrosis as compared to control birds [175]. Similar inhibitory effect on *aflatoxins* production was obtained with other plant-derived antimicrobials, namely *Thymus eriocalyx* [176], *Thymus vulgaris* [177], *thyme, anise, cinnamon,* and *spearmint* [178]. These results suggest that plant compounds could be used as feed/food additives to control *aflatoxins* contamination in food products or animal feed.

5. CONCLUSIONS AND FUTURE DIRECTIONS

Research efforts in the past few decades have significantly improved our understanding of microbial virulence and toxin production. Disruption of toxin production and toxin-mediated pathology in hosts is widely regarded as a potential anti-virulence strategy since this approach, being non-microbicidal, would circumvent the selective pressure for resistance development. The aforementioned information collectively suggests that plant-derived antimicrobials could be used to control toxin production in microorganisms, and toxin mediated pathology in humans and animals. With advancements in nanotechnology and pharmaceutics, future research should focus on improving the drug delivery technology to improve the efficacy of plant compounds at target sites. Moreover, in-depth studies are required to properly characterize the therapeutic/chemical effects of plant compounds in the host for developing reliable and effective drugs. In addition, studies are required to minimize the changes in organoleptic properties of food products when supplemented with plant-derived compounds to further improve their applications, especially in the food industry.

CONFLICTS OF INTEREST

The authors declare no conflict of interest.

REFERENCES

1. Alwan, A. Global Status Report on Non-Communicable Diseases 2010; World Health Organization: Geneva, Switzerland, 2011.
2. Cantas, L.; Shah, S.Q.; Cavaco, L.M.; Manaia, C.M.; Walsh, F.; Popowska, M.; Sørum, H. A brief multi-disciplinary review on antimicrobial resistance in medicine and its linkage to the global environmental microbiota. Front Microbiol. 2013, 4.
3. Van Boeckel, T.P.; Brower, C.; Gilbert, M.; Grenfell, B.T.; Levin, S.A.; Robinson, T.P.; Laxminarayan, R. Global trends in antimicrobial use in food animals. Proc. Natl. Acad. Sci. USA 2015, 112, 5649–5654.
4. Galland, J.C.; Hyatt, D.R.; Crupper, S.S.; Acheson, D.W. Prevalence, antibiotic susceptibility, and diversity of Escherichia coli O157:H7 isolates from a longitudinal study of beef cattle feedlots. Appl. Environ. Microb. 2001, 67, 1619–1627.
5. Zhang, X.; McDaniel, A.D.; Wolf, L.E.; Keusch, G.T.; Waldor, M.K.; Acheson, D.W. Quinolone antibiotics induce Shiga toxin-encoding bacteriophages, toxin production, and death in mice. J. Infect. Dis. 2000, 181, 664–670.
6. Finlay, B.B.; Falkow, S. Common themes in microbial pathogenicity revisited. Microbiol. Mol. Biol. Rev. 1997, 61, 136–169.

7. Taylor, C.M.; Roberts, I.S. Capsular polysaccharides and their role in virulence. In Concepts in Bacterial Virulence; Karger Medical and Scientific Publishers: Basel, Switzerland, 2005; pp. 55–66.
8. Wu, H.J.; Wang, A.H.; Jennings, M.P. Discovery of virulence factors of pathogenic bacteria. Curr. Opin. Chem. Boil.2008, 12, 93–101.
9. Rasko, D.A.; Sperandio, V. Anti-virulence strategies to combat bacteria-mediated disease. Nat. Rev. Drug Discov. 2010, 9, 117–128.
10. Burt, S. Essential oils: Their antibacterial properties and potential applications in foods—A review. Int. J. Food Microbiol. 2004, 94, 223–253.
11. Juneja, V.K.; Dwivedi, H.P.; Yan, X. Novel natural food antimicrobials. Annu. Rev. Food Tech. 2012, 3, 381–403.
12. Negi, P.S. Plant extracts for the control of bacterial growth: Efficacy, stability and safety issues for food application. Int. J. Food Microbiol. 2012, 156, 7–17.
13. Savoia, D. Plant-derived antimicrobial compounds: Alternatives to antibiotics. Future Microbial. 2012, 7, 979–990.
14. Henkel, J.S.; Baldwin, M.R.; Barbieri, J.T. Toxins from bacteria. Mol. Environ. Toxic 2010, 100, 1–29.
15. Schmitt, C.K.; Meysick, K.C.; O'Brien, A.D. Bacterial toxins: Friends or foes? Emerg. Infect. Dis. 1999, 5, 224–234.
16. Hardegree, M.C.; Tu, A.T. Handbook of Natural Toxins: Bacterial Toxins; CRC Press: New York, NY, USA, 1988; Volume 4.
17. Richard, J.L. Some major mycotoxins and their mycotoxicoses—An overview. Int. J. Food Microbiol. 2007, 119, 3–10.
18. Lupski, J.R.; Feigin, R.D. Molecular evolution of pathogenic Escherichia coli. J. Infect. Dis. 1988, 157, 1120–1123.
19. Gyles, C.; Boerlin, P. Horizontally transferred genetic elements and their role in pathogenesis of bacterial disease. Vet. Pathol. 2014, 51, 328–340.
20. Skarin, H.; Segerman, B. Horizontal gene transfer of toxin genes in Clostridium botulinum: Involvement of mobile elements and plasmids. Mob. Genet. Elements 2011, 1, 213–215.
21. Popoff, M.R.; Bouvet, P. Clostridial toxins. Future Microbial. 2009, 4, 1021–1064.
22. Rossetto, O.; Scorzeto, M.; Megighian, A.; Montecucco, C. Tetanus neurotoxin. Toxicon 2013, 66, 59–63.
23. Rossetto, O.; Pirazzini, M.; Montecucco, C. Botulinum neurotoxins: Genetic, structural and mechanistic insights. Nat. Rev. Microbiol. 2014, 12, 535–549.
24. Priatkin, R.G.; Kuzmenko, O.M. Secreted proteins of Staphylococcus aureus. Zh Mikrobiol Epidemiol. Immunobiol. 2010, 4, 118–124.
25. Vandenesch, F.; Lina, G.; Henry, T. Staphylococcus aureus hemolysins, bi-component leukocidins, and cytolytic peptides: A redundant arsenal of membrane-damaging virulence factors? Front Cell Infect. Microbiol. 2012, 2, 12.
26. Los, F.C.; Randis, T.M.; Aroian, R.V.; Ratner, A.J. Role of pore-forming toxins in bacterial infectious diseases. Microbiol. Mol. Biol. Rev. 2013, 77, 173–207.
27. Nair, G.B.; Takeda, Y. The heat-stable enterotoxins. Microb. Pathog. 1998, 24, 123–131.
28. Bergan, J.; Dyve Lingelem, A.B.; Simm, R.; Skotland, T.; Sandvig, K. Shiga toxins. Toxicon 2012, 60, 1085–1107.
29. Deng, Q.; Barbieri, J.T. Molecular mechanisms of the cytotoxicity of ADP-ribosylating toxins. Annu. Rev. Microbiol.2008, 62, 271–288.
30. Kuehne, S.A.; Cartman, S.T.; Minton, N.P. Both, toxin A and toxin B, are important in Clostridium difficile infection. Gut Microbes 2011, 2, 252–255.
31. McCormick, J.K.; Yarwood, J.M.; Schlievert, P.M. Toxic shock syndrome and bacterial superantigens: An update. Annu. Rev. Microbiol. 2001, 55, 77–104.
32. Ramachandran, G. Gram-positive and gram-negative bacterial toxins in sepsis: A brief review. Virulence 2014, 5, 213–218.
33. MacDonald, R.D.; Osei-Ampofo, M. Sepsis and septic shock. Air Med. J. 2015, 34, 71–73.

34. Carter, G.P.; Cheung, J.K.; Larcombe, S.; Lyras, D. Regulation of toxin production in the pathogenic clostridia. Mol. Microbiol. 2014, 91, 221–231.
35. Solecki, R.; Shanidar, I.V. A Neanderthal flower burial in northern Iraq. Science 1975, 190, 880–881.
36. Fabricant, D.S.; Farnsworth, N.R. The value of plants used in traditional medicine for drug discovery. Environ. Health Persp. 2001, 109, 69.
37. Verpoorte, R. Pharmacognosy in the new millennium: leadfinding and biotechnology. J. Pharm. Pharmacol. 2000, 52, 253–262.
38. Holley, R.A.; Patel, D. Improvement in shelf-life and safety of perishable foods by plant essential oils and smoke antimicrobials. Food Microbiol. 2005, 22, 273–292.
39. Newman, D.J.; Cragg, G.M. Natural products as sources of new drugs over the 30 years from 1981 to 2010. J. Nat. Prod. 2012, 75, 311–335.
40. Upadhyay, A.; Upadhyaya, I.; Kollanoor-Johny, A.; Venkitanarayanan, K. Combating pathogenic microorganisms using plant-derived antimicrobials: A Minireview of the mechanistic basis. Biomed. Res. Int. 2014, 2014, 1–18.
41. Dixon, R.A. Natural products and plant disease resistance. Nature 2001, 411, 843–847.
42. Reichling, J. Plant-Microbe Interactions and Secondary Metabolites with Antibacterial, Antifungal and Antiviral Properties. Ann. Plant. Rev. 2010, 39, 214–347.
43. Jones, J.D.; Dangl, J.L. The plant immune system. Nature 2006, 444, 323–329.
44. Stamp, N. Out of the quagmire of plant defense hypotheses. Q. Rev. Biol. 2003, 78, 23–55.
45. Hashemi, S.R.; Davoodi, H. Herbal plants as new immuno-stimulator in poultry industry: A review. Asian J. Anim. Vet Adv. 2012, 7, 105–116.
46. Tsuchiya, H.; Iinuma, M. Reduction of membrane fluidity by antibacterial sophoraflavanone G isolated from Sophora exigua. Phytomedicine 2000, 7, 161–165.
47. Upadhyay, A.; Johny, A.K.; Amalaradjou, M.A.R.; Baskaran, S.A.; Kim, K.S.; Venkitanarayanan, K. Plant-derived antimicrobials reduce Listeria monocytogenes virulence factors in vitro, and down-regulate expression of virulence genes. Int. J. Food Microbiol. 2012, 157, 88–94.
48. Upadhyay, A.; Upadhyaya, I.; Kollanoor-Johny, A.; Venkitanarayanan, K. Antibiofilm effect of plant derived antimicrobials on Listeria monocytogenes. Food Microbiol. 2013, 36, 79–89.
49. Mooyottu, S.; Kollanoor-Johny, A.; Flock, G.; Bouillaut, L.; Upadhyay, A.; Sonenshein, A.L.; Venkitanarayanan, K. Carvacrol and trans-cinnamaldehyde reduce Clostridium difficile toxin production and cytotoxicity in vitro. Int. J. Mol. Sci. 2014, 15, 4415–4430.
50. Upadhyaya, I.; Upadhyay, A.; Kollanoor-Johny, A.; Darre, M.J.; Venkitanarayanan, K. Effect of plant derived antimicrobials on Salmonella Enteritidis adhesion to and invasion of primary chicken oviduct epithelial cells in vitro and virulence gene expression. Int. J. Mol. Sci. 2013, 14, 10608–10625.
51. Yin, H.; Chen, C.; Kollanoor-Johny, A.; Darre, M.J.; Venkitanarayanan, K. Controlling Aspergillus flavus and Aspergillus parasiticus growth and aflatoxin production in poultry feed using carvacrol and trans-cinnamaldehyde. Poul. Sci.2015. in press.
52. Goh, E.B.; Yim, G.; Tsui, W.; McClure, J.; Surette, M.G.; Davies, J. Transcriptional modulation of bacterial gene expression by sub-inhibitory concentrations of antibiotics. Proc. Natl. Acad. Sci. USA 2002, 99, 17025–17030.
53. Qiu, J.; Feng, H.; Lu, J.; Xiang, H.; Wang, D.; Dong, J.; Deng, X. Eugenol reduces the expression of virulence-related exoproteins in Staphylococcus aureus. Appl. Environ. Microb. 2010, 76, 5846–5851.
54. Koh, C.L.; Sam, C.K.; Yin, W.F.; Tan, L.Y.; Krishnan, T.; Chong, Y.M.; Chan, K.G. Plant-derived natural products as sources of anti-quorum sensing compounds. Sensors 2013, 13, 6217–6228.
55. Ahmad, A.; Viljoen, A.M.; Chenia, H.Y. The impact of plant volatiles on bacterial quorum sensing. Lett. Appl. Microbial. 2015, 60, 8–19.

56. Bartlett, J.G. Historical perspectives on studies of Clostridium difficile and C. difficile infection. Clin. Infect. Dis. 2008, 46, S4–S11.
57. Beaugerie, L.; Flahault, A.; Barbut, F.; Atlan, P.; Lalande, V.; Cousin, P.; Petit, J.C. Antibiotic-associated diarrhoea and Clostridium difficile in the community. Aliment Pharm. Therap. 2003, 17, 905–912.
58. Nanwa, N.; Kendzerska, T.; Krahn, M.; Kwong, J.C.; Daneman, N.; Witteman, W.; Sander, B. The economic impact of Clostridium difficile infection: A systematic review. Am. J. Gastroenterol. 2015, 110, 511–519.
59. Hookman, P.; Barkin, J.S. Clostridium difficile associated infection, diarrhea and colitis. World J. Gastroenterol. 2009, 15, 1554.
60. Lucado, J.; Gould, C.; Elixhauser, A. Clostridium difficile infections (CDI) in hospital stays, 2009. HCUP Statistical Brief #124. Agency for Healthcare Research and Quality: Rockville, MD, USA, 2012. Available online: http://www.hcup-us.ahrq.gov/reports/statbriefs/sb124.pdf (accessed on 30 July 2015).
61. Sunenshine, R.H.; McDonald, L.C. Clostridium difficile-associated disease: New challenges from an established pathogen. Clev. Clin. J. Med. 2006, 73, 187.
62. Matamouros, S.; England, P.; Dupuy, B. Clostridium difficile toxin expression is inhibited by the novel regulator TcdC. Mol. Microbiol. 2007, 64, 1274–1288.
63. McDonald, L.C.; Killgore, G.E.; Thompson, A.; Owens, R.C., Jr.; Kazakova, S.V.; Sambol, S.P.; Johnson, S.; Gerding, D.N. An epidemic, toxin gene–variant strain of Clostridium difficile. N. Engl. J. Med. 2005, 353, 2433–2441.
64. Dineen, S.S.; McBride, S.M.; Sonenshein, A.L. Integration of metabolism and virulence by Clostridium difficile CodY. J. Bacterial. 2010, 192, 5350–5362.
65. Lindbäck, T.; Mols, M.; Basset, C.; Granum, P.E.; Kuipers, O.P.; Kovács, Á.T. CodY, a pleiotropic regulator, influences multicellular behaviour and efficient production of virulence factors in Bacillus cereus. Environ. Microbiol. 2012, 14, 2233–2246.
66. Fischer, A.; Nakai, Y.; Eubanks, L.M.; Clancy, C.M.; Tepp, W.H.; Pellett, S.; Dickerson, T.J.; Johnson, E.A.; Janda, K.D.; Montal, M. Bimodal modulation of the botulinum neurotoxin protein- conducting channel. Proc. Natl. Acad. Sci. USA2009, 106, 1330–1335.
67. Shih, Y.L.; Hsu, K. Anti-botulismic effect of toosendanin and its facilitatory action on miniature end-plate potentials. Jpn. J. Physiol. 1983, 33, 677–680.
68. Ultee, A.; Gorris, L.G.M.; Smid, E.J. Bactericidal activity of carvacrol towards the foodborne pathogen. Bacillus cereus1998, 85, 21.
69. Ultee, A.; Kets, E.P.; Alberda, M.; Hoekstra, F.A.; Smid, E.J. Adaptation of the food-borne pathogen Bacillus cereus to carvacrol. Arch. Microbiol. 2000, 174, 233–238.
70. Ultee, A.; Smid, E.J. Influence of carvacrol on growth and toxin production by Bacillus cereus. Int. J. Food Microbiol.2001, 64, 373–378.
71. Ultee, A.; Slump, R.A.; Steging, G.; Smid, E.J. Antimicrobial activity of carvacrol toward Bacillus cereus on rice. J. Food Protect. 2000, 63, 620–624.
72. Chapelsky, S.; Batty, S.; Frost, M.; Mogridge, J. Inhibition of Anthrax Lethal Toxin-Induced Cytolysis of RAW264.7 Cells by Celastrol. PLoS ONE 2008, 3, e1421.
73. Slater, L.H.; Hett, E.C.; Mark, K.; Chumbler, N.M.; Patel, D.; Lacy, D.B.; Hung, D.T. Identification of novel host-targeted compounds that protect from anthrax lethal toxin-induced cell death. ACS Chem. Boil. 2013, 8, 812–822.
74. Smith-Palmer, A.; Stewart, J.; Fyfe, L. Influence of subinhibitory concentrations of plant essential oils on the production of enterotoxins A and B and α-toxin by Staphylococcus aureus. J. Med. Microbiol. 2004, 53, 1023–1027.
75. De Souza, E.L.; de Barros, J.C.; de Oliveira, C.E.V.; da Conceição, M.L. Influence of Origanum vulgare L. essential oil on enterotoxin production, membrane permeability and surface characteristics of Staphylococcus aureus. Int. J. Food Microbiol. 2010, 137, 308–311.
76. Parsaeimehr, M.; Basti, A.A.; Radmehr, B.; Misaghi, A.; Abbasifar, A.; Karim, G.; Khanjari, A. Effect of Zataria multiflora boiss. Essential oil, nisin, and their combination on the production of enterotoxin C and α-hemolysin by Staphylococcus aureus. Foodborne Pathog. Dis. 2010, 7, 299–305.

77. Friedman, M.; Rasooly, R. Review of the inhibition of biological activities of food-related selected toxins by natural compounds. Toxins 2013, 5, 743–775.
78. Filgueiras, C.T.; Vanetti, M.C.D. Effect of eugenol on growth and listeriolysin O production by Listeria monocytogenes. Br. Arch. Biol. Tech. 2006, 49, 405–409.
79. Smith-Palmer, A.; Stewart, J.; Fyfe, L. Inhibition of listeriolysin O and phosphatidylcholine-specific production in Listeria monocytogenes by subinhibitory concentrations of plant essential oils. J. Med. Microbiol. 2002, 51, 567–608.
80. Toda, M.; Okubo, S.; Ikigai, H.; Suzuki, T.; Suzuki, Y.; Hara, Y.; Shimamura, T. The protective activity of tea catechins against experimental infection by Vibrio cholerae O1. Microbiol. Immunol. 1992, 36, 999–1001.
81. Pariwat, P.; Homvisasevongsa, S.; Muanprasat, C.; Chatsudthipong, V. A natural plant-derived dihydroisosteviol prevents cholera toxin-induced intestinal fluid secretion. J. Pharmacol. Exp. Ther. 2008, 324, 798–805.
82. Saito, T.; Miyake, M.; Toba, M.; Okamatsu, H.; Shimizu, S.; Noda, M. Inhibition by Apple Polyphenols of ADP-Ribosyltransferase Activity of Cholera Toxin and Toxin-Induced Fluid Accumulation in Mice. Microbiol. Immunol.2002, 46, 249–255.
83. Morinaga, N.; Iwamaru, Y.; Yahiro, K.; Tagashira, M.; Moss, J.; Noda, M. Differential activities of plant polyphenols on the binding and internalization of cholera toxin in vero cells. J. Biol. Chem. 2005, 280, 23303–23309.
84. Yamasaki, S.; Asakura, M.; Neogi, S.B.; Hinenoya, A.; Iwaoka, E.; Aoki, S. Inhibition of virulence potential of Vibrio cholerae by natural compounds. Indian J. Med. Res. 2011, 133, 232.
85. Chatterjee, S.; Asakura, M.; Chowdhury, N.; Neogi, S.B.; Sugimoto, N.; Haldar, S.; Yamasaki, S. Capsaicin, a potential inhibitor of cholera toxin production in Vibrio cholerae. FEMS Microbiol. Lett. 2010, 306, 54–60.
86. Zhong, Z.; Yu, X.; Zhu, J. Red bayberry extract inhibits growth and virulence gene expression of the human pathogen Vibrio cholerae. J. Antimicrob. Chemother. 2008, 61, 753–754.
87. Bhattaram, V.; Upadhyay, A.; Moyoottu, S.; Kollanoor-Johny, A.; Venkitanarayanan, K. Plant molecules inhibit toxin production, motility and hemolysis in Vibrio cholerae in vitro, and down-regulate virulence gene expression. In Proceedings of the IFT Annual Meeting, Chicago, IL, USA, 12–13 July 2013.
88. Pharmacopoeia Commission of People's Republic of China (ChPC). Pharmacopoeia of People's Republic of China (ChP); Chemical Industry Press: Beijing, China, 2000; Volume 1, pp. 49–50.
89. Sack, R.B.; Froehlich, J.L. Berberine inhibits intestinal secretory response of Vibrio cholerae and Escherichia coli enterotoxins. Infect. Immun. 1982, 35, 471–475.
90. Chen, J.; Ho, T.; Chang, Y.; Wu, S.; Hsiang, C. Anti-diarrheal effect of Galla chinensis on the Escherichia coli heat-labile enterotoxin and ganglioside interaction. J. Ethnopharmacol. 2006, 103, 385–391.
91. Chen, J.; Huang, L.; Wu, S.; Kuo, S.; Ho, T.; Hsiang, C. Ginger and its bioactive component inhibit Enterotoxigenic Escherichia coli heat-labile enterotoxin-induced diarrhea in mice. J. Agric. Food Chem. 2007, 55, 8390–8397.
92. Dubreuil, J.D. Antibacterial and antidiarrheal activities of plant products against Enterotoxinogenic Escherichia coli. Toxins 2013, 5, 2009–2041.
93. Chen, J.; Chang, Y.; Wu, S.; Chao, D.; Chang, C.; Li, C.; Hsiang, C. Inhibition of Escherichia coli heat-labile enterotoxin-induced diarrhea by Chaenomeles speciosa. J. Ethnopharmacol. 2007, 113, 233–239.
94. Gonçalves, C.; Berthiaume, F.; Mourez, M.; Dubreuil, J.D. Escherichia coli STb toxin binding to sulfatide and its inhibition by carragenan. Fems Microbiol. Lett. 2008, 281, 30–35.
95. Duarte, M.C.T.; Leme, E.E.; Delarmelina, C.; Soares, A.A.; Figueira, G.M.; Sartoratto, A. Activity of essential oils from Brazilian medicinal plants on Escherichia coli. J. Ethnopharmacol. 2007, 111, 197–201.

96. Birdi, T.; Daswani, P.; Brijesh, S.; Tetali, P.; Natu, A.; Antia, N. Newer insights into the mechanism of action of Psidium guajava L. leaves in infectious diarrhoea. BMC Compl. Altern. M. 2010, 10, 33.
97. Greenberg, R.N.; Dunn, J.A.; Guerrant, R.L. Reduction of the secretory response to Escherichia coli heat-stable enterotoxin by thiol and disulfide compounds. Infect. Immun. 1983, 41, 174–180.
98. Sakagami, Y.; Murata, H.; Nakanishi, T.; Inatomi, Y.; Watabe, K.; Iinuma, M.; Lang, F.A. Inhibitory effect of plant extracts on production of verotoxin by enterohemorrhagic Escherichia coli O157:H7. J. Health Sci. 2001, 47, 473–477.
99. Takemasa, N.; Ohnishi, S.; Tsuji, M.; Shikata, T.; Yokoigawa, K. Screening and analysis of spices with ability to suppress verocytotoxin production by Escherichia coli O157. J. Food Sci. 2009, 74, M461–M466.
100. Okubo, S.; Sasaki, T.; Hara, Y.; Mori, F.; Shimamura, T. Bactericidal and anti-toxin activities of catechin on enterohemorrhagic Escherichia coli. Kansenshogaku Zasshi 1988, 72, 211–217.
101. Sugita-Konishi, Y.; Hara-Kudo, Y.; Amano, F.; Okubo, T.; Aoi, N.; Iwaki, M.; Kumagai, S. Epigallocatechin gallate and gallocatechin gallate in green tea catechins inhibit extracellular release of Vero toxin from enterohemorrhagic Escherichia coli O157:H7. Biochim. Biophys. Acta (BBA)-Gener. Subj. 1999, 1472, 42–50.
102. Kim, Y.; Lee, J.; Kim, S.; Baek, K.; Lee, J. Cinnamon bark oil and its components inhibit biofilm formation and toxin production. Int. J. Food Microbiol. 2015, 195, 30–39.
103. Lee, K.; Kim, W.; Lim, J.; Nam, S.; Youn, M.; Nam, S.; Park, S. Antipathogenic properties of green tea polyphenol epigallocatechin gallate at concentrations below the MIC against enterohemorrhagic Escherichia coli O157:H7. J. Food Prot. 2009, 72, 325–331.
104. Doughari, J.H.; Ndakidemi, P.A.; Human, I.S.; Benade, S. Antioxidant, antimicrobial and antiverotoxic potentials of extracts of Curtisia dentata. J. Ethnopharmacol. 2012, 141, 1041–1050.
105. Baskaran, S.A. Investigating the Potential of Plant-Derived Antimicrobials for Controlling Enterohemorrhagic Escherichia coli O157:H7. Doctoral Dissertation, University of Connecticut, Storrs, CT, USA, 2012.
106. Quiñones, B.; Massey, S.; Friedman, M.; Swimley, M.S.; Teter, K. Novel cell-based method to detect Shiga toxin 2 from Escherichia coli O157:H7 and inhibitors of toxin activity. Appl. Environ. Microbiol. 2009, 75, 1410–1416.
107. Kulkarni, A.A.; Fuller, C.; Korman, H.; Weiss, A.A.; Iyer, S.S. Glycan encapsulated gold nanoparticles selectively inhibit shiga toxins 1 and 2. Bioconjug. Chem. 2010, 21, 1486–1493.
108. Kitov, P.I.; Sadowska, J.M.; Mulvey, G.; Armstrong, G.D.; Ling, H.; Pannu, N.S.; Bundle, D.R. Shiga-like toxins are neutralized by tailored multivalent carbohydrate ligands. Nature 2000, 403, 669–672.
109. Nishikawa, K.; Watanabe, M.; Kita, E.; Igai, K.; Omata, K.; Yaffe, M.B.; Natori, Y. A multivalent peptide library approach identifies a novel Shiga toxin inhibitor that induces aberrant cellular transport of the toxin. FASEB J. 2006, 20, 2597–2599.
110. Basilico, M.Z.; Basilico, J.C. Inhibitory effects of some spice essential oils on Aspergillus ochraceus NRRL 3174 growth and ochratoxin A production. Lett. Appl. Microbiol. 1999, 29, 238–241.
111. Ikeura, H.; Somsak, N.; Kobayashi, F.; Kanlayanarat, S.; Hayata, Y. Application of selected plant extracts to inhibit growth of Penicillium expansum on apple fruits. Plant Pathol. J. 2011, 10, 79–84.
112. Mossini, S.A.G.; Carla, C.; Kemmelmeier, C. Effect of neem leaf extract and Neem oil on Penicillium growth, sporulation, morphology and ochratoxin A production. Toxins 2009, 1, 3–13.
113. Marí, S.; Velluti, A.; Ramos, A.J.; Sanchis, V. Effect of essential oils on zearalenone and deoxynivalenol production by Fusarium graminearum in non-sterilized maize grain. Food Microbiol. 2004, 21, 313–318.

114. Stępień, Ł.; Waśkiewicz, A.; Wilman, K. Host extract modulates metabolism and fumonisin biosynthesis by the plant-pathogenic fungus Fusarium proliferatum. Int. J. Food Microbiol. 2015, 193, 74–81.
115. Sinha, K.K.; Sinha, A.K.; Prasad, G. The effect of clove and cinnamon oils on growth of and aflatoxin production by Aspergillus flavus. Lett. Appl. Microbiol. 1993, 16, 114–117.
116. Gandomi, H.; Misaghi, A.; Basti, A.A.; Bokaei, S.; Khosravi, A.; Abbasifar, A.; Javan, A.J. Effect of Zataria multiflora Boiss. essential oil on growth and aflatoxin formation by Aspergillus flavus in culture media and cheese. Food Chem. Toxicol. 2009, 47, 2397–2400.
117. Razzaghi-Abyaneh, M.; Shams-Ghahfarokhi, M.; Yoshinari, T.; Rezaee, M.B.; Jaimand, K.; Nagasawa, H.; Sakuda, S. Inhibitory effects of Satureja hortensis L. essential oil on growth and aflatoxin production by Aspergillus parasiticus. Int. J. Food Microl. 2008, 123, 228–233.
118. Stenfors Arnesen, L.P.; Fagerlund, A.; Granum, P.E. From soil to gut: Bacillus cereus and its food poisoning toxins. FEMS Microbiol. 2008, 32, 579–606.
119. Logan, N.A. Bacillus and relatives in foodborne illness. J. Appl. Microbiol. 2012, 112, 417–429.
120. Tewari, A.; Abdullah, S. Bacillus cereus food poisoning: international and Indian perspective. J. Food Sci. Tech. 2015, 52, 2500–2511.
121. Dinges, M.M.; Orwin, P.M.; Schlievert, P.M. Exotoxins of Staphylococcus aureus. Clin. Microbiol. Rev. 2000, 13, 16–34.
122. Geoffroy, C.; Gaillard, J.L.; Alouf, J.E.; Berche, P. Purification, characterization, and toxicity of the sulfhydryl-activated hemolysin listeriolysin O from Listeria monocytogenes. Infect. Immun. 1987, 55, 1641–1646.
123. Gill, D.M. Multiple roles of erythrocyte supernatant in the activation of adenylate cyclase by Vibrio cholerae toxin in vitro. J. Infect. Dis. 1976, 133, S55–S63.
124. Mekalanos, J.J.; Collier, R.J.; Romig, W.R. Enzymic activity of cholera toxin. II. Relationships to proteolytic processing, disulfide bond reduction, and subunit composition. J. Biol. Chem. 1979, 254, 5855–5861.
125. Cassel, D.; Selinger, Z. Mechanism of adenylate cyclase activation by cholera toxin: Inhibition of GTP hydrolysis at the regulatory site. Proc. Natl. Acad. Sci. USA 1977, 74, 3307–3311.
126. Cassel, D.; Pfeuffer, T. Mechanism of cholera toxin action: Covalent modification of the guanyl nucleotide-binding protein of the adenylate cyclase system. Proc. Natl. Acad. Sci. USA 1978, 75, 2669–2673.
127. Galloway, T.S.; van Heyningen, S. Binding of NAD+ by cholera toxin. Biochem. J. 1987, 244, 225–230.
128. Law, D. Virulence factors of enteropathogenic escherichia coli. J. Med. Microbiol. 1988, 26, 1–10.
129. Kaper, J.B.; Nataro, J.P.; Mobley, H.L. Pathogenic Escherichia coli. Nat. Rev. Microbiol. 2004, 2, 123–140.
130. Rangel, J.M.; Sparling, P.H.; Crowe, C.; Griffin, P.M.; Swerdlow, D.L. Epidemiology of Escherichia coli O157:H7 outbreaks, United States 1982–2002. Emerg. Infect. Dis. 2005, 11.
131. Levine, M.M. Escherichia coli that cause diarrhea: Enterotoxigenic, enteropathogenic, enteroinvasive, enterohemorrhagic, and enteroadherent. J. Infect. Dis. 1987, 155, 377–389.
132. O'Brien, A.D.; Holmes, R.K. Shiga and shiga-like toxins. Microbiol. Rev. 1987, 51, 206–220.
133. Welinder-Olsson, C.; Kaijser, B. Enterohemorrhagic Escherichia coli (EHEC). Scand. J. Infect. Dis. 2005, 37, 405–416.
134. Henderson, I.R.; Hicks, S.; Navarro-Garcia, F.; Elias, W.P.; Philips, A.D.; Nataro, J.P. Involvement of the enteroaggregative Escherichia coli plasmid-encoded toxin in causing human intestinal damage. Infect. Immun. 1999, 67, 5338–5344.

135. Eslava, C.; Navarro-Garcia, F.; Czeczulin, J.R.; Henderson, I.R.; Cravioto, A.; Nataro, J.P. Pet, an autotransporter enterotoxin from Enteroaggregative Escherichia coli. Infect. Immun. 1998, 66, 3155–3163.
136. Levine, M.M.; Edelman, R. Enteropathogenic Escherichia coli of classic serotypes associated with infant diarrhea: Epidemiology and pathogenesis. Epidemiol. Rev. 1984, 6, 31–51.
137. Bennett, J.W. Mycotoxins, mycotoxicoses, mycotoxicology and Mycopathologia. Mycopathology 1987, 100, 3–5.
138. Calvo, A.M.; Wilson, R.A.; Bok, J.W.; Keller, N.P. Relationship between secondary metabolism and fungal development. Microbiol. Mol. Biol. Rev. 2002, 66, 447–459.
139. Yu, J.; Chang, P.K.; Ehrlich, K.C.; Cary, J.W.; Bhatnagar, D.; Cleveland, T.E.; Bennett, J.W. Clustered pathway genes in aflatoxin biosynthesis. Appl. Environ. Microbiol. 2004, 70, 1253–1262.
140. Kabak, B. The fate of mycotoxins during thermal food processing. J. Sci. Food Agric. 2009, 89, 549–554.
141. Da Cruz Cabral, L.; Pinto, V.F.; Patriarca, A. Application of plant derived compounds to control fungal spoilage and mycotoxin production in foods. Int. J. Food Microbiol. 2013, 166, 1–14.
142. Abarca, M.L.; Bragulat, M.R.; Castella, G.; Cabanes, F.J. Ochratoxin A production by strains of Aspergillus niger var. niger. Appl. Environ. Microbiol. 1994, 60, 2650–2652.
143. Robens, J.; Cardwell, K. The costs of mycotoxin management to the USA: Management of aflatoxins in the United States. Toxin Rev. 2003, 22, 139–152.
144. Peraica, M.; Radic, B.; Lucic, A.; Pavlovic, M. Toxic effects of mycotoxins in humans. Bull World Health Organ 1999, 77, 754–766.
145. Lang, G.; Buchbauer, G. A review on recent research results (2008–2010) on essential oils as antimicrobials and antifungals. A review. Flavour Fragr. J. 2010, 27, 13–39.
146. Martinez, J.; Patkaniowska, A.; Urlaub, H.; Lührmann, R.; Tuschl, T. Single-stranded antisense siRNAs guide target RNA cleavage in RNAi. Cell 2002, 110, 563–574.
147. Sivakumar, D.; Bautista-Banos, S. A review on the use of essential oils for postharvest decay control and maintenance of fruit quality during storage. Crop Prot. 2014, 64, 27–37.
148. Höhler, D. Ochratoxin A in food and feed: Occurrence, legislation and mode of action. Z Ernahrungswiss 1998, 37, 2–12.
149. Stein, A.F.; Phillips, T.D.; Kubena, L.F.; Harvey, R.B. Renal tubular secretion and reabsorption as factors in ochratoxicosis: Effects of probenecid on nephrotoxicity. J. Toxicol. Environ. Health Part A Curr. Issues 1985, 16, 593–605.
150. Richard, J.L.; Plattner, R.D.; May, J.; Liska, S.L. The occurrence of ochratoxin A in dust collected from a problem household. Mycopathology 1999, 146, 99–103.
151. Roll, R.; Matthiaschk, G.; Korte, A. Embryotoxicity and mutagenicity of mycotoxins. J. Environ. Pathol. Toxicol. Oncol.1989, 10, 1–7.
152. Hopkins, J. The toxicological hazards of patulin. Food Chem. Toxicol. 1993, 31, 455–456.
153. Sharma, R.P. Immunotoxicity of mycotoxins. J. Dairy Sci. 1993, 76, 892–897.
154. Selmanoglu, G.; Koçkaya, E.A. Investigation of the effects of patulin on thyroid and testis, and hormone levels in growing male rats. Food Chem. Toxicol. 2004, 42, 721–727.
155. Mirocha, C.J.; Christensen, C.M.; Nelson, G.H. F-2 (zearalenone) estrogenic mycotoxin from Fusarium. Microb. Toxins2013, 7, 107–138.
156. Blaney, B.J.; Moore, C.J.; Tyler, A.L. Mycotoxins and fungal damage in maize harvested during 1982 in Far North Queensland. Crop Pasture Sci. 1982, 35, 463–471.
157. Kordic, B.; Simic, M.; Lazarevic, D. Effect of low doses of the mycotoxin zearalenone on reproductive characteristics of pigs. Vet Glas (Yugoslavia) 1990, 44, 1.

158. Pestka, J.J.; Smolinski, A.T. Deoxynivalenol: Toxicology and potential effects on humans. J. Toxicol. Environ. Health Sci.2005, 8, 39–69.
159. Liu, J.; Zanardi, S.; Powers, S.; Suman, M. Development and practical application in the cereal food industry of a rapid and quantitative lateral flow immunoassay for deoxynivalenol. Food Cont. 2012, 26, 88–91.
160. Rheeder, J.P.; Marasas, W.F.; Vismer, H.F. Production of fumonisin analogs by Fusarium species. Appl. Environ. Microbiol. 2002, 68, 2101–2105.
161. Desjardins, A.E. Fusarium Mycotoxins: Chemistry, Genetics, and Biology; American Phytopathological Society (APS Press): St Paul, MN, USA, 2006.
162. Stępień, Ł.; Koczyk, G.; Waśkiewicz, A. Genetic and phenotypic variation of Fusarium proliferatum isolates from different host species. J. Appl. Genet. 2011, 52, 487–496.
163. Bargen, S.V.; Martinez, O.; Schadock, I.; Eisold, A.M.; Gossmann, M.; Büttner, C. Genetic variability of phytopathogenic Fusarium proliferatum associated with crown rot in Asparagus officinalis. J. Phytopathol. 2009, 157, 446–456.
164. Stępień, Ł.; Koczyk, G.; Waśkiewicz, A. Diversity of Fusarium species and mycotoxins contaminating pineapple. J. Appl. Genet. 2013, 54, 367–380.
165. Stankovic, S.; Levic, J.; Petrovic, T.; Logrieco, A.; Moretti, A. Pathogenicity and mycotoxin production by Fusarium proliferatum isolated from onion and garlic in Serbia. Eur. J. Plant Pathol. 2007, 118, 165–172.
166. Gong, Y.; Hounsa, A.; Egal, S.; Turner, P.C.; Sutcliffe, A.E.; Hall, A.J.; Wild, C.P. Postweaning exposure to aflatoxin results in impaired child growth: A longitudinal study in Benin, West Africa. Environ. Health Perspect. 2004, 112, 1334–1338.
167. Qureshi, M.A.; Brake, J.; Hamilton, P.B.; Hagler, W.M.; Nesheim, S. Dietary exposure of broiler breeders to aflatoxin results in immune dysfunction in progeny chicks. Poul. Sci. 1998, 77, 812–819.
168. Tessari, E.N.C.; Oliveira, C.A.F.; Cardoso, A.L.S.P.; Ledoux, D.R.; Rottinghaus, G.E. Effects of aflatoxin B1 and fumonisin B1 on body weight, antibody titres and histology of broiler chicks. Br. Poul. Sci. 2006, 47, 357–364.
169. Oguz, H. A review from experimental trials on detoxification of aflatoxin in poultry feed. Euras. J. Vet. Sci. 2011, 27, 1–12.
170. Hussain, Z.; Khan, M.Z.; Khan, A.; Javed, I.; Saleemi, M.K.; Mahmood, S.; Asi, M.R. Residues of aflatoxin B1 in broiler meat: Effect of age and dietary aflatoxin B1 levels. Food Chem. Toxicol. 2010, 48, 3304–3307.
171. IARC. IARC Monographs on the Evaluation of Carcinogenic Risks to Humans: Traditional Herbal Medicines, Some Mycotoxins, Napthalene and Styrene. IARC press: Lyon, France, 2002.
172. Yunus, A.W.; Razzazi-Fazeli, E.; Bohm, J. Aflatoxin B1 in affecting broiler's performance, immunity, and gastrointestinal tract: A review of history and contemporary issues. Toxins 2011, 3, 566–590.
173. Kew, M.C. Synergistic interaction between aflatoxin B1 and hepatitis B virus in hepatocarcinogenesis. Liver Int. 2003, 23, 405–409.
174. FDA. Section 683.100 Action Levels for Aflatoxins in Animal Feeds (CPG 7126.33). Available online: http://www.abvt.org/public/docs/Aflatoxin_In_Animal_Feed.pdf (accessed on 27 July 2015).
175. Yin, H.; Chen, C.; Upadhyaya, I.; Upadhyay, A.; Fancher, S.; Li, J.; Nair, M.S.; Mooyottu, S.; Khan, M.I.; Darre, M.J.; Venkitanarayanan, K. Efficacy of in-feed supplementation of plant-derived antimicrobials in reducing aflatoxicosis in chickens. In Proceeding of the Annual meeting of the Poultry Science Association (PSA), Louisville, KY, USA, 27–30 July 2015.
176. Rasooli, I.; Abyaneh, M.R. Inhibitory effects of Thyme oils on growth and aflatoxin production by Aspergillus parasiticus. Food Cont. 2004, 15, 479–483.
177. Kohiyama, C.Y.; Ribeiro, M.M.Y.; Mossini, S.A.G.; Bando, E.; da Silva Bomfim, N.; Nerilo, S.B.; Machinski, M. Antifungal properties and inhibitory effects upon aflatoxin production of Thymus vulgaris L. by Aspergillus flavus Link. Food Chem. 2015, 173, 1006–1010.

178. Soliman, K.M.; Badeaa, R.I. Effect of oil extracted from some medicinal plants on different mycotoxigenic fungi. Food Chem. Toxicol. 2002, 40, 1669–1675.

CHAPTER 15

Human Health Sciences—From Cultivation to Utilization of Medicinal Plants

Fumio Ikegami*, Yan Wang, Maya Kaneko, Megumi Sumino, Satoru Tsukagoshi

Center for Environment, Health and Field Sciences, Chiba University, Kashiwa, Japan;

ABSTRACT

The aim of this review is to recognize the ability of plants used as food and medicine for our health care. From this point of view, we have studied economical production techniques of medicinal plants and vegetables that have physiological functions such as disease prevention, health maintenance and improvement of physical function. We revealed the suitable cultivation techniques such as the long-term freezing seed storage, and systematic and efficient seedling production of Swertia japonica in the future domestication in Japan. We have also studied the development of a new and friendly product for preparing decoction of Kampo medicine (Japanese traditional medicine) to utilize medicinal plants effectively. It was possible that the decoction of some Kampo medicines in a microwavable container could be carried out in substitution for a common method. This study revealed that the improvement of decoction method of Kampo medicine might contribute to conserve the energic or natural resources, especially medicinal plants of crude drugs in comparison with the conventional way. Moreover, the re-evaluation of some vegetables such as Japanese radish and carrot in terms of the suitability as materials for "Yakuzen", and creating and producing of newly low potassium tomatoes for improving the quality of life (QOL) of dialysis patients and potassium restricted patients were also studied by focusing traditional and local vegetables. Some local cultivars containing stronger flavor and taste with higher amount of functional constituents are suitable for our health care than F_1 (first filial generation) cultivars. Our research will give feedback each other by cross cutting way, and human health science from the cultivation to utilization of medicinal plants and vegetables will be important and needed for our healthy and comfortable life in the future.

Keywords: Kampo Medicine; Yakuzen; Medicinal Plant; Functional Vegetable; Low Potassium Tomato

1. INTRODUCTION

To ensure better health and environment is one of the major issues in the 21st century and then the important challenges have been faced. Efforts are underway to develop a new academic discipline called "human health sciences". The discipline incorporates philosophies of Oriental medicine and aims at maintaining or improving our quality of life (QOL) through a variety of trials. To clarify the effect of not only providing medical care using traditional medicine, traditional food products and functional plants, but also farming or horticultural activity on people's mind and bodily function is our great concern.

In recent years there has been an increase in the number of people in Japan suffering from so-called "lifestyle-related diseases" such as hypertension, diabetes, hyperlipidemia and obesity, as a consequence of changes in people's lifestyles such as adopting Western eating habits and engaging in less exercise. The rapid spread of visceral fat obesity has become an issue in particular, and when this is compounded by risk factors for arteriosclerosis such as hyperlipemia and hypertension, it can easily give rise to serious and potentially fatal illnesses such as angina pectoris, myocardial infarctions and brain infarctions. This condition is known as metabolic syndrome. Therefore, as well as the treatment by early diagnosis (secondly prevention), improvement of the lifestyle itself and the environment concerned with lifestyle (primary prevention) is emphasized nowadays [1].

While the concepts of diseases such as hyperlipemia or obesity do not inhere in Kampo medicine, there are nevertheless many prescriptions for dealing with those symptoms in it [2]. It is possible to maintain our health and prevent lifestyle-related diseases by enhancing our natural healing ability and gastrointestinal function through the skilled use of Kampo medicine. In addition, crude drugs such as ginger, dioscorea rhizome (Dioscorea japonica or D. batatas) and jujube (Ziziphus jujube) are regarded as Yakuzen ingredients [3], and they are as well as popular ones for our daily meal. Thus, improvement of our dietary life is also important by utilizing Yakuzen theory.

Yakuzen is a form of medicinal cooking that combines food and traditional medicines, based on theories of Oriental medicine such as the yin-yang theory and the fivephase theory (**Table 1**) [3]. Yakuzen meals are prepared using in-season ingredients, crude drugs and foodstuffs matched to one's physical constitution and particular symptoms (patient's "sho" in Japanese or "Kampo diagnosis") [2]. The philosophies governing Yakuzen are "food as medicine" and "shindo fuji". In Japan and China, "food as medicine" is derived from the view that since medicine and food share a common origin, food is to be regarded not only as a source of nutrients, but also as a nurturing and restorative element like medicine. Shindo fuji is derived from a belief that human beings ought to eat foods in season from the land where they were born and raised since they cannot be separated from that place. From these backgrounds, the research in economical production techniques of medicinal plants or foods used as medicine and Yakuzen foodstuffs becomes more important.

This paper presents our attempts to produce medicinal plants for Kampo medicine using effective production techniques, and to develop a new decoction method of Kampo medicine for convenient uses. The evaluation of some local cultivars of vegetables, and the approaches to create and produce low potassium tomato fruits as a new functional vegetable for Yakuzen dishes are also described.

2. RESULTS AND DISCUSSION

2.1. Producing Medicinal Plants Using Effective Production Techniques

Kampo is a general term for the unique system of traditional medicine originally developed in Japan under the influence of traditional Chinese medicine (TCM). Recently Kampo medicines (Japanese traditional medicine) have been used by more than 86% of Japanese doctors since

Kampo products were mostly included under the coverage of the public health insurance system in 1976 [2,4]. Kampo treatment places an emphasis on normalizing the distortion of mind and body in patients and improving QOL rather than curing the local disease. Thus, in parallel with the increase of chronic, lifestyle-related and complex diseases such as high blood pressure, lipid abnormality symptom and diabetes, patients have increasingly turned to Kampo treatment.

There are many kinds of medicinal plants that are used as folk medicine or the source of Kampo formulae such as kakkonto and shosaikoto [5-7]. Kampo formulae consist of several crude drugs; most of them are plant origin and have been screened out through many years of use. At present, 172 kinds of crude drugs are listed on the Japanese Pharmacopoeia (16th ed., 2011) [8], and more than 80% of the resources are the imports from China. However, nature resources in China are limited and agricultural production area is gradually declining in recent years [9]. Moreover, Chinese domestic demand on crude drugs is growing and personnel expenses are increasing. These situations influenced the price of crude drugs and it will be difficult to acquire the natural resources that are necessary in Japan. Therefore, the increase of the domestic production of original plant resources for crude drugs becomes a pressing matter, and technical researches to produce them in high quality are in progress. For example of Glycyrrhiza, a fine strain was selected and cultivated by hydroponic production system or in plastic tubes [10-12].

Table 1. Five fundamental elements in nature and their properties.

Phase	Parenchymatous viscera	Main function	Hollow viscera	Flavor	Function of flavor
Wood	Liver	Stabilizing mental activities. Storing blood. Supplying nutrients to the whole body. Detoxicating	Gallbladder	Sour	Contracting the body (lemon, vinegar, *etc*). Too much intake damages sleen and stomach.
Fire	Heart	Circulating blood. Adjusting the rhythm of sleep and waking status	Small intestine	Bitter	Controlling excitation and palpitation (burdock, gourd, *etc*). Too much intake damages lung. Unsuitable for Yin-fluid deficiency.
Earth	Spleen	Digesting and absorbing foods. Maintaining patency of blood vessels	Stomach	Sweet	Nourishing the body (pumpkin, sweet potato, *etc*). Too much intake damages kidney (causes swellen and heat).
Metal	Lung	Inhaling ki by breathing and spreading throughout the body	Large intestine	Pungent	Promoting the circulation of ki (ginger, leek, Japanese radish, *etc*). Too much intake damages liver. Unsuitable for Yin-fluid deficiency.
Water	Kidney	Controlling growth, development and urogenous function. Forming and maintaining bones and teeth	Bladder	Salty	Making the body pliant (salt, sea tangle, etc.). Too much intake damages heart (causes hypertension).

Developing cultivation techniques to increase the yield and raise the medicinal component of medicinal plants, as well as breeding cultivars with genetic superiority in yield and medicinal component, are important for the future development of Kampo medicine and the business of health care.

We currently conduct research on the cultivation techniques and medicinal component of the plants such as purple perilla (Perilla frutescens), bupleurum (Bupleurum falcatum), Japanese green gentian (Swertia japonica) and Japanese angelica (Angelica acutiloba). This research involves using drip fertigation technique on substrate mixed with thermophilic bacteria [13]. The growth and the component content of these plants were affected by kinds of substrate and nutrient solution, and the adequate conditions to produce the plants are gradually clarified.

As part of the research on the cultivation of medicinal plants, the research was conducted on the effective production of S. japonica, a biennial plant widely grown in the wild and used as a traditional stomachic medicine for long time in Japan. The Japanese Pharmacopoeia defines that the principal medicinal component of S. japonica is swertiamarin and the plants must contain it not less than 2% as dry matter basis when the plants are used as a crude drug [8].

S. japonica is currently cultivated only in Nagano and Kouchi prefecture, where the 76% and 24% of the total yield in Japan was obtained, respectively [14]. There are very few previous reports related to the cultivation of S. japonica and most of those are concerned with the reproduction of the plant, such as the promotion of seed germination, the examination of the tissue culture. In other words, there is little information about the relationship of the varieties and growth conditions to the yield and medicinal component.

As stable seed supply is necessary for effective production, the study to clarify the effect of long-term freezing storage of S. japonica seeds on germination, yield and swertiamarin content of the seedling was carried out [15]. The seeds were successively collected for years by the conventional way and stored at –20°C. The germination percentage affected by seed storage period, but there were no correlation between the percentage and the period (**Figure 1**). In addition, there was no correlation between plant dry weight and seed storage period. Although it was revealed that long-term preservation might lead to a decline in swertiamarin content, all plants contained more than 2% swertiamarin and satisfied the criteria as a crude drug by HPLC analysis (**Table 2**). Therefore, the seed viability of S. japonica can be maintained for around 13 years by freezing at –20°C without affecting on dry matter yield and quality as a crude drug. This result would contribute to the constraction of seedling production system and the system can separate the field cultivation from the task of raising seedlings, thereby mitigate the risk from disease in the field.

In order to bring about systematic and efficient seedling production, the rate and percentage of seed germination will need to be improved. To that end the effects of gibberellin treatment on the rate and percentage of germination were investigated. Both the germination percentage and rate increased by 12 - 24 hours soaking in a gibberellin solution of 200 ppm (**Table 3**) [16]. When seeds were soaked for a long period in a high-concentration gibberellin solution, however, both the germination percentage and rate declined. These factors will come to serve as reference for cultivation techniques in the future domestication of medicinal plants in Japan.

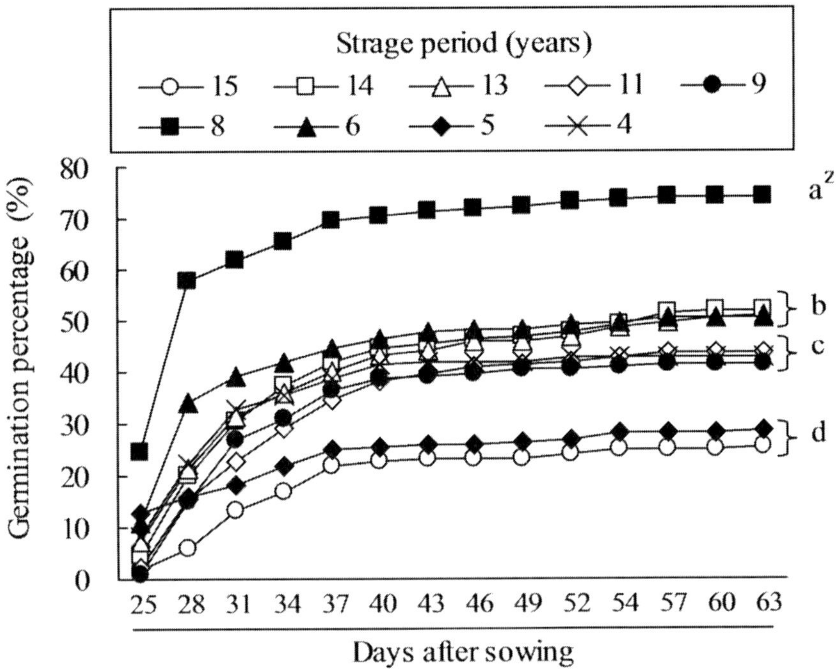

Figure 1. Effect of the period of freeze storage on germination of S. japonica seeds. ᶻDifferent letters indicate significant difference by Tukey's multiple range test at 5% (n = 4).

Table 2. Effect of the period of freeze storage on swertiamarin content of S. japonica seedlings.

Storage period (years)	Content (mg/g DW)
13	33.1 ± 10.3[z]
12	54.4 ± 7.8
11	35.5 ± 8.9
9	45.1 ± 12.7
7	48.5 ± 12.3
6	40.6 ± 6.3
2	48.1 ± 5.3
0 (conventional)	44.9 ± 14.2

[z]Mean ± Standard error (n = 4).

The difference in medicinal component among strains or cultivars was also determined [17]. We could demonstrate that some unutilized strains of S. japonica contained higher swertiamarin than the cultivar used for commercial production (**Table 4**). These strains may contribute to the increase of swertiamarin production or the breeding of new cultivar. This result suggested that there would be some wild genetic resources which have superior characters. The production of vigorous transplant using the transplant culture system with artificial environment control may be effective to promote the growth after transplanting to the field in the second year.

We currently determine the effectiveness of production techniques developed for horticultural crops to increase the plant and component yield of other medicinal plants.

2.2. Improvement of Decoction Method of Kampo Medicine to Conserve Natural Resources

It is one of the great contributions to extract crude drugs of Kampo medicines efficiently for utilizing medicinal plants. The decoction method is described with respect to each Kampo formula in the classic book represented by "Shanghanlun" that was probably completed in the Han dynasty (206 BC-220 AD) and included diagnosis as well as treatment of acute febrile disease called "shanghan" [2]. In current Japan, an electric appliance with glass container or a ceramic one heated directly on fire is commonly used. Though it is difficult for old patients to accept expensive and fragile article, it is necessary to find a new decoction method and a new container.

We have studied the extraction efficiency of Kampo medicine, kakkonto, decocted in different containers in a microwave oven (Hitachi, MRO-DF6) and compared with the common method [18]. Kakkonto is popular and clinically used for the treatment of various feverish diseases such as common cold and inflammation. Daily dose of kakkonto contains 7 precious crude drugs of Pueraria root (8 g), Jujube (4 g), Ephedra herb (4 g), Glycyrrhiza (2 g), Cinnamon bark (3 g), Peony root (3 g) and Ginger (1 g) [2].

We confirmed that authentic components of the decoction were ephedrine and pseudoephedrine in Ephedra herb, puerarin in Pueraria root, paeoniflorin in Peony root and glycyrrhizin in Glycyrrhiza by the HPLC method as well as 6-gingerol in Ginger and cinnamaldehyde in Cinnamon bark by the TLC method described in the Japanese Pharmacopoeia [8]. Tannin content was also determined by the spectroscopic method [19].

When we decocted kakkonto in 600 mL water by a microwavable rice cooker (Skater, MWC-1, polypropylene) or herb teapot (Hario, XCK-1000, glass) at 500 W for 30 min, the yield of powdered extract was within the prescribed amount as well as that obtained by the common method. The all authentic components in the decoction prepared by a microwave oven in 600 mL water at 500 W were almost the same as those obtained by an electric decoction appliance (Hario, HMJ3-1000) in 500 mL water for 30 min (**Table 5**). The evaporation of moisture in a microwave oven was more than that of common method, so we used additional 100 mL water in a microwave oven. Thus, it is possible that the decoction of some Kampo medicines in a microwavable container could be carried out by heating at 500 W for 30 min in substitution for a common method, but we still need to improve a container suitable for the higher extraction efficiency to conserve the energic or natural resources.

Table 3. Effect of soaking in gibberellic acid (GA) on germination of S. japonica seeds.

Treatment		Germination percentage (%)	Germination rate (%)
GA concentration (ppm)	Soaking time (hr)		
	12	84.3 ± 1.1[z] ab[y]	56.3 ± 1.6 bcde
0 (water)	24	82.0 ± 2.1 ab	60.3 ± 2.4 bcd
	48	82.5 ± 1.3 ab	53.0 ± 2.4 bcde
	12	77.5 ± 2.7 abc	64.5 ± 2.4 ab
200	24	87.8 ± 0.7 a	71.5 ± 0.3 a
	48	74.3 ± 3.6 bc	51.0 ± 6.8 cde
	12	74.3 ± 2.2 bc	61.0 ± 1.1 abc
400	24	59.3 ± 4.8 de	48.5 ± 4.0 e
	48	48.8 ± 4.1 e	19.8 ± 4.5 g
	12	67.8 ± 2.4 cd	48.8 ± 2.2 de
600	24	38.3 ± 6.2 f	26.7 ± 3.9 f
	48	51.0 ± 4.6 e	36.5 ± 4.4 f
Anova	GA (G)	34.2[x]***	39.1***
(F value)	Time (H)	14.0***	9.2***
	G × H	9.4***	7.0***

[z]Mean ± Standard error (n = 4); [y]Different letter within column indicate significant difference by Fisher's LSD at 5%; [x]*, ***: Significant at 5% and 0.1%, respectively.

Table 4. Difference of plant dry weight and swertiamarin content among strains and cultivars of S. japonica.

Strain and cultivar		Dry weight (g/plant)	Swertiamarin content	
			(mg/g DW)	(mg/plant)
Strains	4	1.3 ± 0.2^z^	22.5 ± 1.8	28.9 ± 4.6
	14-A	1.5 ± 0.2	18.7 ± 2.8	26.9 ± 4.5
	15-Haya	1.5 ± 0.2	31.3 ± 3.3	46.8 ± 8.9
	S-117	1.0 ± 0.1	33.9 ± 1.5	32.3 ± 3.5
	Kanshoyou	3.0 ± 0.3	14.6 ± 1.7	43.7 ± 7.8
	Kanfuzan	2.2 ± 0.2	21.4 ± 0.5	47.5 ± 4.5
Cultivars	Mimaki-1-go	1.4 ± 0.2	23.2 ± 1.6	31.3 ± 3.6
	Mimaki-2-go	2.0 ± 0.4	21.0 ± 3.8	41.8 ± 2.0
	Mimaki-3-go	2.0 ± 0.5	14.6 ± 2.1	26.8 ± 4.6

[z]Mean ± Standard error (n = 4).

We have also conducted the chemical analysis of the constituents in the residual crude drugs of shosaikoto by decocting twice in an electric decoction appliance (Hario, HMJ3-1000; Tochimoto, EK-SA10) [20]. Shosaikoto is popular and clinically used for the treatment of various feverish diseases such as common cold, pneumonia and liver disorder. Daily dose of shosaikoto contains 7 precious crude drugs of Bupleurum root (7 g), Pinellia tuber (5 g), Scutellaria root (3 g), Jujube (3 g), Ginseng (3 g), Glycyrrhiza (2 g) and Ginger (1 g) [2].

From our study, powdered extract from the second decoction of residual crude drugs of shosaikoto was 26% of that from the first decoction. Saikosaponin b_2 in the second decoction was 2.1 mg comparable to the amount in the first decoction (2.4 mg). Extraction efficiency of the other chemical constituents and factors that influence the extraction efficiency such as decoction time and water volume are now under consideration. Continuous study on the utilization of crude drugs, however, will be essential in the resource-saving environment.

2.3. Evaluation of Some Local Cultivars of Japanese Radish and Carrot in terms of the Suitability as Materials for "Yakuzen" Dishes

We conduct research on medicinal properties of vegetables as Yakuzen ingredients, such as Japanese radish (Raphanus sativus) [21] and carrot (Daucus carota) [22] which are both a medicinal plant in Oriental medicine. To prepare Yakuzen dishes for the constitution or the symptom of each person, materials in season with appropriate properties and tastes play a vital role, and these properties and tastes are called "Sei-Mi" in the Yakuzen theory. "Sei-Mi" is composed of four natures (hot, warm, cool, cold) and five tastes (salty, bitter, sweet, pungent, sour) according to TCM, and each property and/or taste is considered to show its own functionality for human body (**Table 1**). Therefore, the materials for Yakuzen dishes are selected for the physical condition and constitution of each person. If the "Sei-Mi" concept can apply to vegetables, cultivar which has strong inherent flavor and taste with high functional constituents related to the property and taste would be defined to have strong "Sei-Mi," and thus the cultivar can be accounted to have high suitability as materials for Yakuzen dishes.

The inherent flavor and taste of current vegetable cultivars tend to become weak as a result of breeding which gives the priority to the easiness to eat for consumers, or the easiness to grow for growers, etc. In the case of carrot, most of all carrot cultivars nowadays are F_1 (first filial generation) with similar size and shape. In contrast, various cultivars that have various root shapes and colors were used in the time when the carrot production started domestically in Japan. And it was reported that the main constituents in the root varied depending on the cultivar with different root shapes or colors [23,24]. Consequently, it is expected that the flavor, the taste and the content of functional constituents also depends on the cultivars [25,26]. In other words, so-called local cultivars grown for many years may have been keeping the inherent flavor and taste, and some of them may be more suitable for Yakuzen dishes than current ones.

Suitability of Japanese radish cultivars for Yakuzen dishes was evaluated by sensory test, amounts of functional constituents related to the taste and functionality and principal component analysis [21]. One F_1 cultivar and 18 local cultivars belonged to 11 local varietal groups was used (**Table 6**). Cultivars that were considered to possess many aptitudes as Yakuzen ingredients,

Table 5. Authentic components and extract contents in kakkonto decocted by the different methods.

Decoction method			Puerarin (mg)	Paeoniflorin (mg)	Glycyrrhizin (mg)	Ephedrine (mg)	Pseudoephedrine (mg)	Tannin (mg)
Appliance	Water volume	Decoction time						
Microwave oven & Herb teapot	600 mL	30 min	129.4 ± 10.0[z]	59.5 ± 2.4	56.7 ± 2.6	18.0 ± 1.0	6.9 ± 0.6	53.7 ± 3.1
Microwave oven & Rice cooker	600 mL	30 min	121.8 ± 3.5	56.9 ± 4.1	56.9 ± 6.2	16.5 ± 1.0	6.1 ± 0.3	51.9 ± 0.9
Hario, HMJ3-1000	500 mL	30 min	116.5 ± 14.6	57.9 ± 3.7	59.4 ± 6.3	17.2 ± 1.0	6.1 ± 0.5	50.9 ± 3.2

[z]Mean ± SD (n = 4)

Table 6. Local varietal groups and local cultivars of daikon (Japanese radish, Raphanus sativus L.) and carrot (Daucus carota L.).

Daikon Group	Daikon Cultivar	Abbreviation	Carrot Group	Carrot Cultivar	Abbreviation
Miyashige F_1	Taibyo-so-butori	MF_1	F_1	Kouyou-2-gou	K2
Miyashige	Miyashige	MM		Chihama-5-sun	C5
	Aokubi-miyashige-marujiri	MB		Purple Haze	PH
	Shirokubi-miyashige-shirimaru	MW	3-sun	Tokinashi-3-sun	T3
	Gensuke	MG	5-sun	Tokinashi-5-sun	T5
Nerima	Okura	NO		Shin-Kuroda-5-sun	SK5
	Miura	NM		Shin-Ogata-5-sun	SO5
Shogoin	Shogoin	SS		Hakata-Tokinashi-5-sun	HT5
	Kokufu	SK		Super-Harumaki-5-sun	SH5
Horyo	Horyo	H	8-sun	Tokinashi-8-sun	T8
Awa-bansei	Awa-takuan	AA	Long orange	Senko-Futonaga	SeF
	Misono	AM		Senko-Onaga	SeO
Shinshu-ji-daikon	Shinshu-ji-daikon	Si		Kokubun-Senko-Onaga	KSeO
Tohoku-ji-daikon	Kairyo-sendai	TK		Super-Senko-1-Shaku	SSeS
	Akasuji	TA	Danvers	Sapporo-Futo	SaF
Shiro-agari	Wakayama	W	Kintoki	Honbeni-Kintoki	HK
Moriguchi	Moriguti-hosonaga	M	Okinawa zairai	Shima-Ninjin	SN
Ninengo	Suikoun-ninenngo	N			
Minami-kyushu-ji-daikon	Sakurajima	S			

such as high in sweetness, high total soluble solid, amino acids (including glutamine) and isothiocyanate (the pungency component of Japanese radish) content [27], were distributed in the first quadrant of the graph (**Figure 2**). As a result, "Shinshu-ji-daikon" and "Shogoin" was considered to contain higher amount of functional ingredients. "Shinshu-ji-daikon" showed higher contents of sucrose, GABA and total amino acids, and isothiocyanate than the other cultivars. "Shogoin" (SS) also contained higher level of GABA and total amino acid than those of F_1 cultivar. In conclusion, within the range of cultivars we used, "Shinshu-ji-daikon" (Si) was the most suitable material for Yakuzen dishes because it contained stronger taste and higher amount of functional ingredients (**Figure 3**).

Suitability of carrot cultivars for Yakuzen was also evaluated by the same concept as for Japanese radish, especially including volatile compounds content [22]. One F_1 cultivars and 16 local cultivars belonged to seven local varietal groups were used (**Table 6**). As a result of sensory test, three local cultivars ("Hakata-Tokinashi-5- sun"; "HT5", "Kokubun-Senko-Onaga"; "KSeO" and "Sapporo-Futo"; "SaF") were considered to have proper characteristics for Yakuzen because of their strong taste (**Table 7**). And one of the F_1 cultivars ("Chihama-5- sun"; "C5") was also considered to be suitable, object cultivar "Kouyou-2-gou" ("K2") and these four cultivars

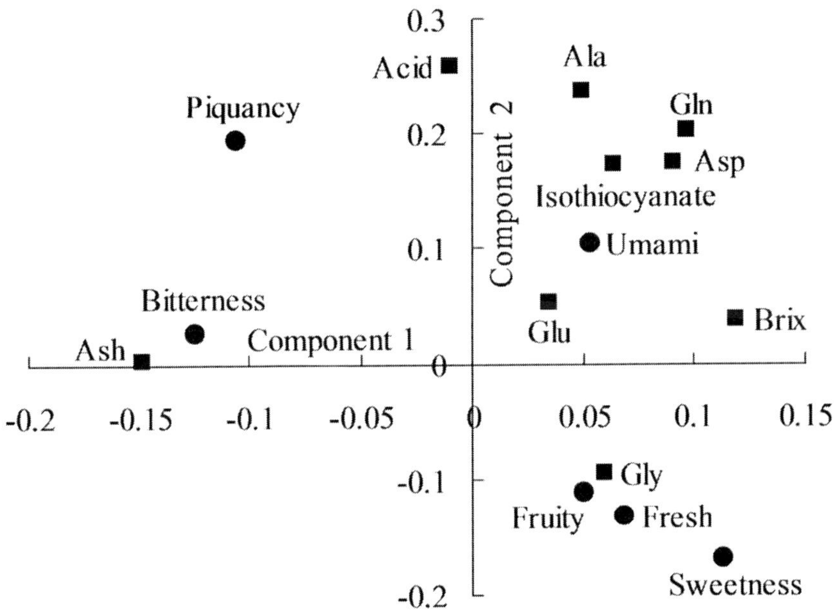

Figure 2. Two-dimensional scatter diagram of factor loading of daikon by principal component analysis. ■, ●: data was derived by sensory test, chemical analysis and growth measurement, respectively.

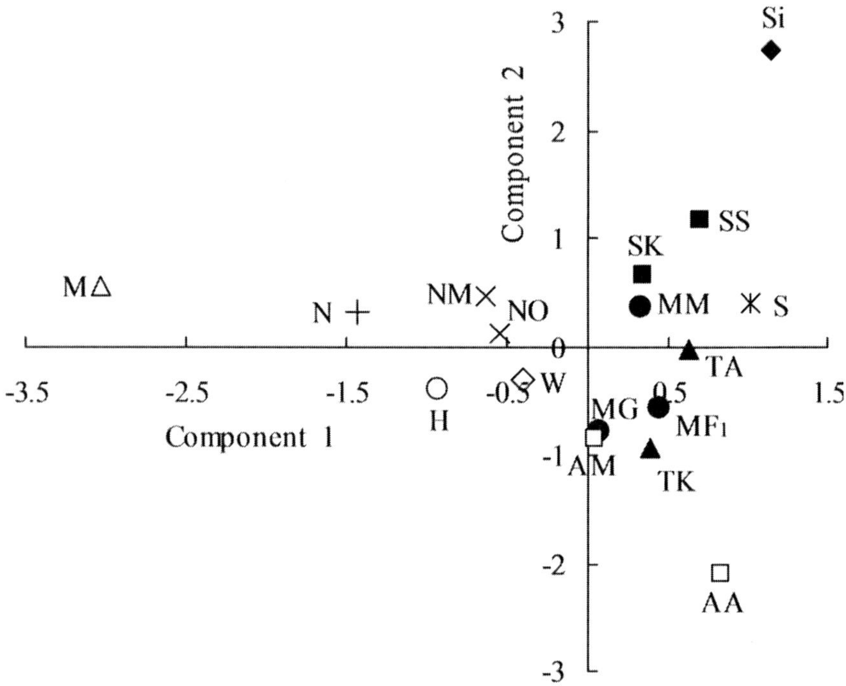

Figure 3. Two-dimensional scatter diagram of the principal component score of daikon cultivars. Abbreviations: See **Table 6**.

were evaluated in detail. Generally, selected four cultivars showed higher sourness, less sugar, higher ash and carotenoid content, and higher anti-oxidant activity than those of "K2" (**Table 8**). Amino acid contents of "KSeO" and "SaF" were ten and six times higher than those of "K2", respectively. On one hand, "K2" and "C5" could characterize to be sweet but less functional. On the other hand, "HT5", "KSeO" and "SaF" had complex taste and much functional constituents, although the factor loadings by principal component analysis were small. Besides, local cultivars had more volatile compounds than F_1 cultivars. It is widely known that smell greatly influences the taste of food. Besides, most of the volatile compounds are terpenoids such as monoterpenes or sesquiterpenes etc., and the composition differs from cultivars, from the data by the GC-MS analysis (**Table 9**). In conclusion, some local cultivars of carrot have stronger taste and more functional constituents than current F1 cultivars. "KSeO" and "SaF" were considered to be the most suitable material for Yakuzen dishes within the range of cultivars, because it showed strong flavor and taste with higher amount of functional constituents.

Table 7. Taste and characteristics of local cultivars evaluated by sensory test[z].

Cultivar	Taste			Other characteristics[y]	Cultivars used for further study
	Strong carrot flavor	Sweet	Bitter, Acrid		
K2		○[x]		Fresh, Fruity, Cosmetic	•
C5	○	○		Fresh and smooth	•
PH			○	Fresh and slightly hard, Potato-like	
T3				Bamboo shoot-like, Weak taste, Porous	
T5		○		Fresh and slightly hard	
SK5				Cosmetic, Cinnamon-like	
SO5	○			Smell of earth, Bamboo shoot-like	
HT5	○			Fresh and slightly hard	•
SH5				Chestnut-like, Bamboo shoot-like	
T8				Raw, Bamboo shoot-like	
SeF	○			Fresh and smooth, Potato-like, Fluffy (not soggy)	
SeO			○	Fresh and slightly hard, Dry, Potato-like	
KSeO	○	○		Kampo medecine-like, Fresh and smooth	•
SSeS				Sweet as dried persimmon, Dry	
SaF	○			Fresh and smooth, Fruity	•
HK			○	Fresh and smooth	
SN			○	Fresh and smooth, Chestnut-like	

[z]Carrot samples were steamed at 100°C for 5 minutes before the test. [y]The description of characteristics was identified by profile interview method. [x]The taste can be clearly felt.

Table 8. Soluble solid, total sugar, carotenoid and ash content, and DPPH radical scavenging capacity of selected carrot cultivars.

Cultivar	Soluble solid (Brix %)	Total sugar (g/100g FW)	Ash (mg/g DW)	Carotenoid[z] (mg/100g FW)	DPPH radical scavenging capacity[y] (%)
K2	7.2 ± 0.1[x] b[w]	7.4 ± 0.2 a	308.2 ± 14.7 bc	11.3 ± 0.8 c	31.9 ± 2.2 b
C5	7.5 ± 0.1 b	5.8 ± 0.1 b	279.3 ± 6.3 c	20.2 ± 1.6 b	42.1 ± 3.1 ab
HT5	7.5 ± 0.3 b	4.9 ± 0.3 c	381.3 ± 19.4 ab	38.1 ± 0.7 a	44.4 ± 4.2 ab
KSeO	8.7 ± 0.1 a	3.9 ± 0.7 c	418.2 ± 6.3 a	8.9 ± 2.8 c	44.6 ± 4.5 ab
SaF	9.2 ± 0.2 a	3.7 ± 0.3 c	418.9 ± 14.0 a	23.0 ± 1.6 b	51.8 ± 4.2 a

[z]Values are expressed as equivalent of β-carotene. [y]Capacity (%) = $(A_{Blank} - A_{Sample})/A_{Blank} \times 100$; [x]Mean ± Standard error (n = 6); [w]Different letter within the column indicates significant difference by Tukey's multiple range test at 5% level.

2.4. Creating and Producing Low Potassium Tomatoes

Consumers today demand higher safety and security for the food. Especially in the field of medical care, the vegetables with lower concentrations of toxic or harmful substances are strongly demanded for patients in the face of a particular disease as well as the safety and security. In order to meet those demands and to raise the QOL of many patients, producing crops with desirable characteristics is considered to be important. An organic coordination with researchers in various scientific fields should be necessary to meet both the patients' and health care professionals' needs and to explore the way for patientcentered medicine.

Table 9. Composition of major volatile compounds of carrot cultivars.

No.	RT (min)	Compound	Peak area %[z]									
			K2	C5	PH	SK5	HT5	SH5	T8	KSeO	SaF	HK
1	9.85	α-pinene				2.2	1.8	0.6	0.6	1.1	0.5	0.5
2	10.79	camphene				0.1			0.3			
3	11.28	sabinene										0.5
4	11.71	β-pinene			0.5	1.4	1.6	0.3	0.4	1.5	2.0	2.3
5	11.93	α-phellandrene				0.1	0.1	0.3	0.2	0.1	nd	
6	12.74	β-myrcene				0.3	0.3	0.2	0.3	0.9	0.4	0.15
7	13.01	α-terpinene				0.1		0.1	0.1	0.1	0.2	
8	13.04	α-thujene									0.2	
9	13.92	limonene			0.6	1.1	0.6	0.7	0.9	0.9	1.2	0.6
10	14.27	β-phellandrene				0.1					0.1	
11	14.61	β-ocimene								3.8	0.0	0.3
12	15.15	γ-terpinene		0.9	1.1	3.5	4.7	3.0	5.8	5.3	10.0	0.8
13	15.99	ρ-cymene				0.5	0.3	0.2	0.3		0.6	
14	16.33	terpinolene	13.0	13.3	12.5	20.6	14.7	22.2	22.5	26.3	31.3	16.6
15	25.29	bornyl acetate	1.1	3.2	2.2	2.5	1.5	0.6	6.2	4.0	2.3	1.8
16	25.96	β-caryophyllene	18.1	8.5	10.7	24.6	24.9	40.7	5.4	7.8	22.0	5.4
17	27.36	neryl acetate	0.1			0.4	0.4	0.3	0.3	6.2	0.3	1.5
18	28.04	α-caryophyllene	1.7	0.8	0.9	1.3	1.3	2.1	0.2	1.0	0.9	0.8
19	29.08	germacrene d	0.5	1.3	0.5			0.3	0.2		0.3	
20	29.09	α-zingiberene				0.5	0.3			1.2		
21	29.24	β-bisabolene	1.3	0.8	0.9	1.1	1.0	1.0	1.5	0.9	1.5	0.6
22	29.37	β-himachalene	1.9	1.7	0.9	0.7	3.3	0.4	0.4	10.0	3.3	2.4
		Others	62.4	69.4	69.2	38.8	43.3	27.0	54.2	29.0	22.9	65.7
		No. of Peaks	10.7	11.0	12.3	23.0	15.5	19.0	23.0	19.0	21.7	14.7

[z]Values are means of 3 samples, and blank means the compound was not detected.

In recent years, the number of patients with diabetes inceases, and they must receive dialysis as a result of development into complications of kidney diseases. Dialysis patients and other patients with kidney dysfunction are given the diet treatment, and they are restricted potassium (K) intake less than 1500 mg/day. Because vegetables and fruits usually contain much potassium, the patients require the fruits and vegetables with low potassium content.

We specifically conduct research to establish the production method for tomato (Solanum lycopersicum) fruits with low potassium content [28]. This research will enable dialysis patients and others with kidney dysfunction to eat fresh tomato and will improve in their QOL.

Table 10. Effect of potassium (K) supply on the fruit K content and total K uptake.

K supply	Fruit weight (g fruit)	K content (mg/100g FW)	Soluble solid (%)	Titratable acid[z] (%)	Total K uptake (g plant)
4 me/L K (control)	42.3 a[y]	202 a	7.4 a	0.56 a	5.74 a
1 me/L K and withdrawal	41.8 a	152 b	6.8 a	0.42 b	0.46 b

[z]Values are equivalent of citric acid. [y]Different letter within the column indicates significant difference by T test at 5% level (n = 6 - 8).

Table 11. Effect of K supply on the fruit K content and fruit quality of 4 medium tomato cultivars grown in DFT in autumn.

K supply	Cultivar	No. of fruit cluster	Fruit weight (g fruit)	K content (mg/100g FW)	K content (mg/fruit)	Soluble solid (%)	Titratable acid[z] (%)
Continuous 4 me/L K (control)	Ai-chan	1	34.9 ± 0.8[y]	249 ± 6	70 ± 4	7.9 ± 0.1	0.72 ± 0.02
		2	39.6 ± 1.3	248 ± 7	83 ± 4	8.1 ± 0.2	0.67 ± 0.02
		3	43.7 ± 1.6	231 ± 7	87 ± 3	7.6 ± 0.1	0.67 ± 0.02
	Frutica	1	38.5 ± 2.5	210 ± 16	77 ± 7	7.4 ± 0.1	0.46 ± 0.06
		2	37.5 ± 2.7	194 ± 10	69 ± 4	7.6 ± 0.1	0.49 ± 0.01
		3	49.3 ± 1.7	174 ± 8	83 ± 5	7.7 ± 0.2	0.44 ± 0.01
1 me/L K and withdrawal	Ai-chan	1	31.3 ± 1.4	105 ± 4	30 ± 1	7.7 ± 0.0	0.38 ± 0.01
		2	40.0 ± 1.2	98 ± 3	34 ± 1	6.9 ± 0.1	0.39 ± 0.01
		3	42.7 ± 1.1	82 ± 5	32 ± 3	7.0 ± 0.2	0.36 ± 0.01
	Frutica	1	38.8 ± 1.1	147 ± 5	54 ± 3	6.9 ± 0.2	0.39 ± 0.01
		2	38.0 ± 0.8	127 ± 6	47 ± 2	7.4 ± 0.2	0.4 ± 0.01
		3	45.2 ± 4.1	126 ± 5	51 ± 4	6.9 ± 0.1	0.39 ± 0.01
ANOVA (P value)	K supply (K)		0.001	<0.001	<0.001	<0.001	<0.001
	Cultivar (C)		<0.001	<0.001	0.011	0.009	<0.001
	No. of fruit cluster (F)		<0.001	0.001	0.014	0.037	<0.001
	Interaction (K × C)		<0.001	<0.001	<0.001	0.006	<0.001
	(K × F)		0.140	0.054	0.297	0.779	<0.001
	(C × F)		0.001	0.019	0.002	<0.001	0.005
	(K × C × F)		0.419	0.017	0.249	<0.001	0.019

[z]Values are equivalent of citric acid. [y]Mean ± Standard error (n = 7).

We first attempted to reduce the fruit potassium content by restricting the amount of potassium supply to the plant. Single cluster tomatoes (cv. Ai-chan, midi plum type) were hydroponically cultivated at 1 me/L of potassium in the nutrient solution. Then, potassium fertilizer was withdrawn from the nutrient solution after the flowering stage. The fruit potassium content decreased, accounting for 75% of the control (continuous 4 me/L of potassium supply), without any adverse effect on fruit weight and soluble solid content (**Table 10**). On the other hand, total potassium uptake was 8% of the control. In this attempt, plant had only 1 fruit cluster, and then potassium might be translocated to the fruit intensively. Thus, plant was allowed to emerge 3 fruit clusters in the next attempt to distribute potassium among 3 clusters and decrease the quantity of translocation to each cluster.

As a result, potassium content was reduced to 40% of control in cv. Ai-chan, but 60% in Frutica. According to the fruit taste, acidity markedly decreased in cv. Ai-chan (**Table 11**). One of the most important problems is that the fruit potassium become lower, the fruit acidity is liable to be lower. We should examine the cultivating method further to improve the sugar content, acidity and the other characteristics concerning to the taste.

3. CONCLUSION

In the ancient past, people used many plants as means for obtaining medical care. They would pick plants out from nature in hope of therapeutic effect to cool a fever, to relieve pain or to suppress inflammation. Through trial and error these plants came to be used more formally as medicines. Although the chemical-based medicine has come to predominate nowadays, they are not necessarily panaceas. Therefore, together with progress being made in medical care and the life sciences, Kampo medicine and other crude drugs made from natural ingredients, and vegetables with highly nutritional functions are into the limelight. Moreover, the functionality of popular foodstuffs such as garlic, buckwheat, soybeans and green tea has being realized again, and thus the use of such plants for lifestyle-related diseases is expected to spread further.

From our study, the possibility of effective production of medicinal plants and foods with high functional ingredients becomes clear by developing cultivation techniques. The results will enhance the effect of treatment by not only Kampo medicines but also Yakuzen dishes. Furthermore, local cultivars that have been grown at restricted area will be spotlighted and re-evaluated for our health maintenance. So as to select the cultivars, growers must obtain the philosophy of Oriental medicine and what property of plant material is desired from professionals. If foods become more functional by using those materials, people can prevent lifestyle-related diseases in particular and disease will be less complicated and easily cured. The effective use of plant materials by considering extraction efficiency will become essential in the future. Thus, the research must be closely coordinated each other and the integration of "food," "agricultural" and "environmental" sciences incorporating philosophies of Oriental medicine is the new academic field of "human health sciences," which embodies humanity's knowledge about "health."

4. ACKNOWLEDGEMENTS

We would like to thank all of our co-workers whose names appear in the references for their dedication, intellectual contributions, and hard

Fumio Ikegami, Yan Wang, Maya Kaneko, Megumi Sumino, Satoru Tsukagoshi work. Our work was partially supported by Grants-in Aid from the Ministry of Education, Culture, Sports and Technology of Japan, and by Research Grants from the Uehara Memorial Foundation, Japan.

REFERENCES

1. Ministry of Health, Labour and Welfare (2009) Health, labour and welfare report 2008-2009, Part 2, No. 62.
2. Japan Society for Oriental Medicine. (2005) Introduction to KAMPO. Elsevier Japan, Tokyo.
3. Tokui, N., Minari, Y., Zhang, Z.L. and Guo, X. (2003) Yakuzen to Chuigaku. Kenpakusha, Tokyo, 2-32.
4. Akiba, T. (2010) History of Kampo extracts for Medical use. Kampo Medicine, 61, 881-888.
5. Ikegami, F. (2001) Herbal drugs in Kampo medicines in present-day Japan. Emerging drugs Vol. 1: Molecular aspects of Asian medicines. PJD Publication Ltd., New Hyde Park, 489-507.
6. Ikegami, F., Fujii, Y., Ishihara, K. and Satoh, T. (2003) Toxicological aspects of Kampo medicines in clinical use. Chemical-Biological Interaction, 145, 235-250.
7. Ikegami, F., Sumino, M., Fujii, Y., Akiba, T. and Satoh, T. (2006) Pharmacology and toxicology of Bupleurum root-containing Kampo medicines in clinical use. Human & Experimental Toxicology, 25, 481-494.
8. The Japanese Pharmacopoeia (2011) The Japanese Pharmacopoeia. 16th Edition, Hirokawa Publishing Co., Tokyo. http://jpdb.nihs.go.jp/jp16e/
9. Kang, D. (2011) Resources of medicinal herbs: Considerations for the necessity of domestic production. Kampo & the Newest Therapy, 20, 131-136.
10. Ozaki, K., Shibano, M., Kusano G. and Watanabe, H. (2007) Aim for production of Glycyrrhizae Radix in Japan (1). A novel cultivation method of Glycyrrhiza uralensis Fisher. Shoyakugaku Zasshi, 61, 89-92.
11. Ozaki, K., Shibano, M., Kusano, G. and Watanabe, H. (2010) Aim for production of Glycyrrhizae Radix in Japan (2). Selection of pharmaceutically fine strains from Glycyrrhiza uralensis Fisher. Shoyakugaku Zasshi, 64, 76-82.
12. Jiho Editorial Department (2011) National Institute of Biomedical Innovation, Kajima Corporation and Chiba University developed the system for the stable domestic production of high-quality Glycyrrhiza uralensis (glabra). Pharmaceutical Technology Japan, 27, 2390.
13. Ikegami, F., Kaneko, M., Tsukagoshi, S., Ito, T., Nagatsuka, T., Matsumoto, J. and Miyamoto H. (2008) Development of the cultivation system for organic herbal medicine. The 128th Annual Meeting of Pharmaceutical Society of Japan, 2, 57.
14. Japan Speciality Agriculture Products Association (2007) Yakuyo-sakumotsu betsu Todoufuken Saibai Gaikyo, 18.
15. Kaneko, M., Tsukagoshi, S., Yanagisawa, K., Yamada, M., Motoki, S., Hagihara, Y. and Ikegami, F. (2011) Effects of freezing of Swertia japonica seeds on germination, seedling yield and swertiamarin content. Shoyakugaku Zasshi, 65, 39-42.

16. Yamada, M., Tsukagoshi, S., Yanagisawa, K., Kaneko, M., Motoki, S., Hagihara, Y. and Ikegami, F. (2011) Effects of strage temperature and gibberellin pretreatment on germination rate of Swertia japonica Seed. Horticultural Research (Japan), 10, 321-324.
17. Kaneko, M., Tsukagoshi, S., Ikegami, F., Yanagisawa, K., Motoki, S. and Hagihara, Y. (2012) Effect of varieties, plant age and altitude of production site on swertiamarin content in Swertia japonica. Acta Horticulture, in press.
18. Wang, Y., Sumino, M., Chino, A. and Ikegami, F. (2011) Study on a new decoction method of Kampo medicine by a microwave oven. Journal of Traditional Medicines, 28, 92.
19. Ikegawa, K., Takayanagi, H. and Anan, T. (1990) Quantitative analysis of tea constituents, 7: Tannin. Tea Research Journal, 71, 52-53.
20. Sumino, M., Saito, Y., Ikegami F. and Namiki T. (2011) Chemical analysis of the constituents in the residual crude drugs of shosaikoto. Journal of Traditional Medicines, 28, 75.
21. Tsukagoshi, S., Inuzuka, S., Hohjo, M., Ikegami, F., Takenaga, S., Nakao, C., Uryu, N., Hagiwara, T., Aoki, H. and Hanamura, T. (2011) Evaluation of Japanese radish cultivars in terms of medicinal properties for "Yakuzen" dishes. HortResearch, 65, 81-86.
22. Tsukagoshi, S., Takano, A., Hohjo, M., Ikegami, F., Hagiwara, T., Nakao, C., Yamada, K., Hanamura, T., Takenaga, S. and Aoki, H. (2011) Evaluation of some local cultivars of carrot in terms of the suitability as materials for "Yakuzen" dishes. Journal of Traditional Medicines, 28, 106-114.
23. Watanabe, K. (1988) Carotenoid pigments in red, orange and yellow root of carrot (Daucus carota L.) cultivars. Japan Journal of Food Engineering, 35, 315-320.
24. Takahashi, A., Okushima, S., Kodama, H. and Yoshida, K. (2001) Constituent and sensory changes of different types of carrots after freezing and frozen storage. Food Preservation Science, 27, 119-132.
25. Kjeldsen, F., Christensen, L.P. and Edelenbos, M. (2001) Quantitative analysis of aroma compounds in carrot (Daucus carota L.) cultivars by capillary gas chromatography using large-volume injection technique. Journal of Agricultural Food Chemistry, 49, 4342-4348.
26. Kreutzmann, S., Thybo, A.K., Edelenbos, M. and Christensen, L.P. (2008) The role of volatile compounds on aroma and flavor perception in coloured raw carrot genotypes. Journal of Food Science Technology, 43, 1619- 1627.
27. Nakamura, Y., Iwahashi, T., Tanaka, A., Koutani, J., Matsuo, T., Okamoto, S., Sato, K. and Ohtsuki, K. (2001) 4-(methylthio)-3-butenylisothiocyanate, a principal antimutagen in Daikon (Raphanus sativus; Japanese white radish). Journal of Agricultural Food Chemistry, 49, 5755- 5760.
28. Hamano, E., Tsukagoshi, S., Hohjo, M. and Ikegami, F. (2009) Production of tomato fruits containing low potassium for kidney disease. 3. Potassium concentration in the fruits of some cultivars as affected by withdrawal of potassium in the nutrient solution. Horticultural Research (Japan), 8, 511.

CHAPTER 16

Aromatic and Medicinal Plants in Mexico

Mariana Palma-Tenango, Ruben San Miguel-Chávez and Ramón Marcos Soto-Hernández*

[1] *Colegio de Postgraduados, Campus Montecillo, Texcoco, Estado de México, México*

ABSTRACT

Medicinal and aromatic plants in Mexico have been studied and explored through history. Day by day there is an increase in ethnobotanical, taxonomic, or phytochemical studies, providing an encouraging picture of research in Mexico and to support its use in traditional medicine. Chemical and biological exploration permit to provide solutions to the treatment of diseases. With this background, the objective of this chapter is to show the potential of endemic medicinal and aromatic plants in Mexico.

Keywords: endemic, medicinal herbs, Mexico

1. INTRODUCTION

Mexico is a country with a wide variety of medicinal and aromatic plants; their use is rooted in its culture, and they are employed to solve health problems in areas with little access to medicines. Although there are several introduced species, Mexico has a high number of endemic plants that are considered medicinal.

Demand for medicinal plants increases with the market, which is broad [1]. They are plants that produce secondary metabolites, each with different active ingredients and different therapeutic properties [2]. However, phytochemical information is generally focused on introduced species. Mexico has medicinal and aromatic species that require scientific research and wider dissemination. The cultivated area in the country is smaller than the demand there, and most of the plants are grown commercially. With this background, it is vital to gather information about endemic species of Mexico that are important for agriculture, pharmaceutics, cosmetics, etc. due to their phytochemical and pharmacological properties.

Mexico has an estimated 30,000 species of plants, where approximately 3,000–to 5,000 are of medicinal use [3]. Most of the medicinal plants are gathered from the wild, and only 15% are cultivated. About 50,000 tons were exported, but there are no precise data about endemic species. The highest percentage of medicinal and aromatic plants is sold in local markets. Their commercialization is economically promising due to their widespread use in herbal medicine, and due to the biological activity, they have shown in some research.

The traditional use of medicinal plants has generated interest to start research to maintain a sustainable use of several species. At least 119 species of plants are used in the empirical treatment of diseases with symptoms similar to those of cancer [4]. Antibacterial activity of 75 species of plants has been reported, and 225 compounds and 140 species (40.57%) had been reported as toxic at least once.

2. HISTORY OF MEXICAN MEDICINAL PLANTS

For several centuries, in what is now known as Mexico, medicinal plants have had an important part in the cultural baggage of indigenous pre-Columbian peoples, continuing through the independent period and reaching the ethnic groups that populate modern-day Mexico. The knowledge of this flora and its use to treat several maladies that affect the population are part of the knowledge that is empirically transmitted from generation to generation [5].

Before the arrival of Spanish conquered in the ancient Tenochtitlan, there was already knowledge of the flora, which was used to cure several illnesses, or during mystical ceremonies. Proof of that were the ancient botanical gardens located near Tenochtitlan, Chapultepec, Huastepec, Ixtapalapa, Peñón, Tetzcoco, and a bit farther, in Atlixco, in the state of Puebla [6]. The deep knowledge of medicinal plants, joined to that of the human body, helped healers to use them to treat illnesses.

It is known that healers used these plants, and depending on the ailment or the patient, the treatment was carried out by a "specialist"; thus, there were physicians, surgeons, midwives, "hueseros" (bone healers), "sobadores" (masseurs), among others [7]. The way they employed the plants was through poultices, concoctions, dry powders, oils, infusions, etc. [7].

When the Spaniards arrived in the sixteenth century, they were marveled by the knowledge, and the use of medicinal plants that the indigenous peoples had to treat illnesses [8]. But it was at that moment when the European medicine had a negative influence in the continuity of the use of medicinal plants native of colonial Mexico.

As the Spanish control increased in New Spain, the influence of the Catholic Church became evident, as it forbidden the treatment of illnesses using the knowledge and traditions of the ancient Mexicans because it considered them magical and superstitious; this led to the punishment of many people, and it resulted in the practice of these traditions in secret [9].

Also, the European culture, represented by the Spaniards, brought new ways of healing and it introduced plants such as chamomile, rosemary, basil, thyme, marjoram, wormwood, English marigold, fennel, mint, peppermint, among others [4]. As a result of this interbreeding, the "botica" was introduced; this was the place where remedies containing plant extracts, oils, essences, powders, etc., were prepared. The botica was overseen by a physician who had to take a quite rigorous test to be able to treat patients. He diagnosed, wrote prescriptions, and prepared the "medicines" for the patient [10].

Despite the prohibitions that the Spaniards had established at the beginning, the interest to commercialize the native plants of New Spain in Europe continued, and thus, in the sixteenth century, the codex de la Cruz Badiano appeared; this codex had been created for the most part for the purpose of this trade [11]. This compendium documented the vast medicinal flora of the century, with at least 41 illustrations of plants, and which practically constituted the first document and list of medicinal plants of Mexico.

The Spanish government promoted the Royal Botanic Expedition, which was carried out towards the end of the eighteenth century in New Spain. This expedition was an extremely important contribution to the knowledge of the botany in Mexico. José Mariano Mociño, who is considered the first Mexican botanist, took part in this expedition and in the creation of the manuscripts of the great Flora Mexicana [12].

Through the years, the pharmaceutical industry appeared, and the traditional medicine and the use of medicinal plants diminished, and they were relegated, and even considered illegal.

Nowadays, the use of medicinal plants still persists in the Mexican traditional medicine, especially in rural zones, where it may be the only available resource to treat illnesses and diseases. Unfortunately, there is too an inappropriate use of the technique, lack of asepsis, or the prescription of ingredients by people without proper training; this contributes to the discredit of this resource.

In spite of all the negative publicity that this technique attracted for many years, nowadays researchers of several institutions, including universities, health institutes, and the pharmaceutical industry have started paying attention to studies with scientific evidence that show that it is a useful alternative to solve health problems.

3. ENDEMIC MEDICINAL AND AROMATIC PLANTS OF MEXICO

Mexico is a country with the fourth largest floristic richness in the world; there are reported 23,314 native vascular plants, of which 11,600 are endemic; although it is reported that the percentage of endemism is actually 50% [3]. The family Lamiaceae is one of the most diverse in the country.

There is no exact number of endemic species with medicinal and aromatic uses, but 3,000 species are reported with medicinal uses in Mexican traditional medicine. It is mentioned that 1,549 are used in the Mayan culture, 816 in the Nahuas, and 3,059 in the Zapotecs. [13].

In Mexican culture, plants are used with different medicinal objectives. Every day, new phytochemical, ethnobotanical, or biodiversity reports arise. This has allowed exploring their potential and supporting their traditional use.

Several authors have reported studies about endemic plants that are considered medicinal or aromatic (**Table 1**). They show a promising future due to their compounds or biological activities. These species are sold all over the country, and some compounds are already being semi-synthesized, as is the case of taxol, to be commercialized and to be used in cancer treatments all over the world.

Table 1. Plants aromatic and medicinal endemic reported in Mexico, with economic potentiall.

Scientific name*	Family	Composition	Activity	Reference
Agastache mexicana subsp. *xolocotziana* Bye, E.L. Linares and Ramamoorthy	Lamiaceae	Essential oils – methyl eugenol – Estragole – Limonene – Estragole – Menthone – Pulegone	Antifungal	[14, 15]
Agastache mexicana subsp. *mexicana*	Lamiaceae	Essential oils – Limonene – Linalool – Methyl Clavicol	Antifungal	[15]
Montanoa tomentosa Cerv.	Asteraceae	Volatile compounds – Sabinene – α-pinene – α-tujene Diterpenes Sequiterpene lactones	Uterotonic, antibacterial, and antifungal	[16, 17]
Lippia graveolens Kunth	Verbenaceae	Essential oils – α-Pinene – α-Thujene – Camphene– -3-Carene – α-Terpinene– Limonene – p-Cymene– Thymol – Carvacrol	Acaricidal, antibacterial	[18, 19]
Erythrina americana Mill. *Erythrina coralloides* DC. *Erythrina lepthoriza* A. DC. *Erythrina mexicana* Krukoff *Erythrina oaxacana* Krukoff *Erythrina sousae* Krukoff	Fabaceae	Alkaloids	Sedative, antifungal	[20, 21]
Hintonia latiflora (Sessé and Moc. ex DC.) Bullock	Rubiaceae	Alkaloids, flavonoids, phenylcoumarin and glucocucurbitacins	Malaria, gastroprotector	[22]
Salvia hispanica L.	Lamiaceae	Fatty acids, phenolic compounds	Antioxidant activity	[23, 24]
Bursera fagaroides var. *fagaroides*	Burceraceae	Flavonoids, lignanans	Citotoxy activity, antitumoral	[25]
Jatropha neopauciflora Pax	Euphorbiaceae	Uncommon Sesquiterpenoids and New Triterpenoids	Antimicrobial and insecticide, antibacterial activities, Cytotoxic	[26, 27]
Eysenhardtia platycarpa Pennell and Saff.	Fabaceae	Flavones, oleanolic acid, lupeol, betulinic acid, β-sitosterol, β-sitosteryl β-D-glucopyranoside, β-sitosteryl palmitate, and 3-O-methyl-myo-inositol	Antihyperglycemic	[28]

*The botanical names were corroborated at Missouri Botanical Garden (2016)-TROPICOS.

4. SECONDARY METABOLITES IN MEXICAN PLANTS

Research about natural products from endemic, native, and introduced medicinal plants is based on phytochemical studies. It is important to discover new molecules and evaluate their biological activity. Worldwide, 70% of anti-cancer compounds [29] and 75% of drugs destined to treat infectious diseases come from natural products [30]. Secondary metabolites are used as pigments, fibers, glue, oils, waxes, scent agents, perfumes, and drugs.

The acknowledgment of the biological properties of natural products has raised interest not only in their therapeutic activities, but also in their possible uses as antibiotics, to control pests and diseases, and as cosmetics [30].

A clear example is Taxus: there have been several phytochemical studies about this genus and one of its species is endemic to Mexico. In Mexican yew (*Taxus globosa* Schltdl.), the presence of taxol as the main component was documented [31]. Taxol was approved in 1992 to be commercialized as a therapeutic agent used in chemotherapy against a type of cancer.

Secondary metabolites are classified by their chemical structure, biosynthetic approach, or chemosystematic composition. The basic classification includes flavonoids, phenylpropanoids, phenolic compounds, tannins, saponins, essential oils, alkaloids, terpenoids, and glucosinolates [32].

The main secondary metabolites in endemic medicinal and aromatic plants of Mexico that have been investigated are phenolic compounds, flavonoids, terpenoids, and essential oils. Every day the list of groups of compounds or specific molecules and their biological evaluation grows. Studies about medicinal plants report several properties and the presence of secondary metabolites in several plant parts such as barks, leaves, fruits, flowers, roots, stems, wood, and whole plants.

5. EXTRACTION METHODS AND ANALYSES

Phytochemical studies of Mexican medicinal and aromatic plants have been carried out using traditional extraction methods (**Figure 1**). They are used because they are simple, low cost, and easy to obtain. However, these processes can be slow, and there is a constant search to find techniques that can optimize extractions, that are cheap, fast, and that extract contains the highest number of compounds.

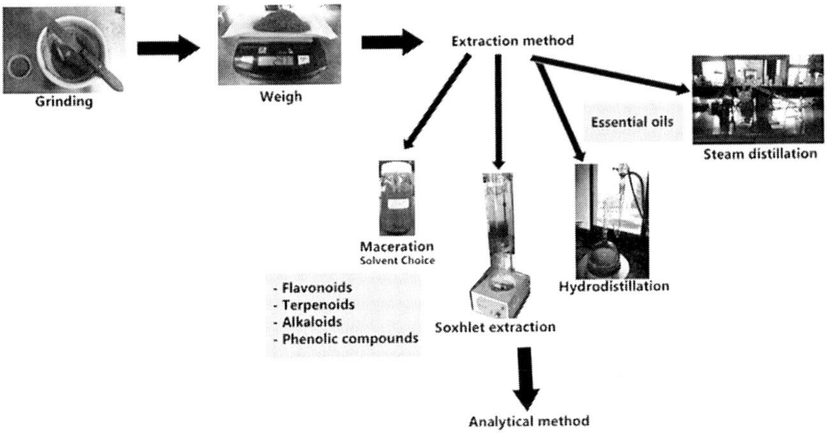

Figure 1. Main extraction methods of secondary metabolites in Mexican plants.

The techniques that have been reported to analyze and characterize chemical compounds from Mexican plants have been (LC), (GC-MS), (NMR), (IR), (UV), and (MS).

Several phytochemical methods describe details to extract and analyze compounds from Mexican plants, but the most used in endemic plants (**Table 2**) are maceration and hydrodistillation, and the most used solvents are hexane, methanol, dichloromethane, acetone, petroleum ether, ethanol, and water.

6. COMMERCIALIZATION OF MEXICAN MEDICINAL PLANTS

6.1. National Market

The sale of the plants in Mexico occurs mainly in regional markets. These plants are concentrated in specific locations, from where they are distributed to several parts of the country. Important zones of trade of medicinal plants and aromatic herbs are the market of Ozumba, in the state of Mexico, and Atlixco and San Martín Texmelucan, in the state of Puebla.

Table 2. Methods of extraction and analysis of secondary metabolites in Mexican endemic plants.

Scientific name*	Family	Plant parts use	Method extraction	Compounds	Analysis	Ref.
Agastache mexicana subsp. *xolocotziana* Bye, E.L. Linares and Ramamoorthy	Lamiaceae	Inflorescence, leaves and stems	Hydrodistillation Maceration	Essential oils Flavonoids	GC-MS HPLC-MS	[14, 33]
Montanoa tomentosa Cerv.	Asteraceae	Leaves	Maceration with sonicated	Terpenoids	HPLC, GC–MS	[16]
Salvia hispanica L.	Lamiaceae	Seeds	Soxhlet	Fatty acids composition, Total phenolic compounds	GC-MS Spectrophotometry maceration	[23, 24]
Heterotheca inuloides Cass.	Asteraceae	Inflorescence	SC-CO2	Volatile Fatty acids	GC-MS	[34]
Tagetes lucida Cav.	Asteraceae	Inflorescence	Maceration	Coumarinic constituents	UHPLC	[35, 36]
Bursera fagaroides var. *Fagaroides* *Bursera fagaroides* var. *elongata* McVaugh and Rzed *Bursera fagaroides* var. *purpusii* (Brandegee) McVaugh and Rzed.	Burseraceae	Leaves	Chromatographic fractionation	Flavonoids, lignans, volatile compounds	HPLC, 1H-NMR	[25, 37]
Jatropha neopauciflora Pax	Euphorbiaceae	Bark	Maceration	Sesquiterpenoids	CC, HPLC, 1H-NMR	[27]
Lippia graveolens Kunth	Verbenaceae	Leaves and stems	Hydrodistillation Maceration	Essential oils Phenolic compounds	GC-MS UHLPC-MS	[38, 39]
Amphipterygium adstringens (Schltdl.) Standl.	Anacardiaceae	Bark	Maceration	Terpenoids and maticadienoic acid	Spectroscopic and RNM methods	[40, 41]

*The botanical names were corroborated at Missouri Botanical Garden (2016)-TROPICOS.

The plants that are mostly gathered from the wild are sold in large quantities in the "Mercado de Sonora" (market of Sonora). This market, located in Mexico City, is the most famous market due to the trade of dry and fresh plants [42]; approximately 15 tons of each medicinal plant is sold every year in this market.

The "Mercado de Ozumba" (market of Ozumba) is another place where one can get plants (**Figure 2**) that is closest to the market of Sonora. Here, local useful and medicinal plants, which have an important role for the economic and ceremonial life of the inhabitants of this region, are sold [42].

Figure 2. Trade of medicinal plants in Ozumba, state of Mexico.

There are no precise reports about the species that are sold in regional markets, or about the total amount of income that is generated. Prices can vary depending on the species, part of the plant, or the location. The lack of information is related to illegal collection of plants. Plants collected from the wild are an important source of income for the inhabitants of the zones where they grow.

The species that are known in the country as having medicinal properties, and that have been collected for years (endangering their presence or distribution) are peyote (*Lophophora williamsii* (Lem. Ex Salm-Dyck) J.M Coult.), Mexican valerian (*Valeriana edulis subsp. procera* (Kunth) F.G. Mey.), cuachalalate (*Amphipterygium adstringens* (Schltdl.) Standl.), tepezcohuite (*Mimosa tenuiflora* Benth), and probably other species for which reports are lacking.

Plants sold in markets can be endemic or introduced. Trade conditions vary depending on the state of the plant: when dry, they are sold by weight, and when fresh, they are sold in bundles. They can be sold wholesale or retail, depending on the location, season of the year, and plant availability.

Some plants have the same common name and the same use all over the country. For example, Mexican arnica (*Heterotheca inuloides* Cass.) is sold dry or fresh (**Figure 3**), and also as part of different types of products including ointments, gels, shampoo, soap, etc.

Figure 3. Trade of Mexican arnica in the market of Ozumba, state of Mexico. (Photography: Mariana Palma Tenango).

6.2. International Market

The international markets demand Mexican medicinal plants [43]. For example, damiana (*Chrysactinia mexicana* A. Gray), sarsaparilla (*Smilax aristolochiifolia* Mill.), Mexican arnica (*H. inuloides*), Mexican oregano (*Lippia graveolens* Kunth), cuachalalate (*A. adstringens* (Schltdl.) Standl.), mexican valerian (*Valeriana edulis* subsp. *procera* (Kunth) F.G. Mey.) are exported to the United States, Japan, and Germany.

Mexico exports essential oils and resinoids to several countries, especially the United States (**Figure 4**). There are no specific records on exports of essential oils of endemic plants in the country. But it is known that the most popular essential oils are Mexican oregano, sweet lime, and Mexican lime. [44].

Some of the plants with medicinal and gastronomic importance are cacao (*Theobroma cacao* L.) and vanilla (*Vanilla planifolia* Jacks.), and the countries with the highest percentage of their consumption are the United States, France, and Germany. However, vanilla from Madagascar competes against Mexican vanilla, which has diminished the market for the latter.

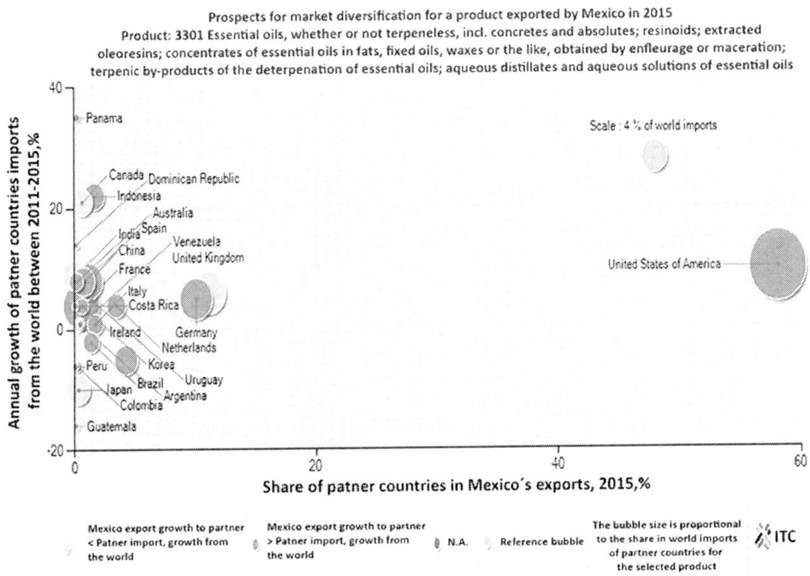

Figure 4. Prospect for market diversification for product exported by Mexico in 2015. Product 3301. Essential oils (terpenes or not), resinoids, and extracted oleoresins [**44**].

6.3. Future Prospects

Currently, there are several challenges; the most important is the sustainable use of plants with medicinal, cosmetologic, culinary or ritual applications, and their preservation through germoplasm banks. The domestic and international demand for medicinal plants is growing. About 90% of the plants are collected from the wild, and the remaining 10% is cultivated. Due to this, overexploitation is a real danger, because the species are collected without thinking about their recovery or the environmental damage. In Mexico, 83 species are considered endangered, 206 threatened, and 175 vulnerable [3]. One example of this is cuachalalate (*A. adstringens*), which is exploited in the state of Guerrero; when too much of its bark is removed, the tree is unable to recover and dies. Something similar occurs with mexican valerian (*V. edulis* subsp. *procera*), whose roots were collected relatively easily in the past in the wooded area of the boroughs of Milpa Alta and Xochimilco in Mexico City; nowadays it is almost impossible to find the plant. To avoid these situations, it is desirable that secretariats and institutes in charge of the care and promotion of the diversity of Mexican plant species enforce the laws and regulations that are already in place, to preserve these natural resources.

Another important aspect to take into account is the necessity to encourage the culture of plants that are collected from the wild, and to create technology packages to improve culturing, harvesting, and drying of the plants to monitor the quality of the secondary metabolites that are being produced, because in these plants what is sought is precisely this mixture of secondary metabolites that give, in the end, the beneficial effect to health and/or cosmetology that is desired by people.

The pharmaceutical industry, which in the beginning turned its back on the traditional medicine, is starting to see medicinal plants and their uses and the traditions of the indigenous peoples that know and use them on a daily basis, as a source of active ingredients to be incorporated into its drugs. Today it is not surprising anymore to see vitamin supplements with Ginseng extracts, cough syrups with propolis or great mullein, or pills with cuachalalate extracts to cure gastritis, etc.

Medicinal plants used in traditional medicine may represent alternative sources for new compounds to treat various diseases. For example, plants used for gastrointestinal disorders [35], used by antibacterial activities [45], anxiolytic or antidepressant activity [46], or for the treatment of colorectal cancer [4]. In Mexican arnica flowers (*H. inuloides*) were actives as antidiarrhoeals or antiparasitics [47].

The list increases, providing important data of compounds and biological activities. But it is necessary to increase developing phytochemical studies associated to the agricultural sector to produce medicinal plants, natural products, extracts, and other subproducts in a sustainable way.

Aromatherapy is another source of demand for aromatic plants. In Mexico, the demand of essential oils is higher than their production, which has led to the import of these substances.

The diversity in Mexico is large, and despite the scientific technological developments, there remains a lot to be researched about all the species with medicinal potential. We can say that the table is set and waiting for people to research, rationally exploit, and care for all this exquisite variety of medicinal and aromatic plants.

7. CONCLUSIONS

Endemic plants are known and commercialized in Mexico, but there are no precise numbers about their cultured surface. Phytochemical and biological activity research requires a previous botanical and taxonomical classification, and thus, their research is slow, but necessary to get scientific bases about their use.

REFERENCES

1. Rosete J C R, Aguilar C J A, Juan-Rosete R E, Bugarin-Montoya R, Juan-Lopez P, Cruz-Crespo E. Herbs and medicinal plants in México: Tradition. Revista Biociencias. 2013;52b:119–129.
2. Bourgaud F, Gravot A, Milesi S, Contier E. Production of plant secondary metabolites: A historical perspective. Plant Science. 2001;161:839–851.
3. Villaseñor J L. Checklist of native vascular plants of México. Revista mexicana de Biodivesidad. 2016;87b:559–902.
4. Jacobo-Herrera N J, Jacobo-Herrera F E, Zentella-Dehesa A, Andrade-Cetto A, Heinrich M, Pérez-Plasencia C. Medicinal plants used in Mexican traditional medicine for the treatment of colorectal cancer. Journal of Ethnopharmacology, 2016;179:391–402.
5. Estrada L E, Reyes T B, Navarro C, editors. Medicina tradicional y herbolaria a 500. México: Universidad Autónoma Chapingo; 1995. 439 p.
6. Granziera P. Concept of the garden in Pre-Hispanic Mexico. Garden History. 2001;29:185–213.
7. Pijoan P. Aztec Medicine and ethnobotany. Offarm. 2003;22(9):128–136
8. de Sahagun F B, editor. Historia general de las cosas de la Nueva España. 1st ed. México: Editorial Porrúa; 2006. 1076 p.
9. Aguirre B G, editor. The activity of holly office of the inquisition of the New Spain. 1st ed. México: INAH; 1981. 320 p.
10. Ortiz R M, Puerto S F J, Aceves P P E. The rules of pharmaceutical exercise in Mexico Parte I. Revista Mexicana de Ciencias Farmacéuticas. 2008;39(1):12–19.
11. Hersch M P, Fierro A A, editors. The trade of medicinal plants, some significant features in the centre of Mexico. 1st ed. México: Universidad Autónoma Metropolitana; 2001. 78 p.
12. Argueta A, Gallardo V M. Handbook of plants of the traditional mexican medicine. 1st ed. México: Instituto Nacional Indigenista; 1994. 1786 p.
13. Heinrich m, Ankle A, Frei B, Weimann C, Sticher O. Medicinal plants in México: Healers consensus and cultural importance. Social Science and medicine. 1998;47:1863–1875.
14. Juarez Z N, Hernández L R, Bach H, Sanchez A E, Bach H. Antifungal activity of essential oils extracted from *Agastache mexicana* ssp. Xolocotziana and *Porophyllum linaria* against post-harvest pathogen. Industrial Crops and Products. 2015;74:178–182.
15. Estrada-Reyes R, Aguirre-Hernández E, García-Argáez A, Soto-Hernández M, Linares E, Bye R, Martinez-Vazquez M. Comparative chemical composition of *Agastache mexicana* subsp. Mexicana and *A. mexicana* subsp. Xolocotziana. Biochemical Systematics and Ecology. 2004;32(7):685–694.
16. Villa-Ruano N, Betancourt-Jiménez M G, Lozoya-Gloria E. Biosynthesis of uterotonic diterpenes from *Montanoa tomentosa* (Zoapatle). Journal of Plant Physiology. 2009;166(18):1961–1967.
17. Cotoras M, Folch C, Mendoza L. Characterization of the antifungal activity on *Botrytis cinerea* of the natural diterpenoids kaurenoic acid and 3? Hydroxy-kaurenoic acid. Journal of Agricultural and Food Chemistry. 2004;52(10):2821–2826.
18. Martinez-Velazquez M, Rosario-Cruz R, Castillo-Herrera G, Flores-Fernandez J M, Alvarez A H, Lugo-Cervantes E. Acaricidal effect of essential oils from *Lippia graveolens* (Lamiales: Verbenaceae), *Rosmarinus officinalis* (Lamiales: Lamiaceae),

and *Allium sativum* (Liliales: Liliaceae) *Against Rhipicephalus* (Boophilus) microplus (Acari: Ixodidae). Journal of Medical Entomology. 2011;48(4):822–827.

19. Hernández T, Canales M, Avila J G, Duran A, Caballero J, De Vivar A R, Lira R. Ethnobotany and antibacterial activity of some plants used in traditional medicine of Zapotitlán de las Salinas, Puebla (México). Journal of Ethnopharmacology. 2003;88:181–188.
20. García-mateos R, Soto-Hernández M, Kelly D. Alkaloids from six Erythrina species endemic to Mexico. Biochemical Systematics and Ecology. 1998;26:545–551.
21. Garín-Aguilar ME, Rámirez L J, Soto-Hernández M, Valencia del T, Martinez V M. Effect of crude extracts of *Erythrina americana* Mill. on aggressive behavior in rats. Journal of Ethnopharmacology. 2000;69:189–196
22. Beltrán-Rodriguez L, Romero-Manzanares A, Luna-Cavazos M, Vibrans, H, Manzo-Ramos F, Cuevas-Sanchez J, García-Moya E. Natural History and harvest of the bark of yellow quina *Hintonia latiflora* (Rubiaceae). Botanical Sciences. 2015;93(2):1–12.
23. Porras-Loaiza P, Jiménez-Munguía M T, Sosa-Morales M E, Palou E, López-Malo A. Physical properties, chemical characterization and fatty acid composition of Mexican chia (*Salvia hispanica* L.) seeds. International Journal of Food Science and Technology. 2014;49(2):571–577.
24. Reyes-Caudillo E, Tecante A, Valdivia-López M A. Dietary fibre content and antioxidant activity of phenolic compounds present in Mexican chia (*Salvia hispanica* L.) seeds. Food Chemistry. 2008;107(2):656–663.
25. Rojas-Sepúlveda A M, Mendieta-Serrano M, Mojica M Y A, Salas-Vidal E, Marquina S, Villarreal M L, Alvarez L. Cytotoxic podophyllotoxin type-lignans from the steam bark of *Bursera fagaroides* var. fagaroides. Molecules. 2012;17(8):9506–9519.
26. Garcia A, Delgado G. Uncommon sesquiterpenoids and new triterpenoids from *Jatropha neopauciflora* (Euphorbiaceae). Helvetica Chimica Acta. 2006;89(1):16–29.
27. García A, Delgado G. Cytotoxic cis-fused bicyclic sesquiterpenoids from *Jatropha neopauciflora*. Journal of Natural Products. 2006;69(11):1618–1621.
28. Narváez-Mastache J M, Garduño-Ramírez M L, Alvarez L, Delgado G. Antihyperglycemic activity and chemical constituents of *Eysenhardtia platycarpa*. Journal of Natural Products. 2006;69(12):1687–1691.
29. Jacobo-Herrera, N J, Jacobo-Herrera F E, Zentella-Dehesa A, Andrade-Cetto A, Heinrich M, Pérez-Plasencia C. Medicinal plants used in Mexican traditional medicine for the treatment of colorectal cancer. Journal of Ethnopharmacology. 2016;179:391–402.
30. McChesney J D, Venkataraman S K,Henri J T. Plant natural products: Back to the future or into extinction?. Phytochemistry. 2007;68(14):2015–2022.
31. Soto M, Sanjurjo M, Gonzalez M T, Cruz D, Giral F. The mexican yew (*Taxus Globosa* Sch.). potential of its exploit in taxol. Ciencia Ergo Sum. 2000;73(3):277–279.
32. Croteau R, Kutchan T M, Lewis N G. Natural products secondary metabolites. In: American Society of Plant Physiologists, editor. Biochemistry molecular biology of plants. Wiley-Blackwell, Somerset New Jersey, 1st ed. 7; 2000. pp. 1250–1318.
33. Estrada-Reyes R, López-Rubalcava C, Ferreyra-Cruz O A, Dorantes-Barrón A M, Heinze G, Moreno Aguilar J, Martínez-Vázquez M. Central nervous system effects and chemical composition of two subspecies of *Agastache mexicana:* An ethnomedicine of Mexico. Journal of Ethnopharmacology. 2014;153(1):98–110.
34. García-Pérez J, Cuéllar-Bermúdez S, Arévalo-Gallegos A, Rodríguez-Rodríguez J, Iqbal H, Parra-Saldivar R. Identification of bioactivity, volatile and fatty acid profile in supercritical fluid extracts of Mexican arnica. International Journal of Molecular Sciences. 2016;17(9):1528.
35. Castillo-Juárez I, González V, Jaime-Aguilar H, Martínez G, Linares E, Bye R, Romero I. Anti-*Helicobacter pylori* activity of plants used in Mexican traditional medicine for gastrointestinal disorders. Journal of Ethnopharmacology. 2009;122(3):402–405.

36. Pérez-Ortega G, González-Trujano M E, ángeles-López G E, Brindis F, Vibrans H, Reyes-Chilpa R. *Tagetes lucida* Cav.: Ethnobotany, phytochemistry and pharmacology of its tranquilizing properties. Journal of Ethnopharmacology. 2016;181:221–228.
37. Noge K, Becerra J X, Germacrene D A. Common sesquiterpene in the genus Bursera (Burse-raceae). Molecules. 2009;14(12):5289–5297.
38. Castillo-Herrera G A. Extraction method that enriches phenolic content in oregano. Journal of Food Process Engineering. 2007;30:661–669.
39. Mammo F, Endale M. Recent trends in rapid dereplication of natural product extracts: An update. Journal of Coastal Life Medicine. 2015;3(3):178–182.
40. Olivera Ortega A G, Soto-Hernández M, Martínez-Vázquez M, Terrazas-Salgado T, Solares-Arenas F. Phytochemical study of cuachalalate (*Amphiptherygium adstringens*, Schiede ex Schlecht). Journal of Ethnopharmacology. 1999;68:109–113.
41. Andrade-cetto A, Heinrich M. Mexican plants with hypoglycaemic effect used in the treatment of diabetes. Journal of Ethnopharmacology. 2005;99:325–348.
42. Linares E, Bye R. Traditional markets in Mesoamerica: A mosaic of history and traditions. In: Lira R, Casas A, Blancas J, editors. Ethnobotany of Mexico. 1st ed. New York: Springer; 2016. pp. 151–178.
43. Jiménez A, editor. Herbolaria mexicana. 2nd ed. México: Biblioteca básica de Agricultura; 2012. 531 p.
44. COMTRADE. Trade statistics for international business development [Internet]. Product 3301. Essential oils, whether or not terpeneless, incl. concretes and absolutes; resinoids; extracted. Available from: http://www.trademap.org/Product_SelProductCountry.aspx?nvpm=1|484||||33|||4|1|1|2|1|1|1|1|1 [Accessed: October, 2016]
45. Sharma A, Flores-Vallejo R del C, Cardoso-Taketa A, Villarreal M L. Antibacterial activities of medicinal plants used in Mexican traditional medicine. Journal of Ethnopharmacology. 2016;2(1):62–68.
46. López-Rubalcava C, Estrada-Camarena E. Mexican medicinal plants with anxiolytic or antidepressant activity: Focus on preclinical research. Journal of Ethnopharmacology. 2016;186:377–391.
47. Rodriguez-Chavez J L, Rufino-Gonzalez Y, Ponce-Macotela M, Delgado G. In vitro activity of "Mexican Arnica" *Heterotheca inuloides* Cass natural products and some derivatives against *Giardia intestinalis*. Parasitology. 2015;142(4):576–584.

CHAPTER 17

Culture, History and Applications of Medicinal and Aromatic Plants in Japan

Maiko Inoue[1*], Shinichiro Hayashi[2] and Lyle E. Craker[3]

[1] *Specified Nonprofit Corporation of Horticulture Therapy and Zen, Hamamatsu, Japan*

[2] *GREEN FLASK Inc., Tokyo, Japan*

[3] *University of Massachusetts, Amherst, MA, USA*

ABSTRACT

Historically, the Japanese began to use aromatic and medicinal plants for ritual activities, food flavor, and treatment of their bodies. The exotic plants, new ideas, and culture associated with medicinal and aromatic plants were introduced to Japan from other countries, primarily via Korea. In this way, experience and knowledge of uses were accumulated, and applications of aromatic and medicinal plants were expanded. The oldest Japanese medicine "Wa ho" leads the way to folk medicine today, and traditional Japanese medicine (Kampo) has spread into modern use. The elegance tradition of "Kodo," an incense ceremony of Japan, was developed because of the use of aromatic incensed wood in sixteenth century as recreation. Paired along with this ceremony is the Japanese sa-do tea ceremony that the spirituality and esthetic sense are inherited to Japanese today. Japanese green tea is becoming popular in many countries due to the constituent, catechins, that medically treats vascular disease, several cancers, and type II diabetes. Today, the Japanese medical system has new direction, integrating medicine with the adoption of modern western and alternative medicine. Scientific data must continue to be collected for interactions between the two medicinal systems for integrative medicine to be ideal for body, mind, and spirit of humans and nature.

Keywords: Kampo (traditional Japanese medicine), green tea, integrative medicine, body and spirit, phytotherapy

1. INTRODUCTION

Aromatic and medicinal plants fill a significant role in human societies that have helped improve the lives of people since ancient times. Ancient people became aware of the value and attractiveness of aromatic and medicinal plants, and the significance of historical books is a guide for the use of the plant material (**Table 1**) [1]. Initial books on use on medicinal and aromatic plants were sourced in various parts of the world, such as the Middle East, Greece, China, and India, indicating that these ancient civilizations used indigenous aromatic and medicinal plants to improve lives in their own separate ways before ideas were shared. Japan is no exception; names of some local aromatic and medicinal plants were recorded in the oldest Japanese history book "Kojiki" written in 712 A.D. Aromatic and medicinal plants, however, continue to influence human life, culture, and history. Currently, an estimated 70,000 plant species are used in traditional medicine [2].

In Japan, most aromatic and medicinal plants have been used in the crude drug system, Kampo, the traditional Japanese medicine system, and in herbal tea as alternative or complemental medicines. The production value of Kampo was $1.38 billion, accounting for only two percent of all Japanese pharmaceuticals in 2011 [3], and although a total of 83.5% of 684 medical doctors were using Kampo for their prescriptions, according to web survey by the Japan Kampo Medicines Manufacturers Association (JKMA) in 2008 [4], the market for Kampo medicines increased by 23% over 5 years (2007–2011). In 2008, the usage of crude medicine for Kampo was 20,000 metric tons, of which a total of 83 percent was imported from China [5]. Almost 250 kinds of crude (not processed) medicines are used as Kampo materials in 2008, and the top 20 are listed as example (**Table 2**) [5].

Table 1. Significant materials and books in history of medicinal plants and medicine [1].

Timeline	Book and author	Book about	Country
~500–300 BC	*Hippocratic corpus* written by Hippocrates; *De Causis Plantarum* and *Historia Plantarum* written by Theophrastis	Formalized medicine practices in diagnosis and treatments; list of medicinal plants and application	Greek
~150–100 BC	Huangdi Neijing (author unknown)	Theoretical foundation of Chinese medicine, diagnostic methods, and acupuncture	China
~100 BC-100 AC	Celsus wrote *De Medicina*; Dioscorides wrote *De Materia Medica*	Alexandrian medicine; pharmacopoeia of herbs and the medicines	Greek
~200–300 AC	*Shennong Bencaojing* (author unknown); *Shanghan Lun* written by Shang Zhongjing	Agriculture and medicinal plants; 113 herbal prescriptions and six stages of disease	China
~750–800 AC	Lu Yu wrote *The Classic of Tea*	Tea tree, making tea, and tea ceremony	China
~800–1000 AC	Avicenna writes *Book of Healing* and *Canon of Medicine*; Albucasis wrote *Kitab al-Tasrif*; Hildegard of Bingen wrote *Ohysica*; Averroes wrote *Kulliyat*	Clinical trials on medicines; Encyclopedia of medical practices; Scientific and medicinal properties of various plants; Medical encyclopedia	Arabic countries and Germany
~1000–1500 AC	Ibn al-Baitar wrote *Compendium on Simple Medicaments and Foods*	Pharmacopoeia listing 1400 plants	Arabic countries
~1500–1600 AC	John Gerard wrote *Herball, or Generall Historie of Plantes*	Heavily illustrated of 1000 plants	England
~1900–2000 AC	René-Maurice Gattefossé wrote *Aromathérapie*	Aroma and essential oil for Medicine	France

2. INTRODUCTION AND OPENING: MEDICAL SYSTEMS AND RECREATIONS

Several of the plants were discovered to have medicinal effects and were spread throughout the world along with traditional medicines, and some of were being as indigenous. At first, traditional medicines were introduced to Japan from China via Korea from the seventh to ninth centuries which is called traditional Chinese medicine (TCM). Before TCM was introduced to Japan, an older medicine called "Wa ho" existed. As TCM started flourishing, Wa ho declined and will likely remain as a folk medicine. TCM influenced traditional Japanese medicine (TJM) known as Kampo, in terms of fundamental principle and diagnosis, such as Yin-Yang (two opposites form the whole), Qi (life energy), and the Five Elements (interactions and relationships). Many ancient thin wooden strips which were excavated from a 7th century ruin recorded Chinese medicinal plant names that indicated their use as medicines In 756, a total of 60 crude medicines, for example, pepper, cinnamon, licorice, rhubarb, betel nut palm, and croton, were contributed to Shosoin, where the treasure house that belongs to Todai-ji, Nara, and 38 of them remained (**Figure 1**). At Shosoin, seven scent bags, included Borneo camphor, musk, and other fragrances from China also remained.

Table 2. Usage of top 20 crude medicines for Kampo in 2008 [5].

Ranking	Common name	Spices	Family	Parts used	usage (kg)
1	Licorice	Glycyrrhiza uralinsis	Leguminosae	Root	1,267,395
2	Chinese Peony	Paeonia lactiflora	Paeoniaceae	Root	1,164,126
3	Cinnamon	Cinnamomum cassia	Lauraceae	Bark	1,033,793
4	Indian bread, tuckahoe	Poria cocos	Polyporaceae	Fungus	996,311
5	Jujube	Ziziphus jujuba	Rhamnaceae	Fruit	675,997
6	Pinellia	Pinellia ternata	Araceae	Tuber	629,063
7	Oriental Ginseng	Panax ginseng	Araliaceae	Root	610,092
8	Angelica acutiloba	Angelica acutiloba	Apiaceae	Root	580,607
9	Ephedra	Ephedra sinica	Ephedraceae	Stem	568,686
10	glutinous starch syrup	Oryza sativa	Gramineae	Seed	555,718
11	Kudzu vine, Japanese arrowroot	Pueraria lobata	Papilionaceae	Root	553,999
12	Atractylodes lancea	Atractylodes lancea	Asteraceae	Rhizome	501,647
13	Job's Tears	Coix lacryma-jobi	Gramineae	Seed	449,253
14	Sickle Hare's Ear	Bupleurum falcatum	Umbelliferae	Root	443,811
15	Rhubarb	Rheum palmatum	Polygonaceae	Rhizome	439,590
16	Atractylodes	Atractylodes japonica	Asteraceae	Rhizome	427,357
17	Alexandrian Senna	Senna alexandrina	Caesalpiniaceae	Leaves, pods	426,230
18	Chinese foxglove	Rehmannia glutinosa	Scrophulariaceae	Root	397,512
19	Baikal skullcap	Scutellaria baicalensis	Labiatae	Root	383,969
20	Gypsum				380,348

Figure 1. Shosoin treasure house that belongs to Todai-ji, Nara (Wikipedia).

In 894, Japanese missions to Tang China were abolished by the imperial court of Japan, and the exchange of crude medicines, TCM, and culture was stopped. Thereby, medicine in Japan was developed uniquely until Sanki Tashiro (1465–1537) introduced Chinese medicine in 1498. Dosan Manase (1507–1594), Tashiro's apprentice, encouraged the more formal Chinese medicine, known as "gosei-ho," while Geni Nagoya (1625–1694), Gonzan Goto (1659–1733), Toyo Yamawaki (1705–1762), and Todo Yoshimasu (1702–1773) regarded "ko-ho" in which medicine should be clinical and adopted "*Shanghan Lun*" (*Treatise on Cold Damage Disorders*) for acute fever symptoms and "*Jin Gui Yao Lue*" (*Essential Prescriptions from the Golden Cabinet*) for pathology and treatments of various diseases which are the oldest complete clinical textbooks written by Zhang Zhongjing (150–219) [4, 6, 7]. Ko-ho put emphasis on abdominal palpation and is main stream of Kampo [8]. The word "Kampo" is used for comparison with "Rampo," a conventional western medicine introduced from the Netherlands in the 17th to 19th centuries. Kampo, however, was the main medicine in Japan until the end of the Edo era. In 1869, the Meiji administration decided that the national medicine system would be the German system, and then, in 1876, the government declared a regulation that passing a western medicine examination was necessary to become a medical doctor. This issue made Kampo usage decline rapidly, only a part of doctors, pharmacists, and medicine traders carried Kampo. After World War II, however, the side effects from using synthetic medicines happened frequently, thereby people returned to the idea of Kampo. At the same time, the Japan Society for Oriental Medicine (JSOM) was established to promote oriental medicine and contribute to the progress and dissemination of oriental medicine in 1950. The national health insurance started to cover Kampo in 1967, and therefore, the demands for Kampo increased. Finally, the medical specialist system of Kampo and pharmacist system were started in 1990 and 2000, respectively. Kampo was added to the core curriculum in medical education system in 2001. Thus, history indicates a U-turn movement for Kampo (**Figure 2**) [6, 8, 9].

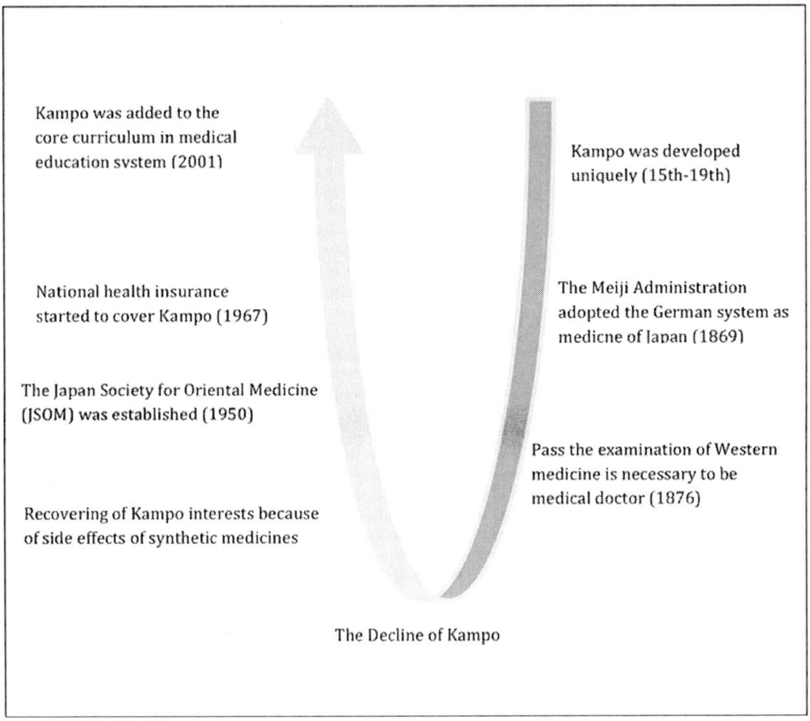

Figure 2. U-turn movement of Kampo in Japan.

According to "*Nihon Shoki*," the second oldest historical book in Japan, aloeswood (incense wood) drifted ashore on Awaji island of Japan in 595 A.D. Thus, incense in Japan started in the sixth century, with more brought from China, along with Buddhism. At first, the Japanese used incense wood in religious practices. *Ganjin*, the Chinese monk that arrived in Japan to propagate Buddhism, taught the people to make and use incense. Later, the Japanese started for transferring the elegant aroma to the robes and sutra books on Nara period. In the Heian period, aristocrats enjoyed mixing incense into their daily life with poems written about the incense, such as *The Tale of Genji* in the eleventh century [10].

In the late Muromachi period during the sixteenth century, samurai started to pursue a holistic approach, including the senses, human spirit, and nature. This new approach was the start of "*Ko-do*" (a way of incense or an elegant incense ceremony). In addition, *Ka-do* (*Ikebana*, Japanese flower arrangement) and *Sa-do* (Japanese tea ceremony). The incense, aloeswood, was made over time by the formation and ripeness of an aromatic resin. Usually, the incense wood consists of three kinds, *Kyara*, *Jinko*, and *Byakudan* (**Figure 3a**). Raw materials such as aloes wood are costly because of the limited amount available and expensive equipment (**Figure 3b**). *Ko-do* has not spread widely due to the difficulty in obtaining raw materials as compared with the materials for *Ka-do* and *Sa-do*.

Figure 3. (a) Incense wood of Jinko; (b) Incense burner (Koro).

3. CONVENTIONAL AND MODERN USES: SPIRIT AND DAILY LIFE

In the Edo era, Kampo was a medicine for wealthy people only as Kampo doctors visited a home by basket palanquin (a litter for one passenger) and treated patients using expensive Kampo materials. Therefore, the common people had to use Japanese folk medicines developed in earlier times. Such folk medicines exist today, but are not as popular as compared with modern western medicines and Kampo. Folk medicine, however, is an important cultural heritage that the Japanese ancestors left to future generations, including recipes of folk medicines, remedies for home use (**Table 3**). Some manufactured folk medicines are sold at pharmacy as quasi-drugs such as medicinal teas, cosmetics, medicinal bath, and supplements (**Figure 4a**).

Table 3. Plant materials, medicinal effects, and uses for folk medicines.

Japanese and common name	Scientific name	Medicinal effects	Uses	Part used
Hechima, Luffa	*Luffa cylindrica*	Heat rash, chapped skin, sunburn	Cosmetic	Water from fruit
Yomogi, Japanese mugwort	*Artemisia indica*	Dry skin, acne	Cosmetic	Leaf
Biwa, Loquat leaf	*Eriobotrya japonica*	Heat rash, eczema, insect bite, rash	Cosmetic	Leaf
Shiso, Perilla herb	*Perilla frutescens*	Eczema, acne, oily skin	Cosmetic	Leaf
Obako, Plantago seed	*Plantago asiatica*	Water metabolism, poor visibility, phlegm	Tea	Seed
Gobo, Burdock fruit	*Arctium lappa*	Heat diffusion, inflammation, promote eruption	Tea	Fruit
Dokudami, Houttuynia herb	*Houttuynia cordata*	Laxative, diuretic	Tea	Upper ground (flowering time)
Senna, Senna leaf	*Senna alexandrina*	Laxative	Tea	Leaf
Kuko, Lycium fruit	*Lycium chinense*	Weak constitution, fatigue, cold hands and feet, insomnia	Liquor	Fruit
Ninjin, Ginseng	*Panax ginseng*	Tonic	Liquor	Root
Toki, Japanese angelica root	*Angelica acutiloba*	Tonic, circulatory disorder, analgesic, sedative, compensate blood	Bath	Root
Senkyu, Cnidium Rhizome	*Cnidium officinale*	Anemia, cold hands and feet, menstrual irregularity	Bath	Rhizome
Kamitsure, German Chamomile	*Matricaria recutita*	Inflammation, sedative, calm spasm, excrete gas	Bath	Flower
Syobu, Calamus	*Acorus calamus*	Backache, neuralgia	Bath	Root, leaf
Yuzu, Yuzu	*Citrus junos*	Circulation improvement, cold prevention, backache, neuralgia	Bath	Fruit

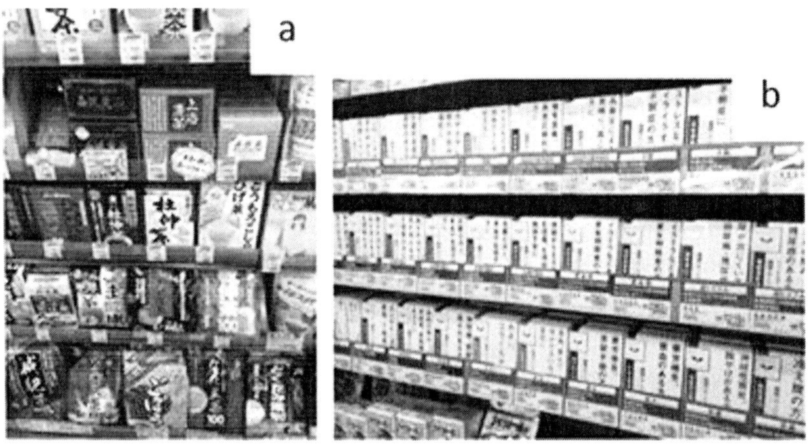

Figure 4. (a) Medicinal teas of folk medicine are sold at pharmacy; (b) Kampo medicines are sold at pharmacy.

The latest Japanese Pharmacopoeia (No. 17 issued 2016) included 324 crude medicines for Kampo [11]. Usually, Kampo medicines are a mix of several kinds of crude medicine, and each prescription has a name. Recently, many medical doctors prescribe Kampo medicines because of an increase in patients who prefer crude medicine rather than synthetic medicine. Patients can get Kampo medicines easily even without prescription (**Figure 4b**).

Aromatic and medicinal plants have spread broadly into Japanese spirit and daily life. The tea and aroma activities were simply an amusement for people until the Muromachi period that was started before the Japanese "sa-do" (tea ceremony) and "ko-do." At the end of the Muromachi period, Buddhist monks reformed the philosophy of tea to express conjointly with ethics and religion, changing the societal view about man and nature [12]. This innovative philosophy reflects the esthetics and world view of Japanese, for example, "wabi" and "sabi." Wabi expresses an esthetic of beauty of deficiency, in other words, senses which try to find sufficiency of mind from humble and deficiency. Sabi is beauty which could feel profoundness and richness inside of tranquility naturally. Subsequently, *Sen no rikyu* who is a famous tea master completed the foundation of present sa-do. The sense of wabi and sabi connects with Japanese spirit today, and sa-do is traditional practice for wholehearted hospitality for our daily life (**Figure 5**). A book "ALL ABOUT TEA" known as the encyclopedia of tea describes the commercial side of tea including the production, cultivation technique, chemical constituent, distribution, trade, consumption, and cultural side related to the as history of drinking tea and the relationships among drinking tea, the literature, and art [13]. The book helps readers to comprehend sa-do as well.

In Japan, Buddhists worship ancestors with incense sticks as one of the traditional manners in appreciating all generations of family in daily life. The incense sticks are made from mixed lots of materials, including charcoals, incensed woods, crude medicines, and fragrances. Such incense sticks are used mostly in homes and temples (**Figure 6**).

4. REPRESENTATIVE PLANT: JAPANESE GREEN TEA

Tea is the second consumed drink in the world after water. Tea is called by different names in different countries or areas. The origins, however, were split into two generally, "cha or chai" and "te or tea" depending the proximity to either China or Europe, respectively [14]. Almost all teas such as green, white, oolong, black, and pu'erh except maté and rooibos are made

Figure 5. One of the sa-do practice called "otemae.".

Figure 6. Incense sticks are offered to any Buddhist ceremony at temple.

from *Camellia sinensis*, and the significant differences come from the processing of the leaves. Oxidation in leaves starts immediately after leaves are picked. Black tea is fully oxidized; oolong tea is half of oxidized; and green and white teas are barely oxidized at all.

As well-known, drinking tea was introduced from China by a Japanese Buddhist monk in seventh to eighth century. The manner became common among aristocrats and monks in ninth century at the time, and tea was hardened tea leaves like ball named "*dan cha*." In Song dynasty of China, the way of drinking tea was changed to use matcha and introduced to Japan. Buddhist monks drank matcha to avoid sleepiness and to concentrate their mentality for Zen practices. Chinese monk *Eisai* wrote "Kissa yojoki" was written about the medical effects of tea in 1211 [15]. In early fourteenth century, a game called "*To-cha*" was in fashion, people would first drink tea and then guess the area in which the tea was grown. In the late fifteenth century, *Jyuko Murata* (1423–1502) created "wabi-cha," based on the simplicity of Zen spirit and spirituality of "wabi." Subsequently, *Sen no rikyu* established sa-do. In 1830s, the finest quality green tea was "gyokuro," which tasted mellow, sweet, and full bodied, came into being in *Uji* area.

The most famous tea production area is Sizuoka prefecture, where tea is exposed to a strong wind and rain and cultivated on the hillside or riverside where other crops could not survive (**Figure 7**). The different quality of soil produces a different color and quality of tea leaf (**Figure 7**). In 2015, the total area of picked green tea was 35,600 ha, a one percent decline The yield of fresh leaf, 357,800 metric tons in 2015, declined by four percent from the previous year in Japan [16]. According to survey, more than 82% Japanese drink green tea every day [17]. In Japan, green tea is drunk in the morning after waking, after or during every meal, for break in the afternoon, to show hospitality whenever guests visit. Usually, Japanese enjoy many kinds of flavors and tastes, for example, "*sencha*," "*gyokuro*," "*matcha*," "*hojicha*," "*genmaicha*," and "bancha" have differences that come from the processing methods of green tea (**Table 4**). The taste is a little bitterness and astringency, and it is perfectly refreshing your mouth after meals. Most distributed Japanese green tea is Sen-cha which accounts for 60% of all kinds of green tea [16].

Processing green tea is delicate and has several significant steps that require specialized skills. After the leaves are picked from the tree, a total of eight work sequences are necessary for ready to drink tea (**Table 5**).

Table 4. Different kinds of green tea.

Table 4. Different kinds of green tea.

green tea		Name	Characteristic
Leaf	color		
		sen cha	young and soft leaves are hand-rolled, commonly drunk in Japan
		gyokuro	top quality of green tea, grown under tha shade for 20-30 days before harvested
		genmai cha	green tea with roasted brown rice, popularly as an alternative standard green tea
		hoji cha	roasted green tea, good flavor of roased
		matcha	the finest tea buds of shade-grown grinded by mill stones, used in sa-do

Figure 7. Green tea farms at Kawane honcho, Shizuoka.

Table 5. Processing sequences, action, and water contents for green tea manufacture.

Name of task	Action	Time (min)	Water contents (%)
Sassei (steam)	Fresh leaves are steamed at 100°C, and oxidation is stopped	0.5–1	100
Reikyaku (cool down)	Wait until room temperature and some water is removed from leaf surface	5	–
Sojyu (first hand-rolling)	Wield leaves and some mass of leaves come up by light hand-rolling at 35°C	60	40
Jyunen (second hand-rolling)	Push and make a mass bigger by knead hand-rolling and remove some water from inside of leaf at 100°C	20	–
Chujyu (third hand-rolling)	Loosen a mass up and twist leaf by hand-rolling at 95°C	30	30
Seijyu (forth hand-rolling)	Shape like needle and polish by hand-rolling at 85°C	60	10
Kansou (dry)	Dry at 65°C	120	5
Shiage (finishing)	Removed leaf powder and piece of stem with some heat for dry	–	4

Manufacturing Japanese green tea takes more than 5 h with much labor if all sequences are done by hand. Recently, most of the processing steps have become automated. Hand making tea, however, is still practiced by some processors as the hand-rolling technique is a cultural heritage (**Figure 8a–e**).

Recently, a number of scientific studies have suggested that green tea has medicinal effects, reducing high cholesterol and treating cancer, diabetes, and liver disease [18]. In the natural elements and nutrients in Japanese green teas, the most consumed *sencha* has the highest tannin (catechin), carbohydrate, and vitamin C, E, and B1 levels among *gyokuro, sencha, hojicha*, and *matcha* (**Table 6**) [19]. Catechins are one of the major polyphenolic compounds in green tea, which works on preventing various vascular diseases (**Figure 9**). Usually, catechin is very widespread, especially in woody plants and some medicinal plants, such as Agrimonia *eupatoria* (agrimony), *Crataegus laevigata* (hawthorn), *Salix alba* (white willow), and *Vaccinium myrtillus* (bilberry) contain catechin [20]. Significant effects of the catechins are antioxidative, antihypertensive, anti-inflammatory, antiproliferative, antithrombogenic, and lipid lowering actions that are important for maintaining vascular health [21]. The major catechin, epigallocatechin-3-gallate (EGCG), has been suggested to have great potential as a cancer preventative agent for liver, stomach, skin, lung, mammary gland, and colon cancer [22]. EGCG functions not only as a powerful antioxidant and preventing oxidative damage in healthy cells, but also an antiangiogenic and antitumor agent. EGCG stimulates telomere fragmentation by inhibiting telomerase activity. Some scientific articles suggest that green tea catechin may help in controlling type II diabetes by daily consumption [23–25]. EGCG blocks sodium-dependent glucose transporter 1 (SGLY1) and lipid micelle formation in the intestine.

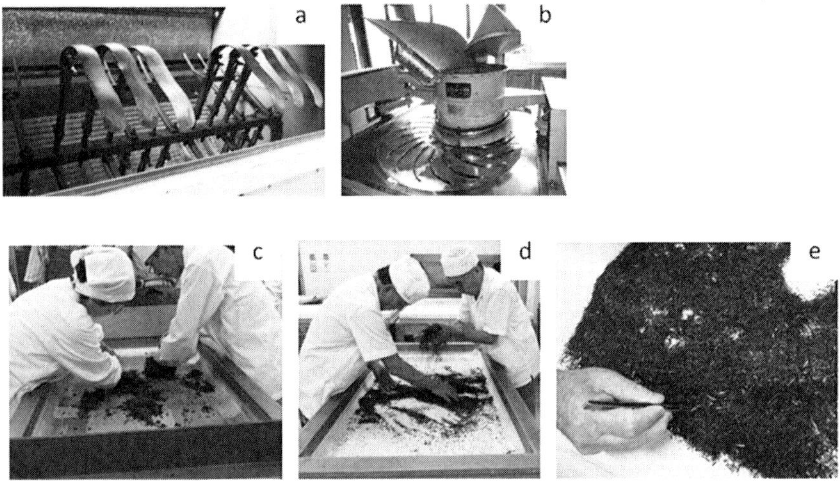

Figure 8. Processing machines, at Kawanehoncho, Shizuoka, (a) sojyu-ki, (b) jyunen-ki. Processing by hands, at Kawanehoncho, Shizuoka, (c) sojyu, (d) seijyu, and (e) shiage.

Table 6. Natural elements and nutrients in Japanese green tea [19].

	Gyokuro	Sencha	Houjicha	Matcha
Tannin (Catechin)	10.0	13.0	9.5	10.0
Caffeine	3.5	2.3	1.5	3.2
Theanine/Amino Acid	29.1	24.5	18.2	30.6
Fat	4.1	4.7	4.8	5.3
Carbohydrate	43.9	47.7	39.2	38.5
Fiber	11.1	10.6	18.7	10.0
Ash content	6.3	5.0	5.5	7.4
Vitamin A (μg)	21,000	13,000	6,700	29,000
C (mg)	110	260	44	60
E (mg)	16.6	68.1	-	28.1
B1 (mg)	0.30	0.36	0.10	0.60
B2 (mg)	1.16	1.43	0.82	1.35
Niacin (mg)	6.0	4.1	5.6	4.0

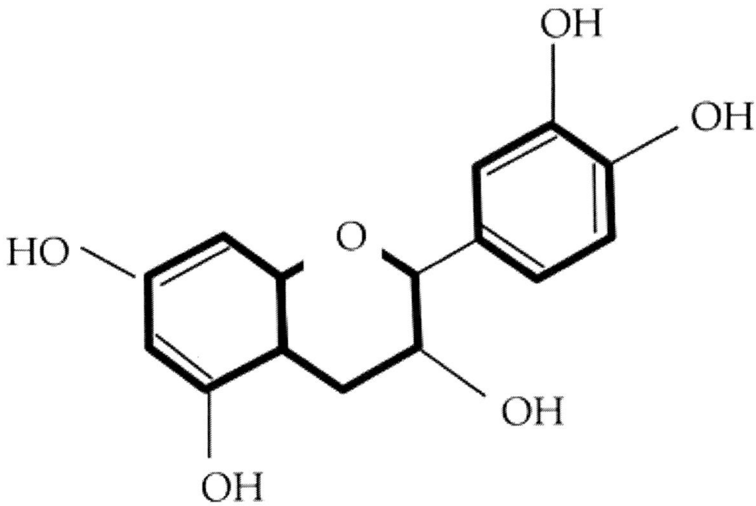

Figure 9. Catechin.

Most convenience stores and supermarkets in Japan sell many kinds of bottled green tea. Some brands put a mark "food for specified health uses" indicating healthy effects, such as cholesterol reduction or gentle rising of blood sugar after meals, and have received approval for special marking from the Consumer Affairs Agency (**Figure 10**) [26]. These types of foods are used for sustaining health and preventing pre-symptomatic disease.

Figure 10. Food mark specifies green tea health uses.

5. APPLICATION

5.1. Integrative Medicine

Integrative medicine is a person-centered care system that uses both modern western medicine, to take advantage of pharmaceuticals, operations, radiology, and other complementary and alternative medicines have become popular among advanced countries to change the structure of disease control. Many complementary and alternative medicine systems exist. Phytotherapy has been actively adapted into integrative medicine in United States because herbs and the origin of several pharmaceuticals and the accumulated scientific evidence are comprehended easily. For example, Dr. Andrew Weil of University of Arizona, a leading figure of integrative medicine, is an herbalist and prescribes medicinal plants for treatment [27].

Aromatherapy was introduced to Japan from the UK in 1985, and became popular as a relaxation technique. Subsequently, aromatherapy has become better known by medical professionals and researchers. This activity advanced clinical application in psychosomatic medicine, obstetrics and gynecology, and palliative care coupled with the spread of integrative medicine. In 2008, the Japanese Society of Integrative Medicine was established, a project team of integrative medicine in Ministry of Health, Labor, and Welfare was launched in 2010, and the Japanese Society of Phytotherapy was established in 2012 for study of medicinal herbs.

Although still controversial, functional mechanisms of complementary and alternative medicines are thought to have the improved the spontaneous healing power of humans. Currently, this understanding is compatible with oriental thought in a regimen and a balanced diet lead to a healthy body. Accumulation of scientific evidence, culture, and history exists behind the rapid spread of aromatherapy and medicinal herbs in Japan. The next generation of integrative medicine that creates inclusive correlation between body, mind, spirituality, natural environment, and local community can be expected to help those suffering from aging and improve the health of a maturing society.

5.2. Dietary Supplements

In recent years, diffusion of dietary supplements is remarkably increasing worldwide, and the trend to use natural substances in healthcare and wellness will continue to rapidly expand dietary supplement markets. In one of the highest dietary supplement consuming countries, 53% of the adults in the United States used at least one dietary supplement each day in 2003–2006 [28]. This number was more than two times that of Japan [29], and comes from different systems of health insurance. To enter health insurance is not necessary for USA citizens have a relatively expensive medical care system, making individuals concerned about their health and wellness. In contrast, Japan citizens are basically mandated to enter a health insurance. According to various reports, some consumers in Japan have the wrong knowledge about dietary supplements. For example, an inappropriate meal is no problem if you take dietary supplements [30]. Nevertheless, the Japanese market of dietary supplement keeps growing rapidly, and the prospect of dietary supplement in Japan will continue to shift more to prevent disease and maintain health and wellness.

6. CONCLUSIONS

Medicinal and aromatic plants help people remain healthy and have influenced culture, nature, and history of humans from ancient times. Further study of medicinal and aromatic plants may discover new constituents that become future medicines. In future research, more clinical trial and interaction between medicinal plants and pharmaceuticals need to be examined and useful information shared throughout the world. The important thing is to appreciate the blessings of nature and sustain all of genetic resources. Today, the society of severe aging and maturity needs to shift from animalistic world view, such as the stronger prey upon the weaker in a high-growth period, to a cooperative vegetative world view with the key factor being medicinal and aromatic plants.

REFERENCES

1. Mamedov NA, Craker LE. Man and medicinal plants: a short review. Acta Hortic (ISHS). 2012;964:181–190.
2. Farnsworth NR, Soejarto DD. Global importance of medicinal plants. In: Akereb O, Heywood V, Synge H editors. Conservation of Medicinal Plants. Cambridge University Press: Cambridge; 1991. p. 362. ISBN 0521392063, 9780521392068
3. Ministry of Agriculture, Forestry, Fisheries. Medicinal crops program of Ministry of Agriculture, Forestry, Fisheries (in Japanese). [Internet]. 2013. Available from: http://www.maff.go.jp/j/keikaku/pdf/yakuyou_sesaku1.pdf [Accessed: 2016-08-10].
4. Motoo Y, Seki T, Tsutani K. Traditional Japanese Medicine, Kampo: Its history and current status. The Chinese Journal of Integrated Traditional and Western Medicine. 2011;17(2):85–87.
5. Japan Kampo Medicines Manufacturers Association. Report of usage of crude materials (in Japanese). [internet]. 2011. Available from http://www.nikkankyo.org/topix/news/111001/shiyouryou-chousa.pdf [Accessed: 2016-08-11].

6. Sashida Y. Pharmacognocy. (in Japanese) Nankodo: Tokyo; 2002. p 394. ISBN978-4-524-40224-3
7. Hatano T. Herbal drugs in traditional Japanese medicine. In: Sakagami H, editors. Alternative Medicine. InTech: Rijeka; 2012. p 306.
8. Watanabe K, Matsuura K, Gao P, Hottenbacher L, Tokunaga H, Nishimura K, Imazu Y, Reissenweber H, Witt CM. Traditional Japanese Kampo medicine: Clinical research between modernity and traditional medicine – The state of research and methodological suggestions for the future. Evidence-Based Complementary and Alternative Medicine, Volume 2011. Hindawi Publishing Corporation: Cairo; 2011. ID 513842
9. Irie Y. Kampo Medicine for Beginners (in Japanese). Sogensha: Osaka; 2008. p. 269. ISBN978-4-422-41078-4
10. Pybus D. Kodo: The Way of Incense. Tuttle Publishing: North Clarendon; 2001. 96 p. ISBN13: 9780804832861
11. The Japanese Pharmacopeia, 17th edition. Ministry of Health, Labour and Welfare [Internet]. 2016. Available from: http://jpdb.nihs.go.jp/jp17e/ [Accessed: 2016-08-25].
12. Okakura K. The Book of Tea. Duffield & Company: New York; 1906. p. 160.
13. Ukers WH. All About Tea. The Tea and Coffee Trade Journal Company. New York; 1935. p. 559.
14. Yamada S, Tsunoyama S. Cha 108 Chapters-Culture Health Amenity (in Japanese). Fukujyuen: Kizugawa; 1990. p. 223.
15. History of green tea. HIBIKI-AN [Internet]. 2004–2016. Available from: http://www.hibiki-an.com/contents.php/cnID/17 [Accessed: 2016-09-15].
16. The area of green tea picked and the yield of fresh and *aracha* in 2015 (in Japanese). Statistics of Agriculture, Forestry and Fisheries [Internet]. 2015. Available from: http://www.maff.go.jp/j/tokei/kouhyou/sakumotu/sakkyou_kome/pdf/syukaku_1tya_15.pdf [Accessed: 2016-08-27].
17. The actual consumed condition of green tea. (in Japanese) Ministry of Agriculture, Forestry, Fisheries [Internet]. 2005. Available from: http://www.maff.go.jp/j/heya/h_moniter/pdf/h1702.pdf [Accessed: 2016-09-15].
18. Ehrlich S. Green tea. University of Maryland Medical Center [Internet]. 2015. Available from: http://umm.edu/health/medical/altmed/herb/green-tea [Accessed: 2016-08-25].
19. Green tea for health. HIBIKI-AN [Internet]. 2004-2016. Available from: http://www.hibiki-an.com/contents.php/cnID/10 [Accessed: 2016-09-15].
20. Hoffmann D, Ahgv F. Medical Herbalism: The Science and Practice of Herbal Medicine. Healing Arts Press: Rochester; 2003. p. 666. ISBN-978-089281749-8
21. Velayutham P, Babu A, Liu D. Green tea catechins and cardiovascular health: An update. Current Medicinal Chemistry. 2008;15(18):1840–1850.
22. Singh BN, Shankar S, Srivastava RK. Green tea catechin, epigallocatechin-3-gallate (EGCG): mechanisms, perspectives and clinical applications. Biochemical Pharmacology. 2011;15;82(12):1807–1821.
23. Park JH, Bae JH, Im SS, Song DK. Green tea and type 2 diabetes. Integrative Medicine Research. 2014;3:4–10
24. Fukino Y, Shimbo M, Aoki N, Okubo T, Iso H. Randomized controlled trial for an effect of green tea consumption on insulin resistance and inflammation markers. Journal of Nutritional Science and Vitaminology (Tokyo). 2005;51:335–42.
25. Iso H, Date C, Wakai K, Fukui M, Tamakoshi A, JACC Study Group. The relationship between green tea and total caffeine intake and risk for self-reported type 2 diabetes among Japanese adults. Annals of Internal Medicine. 2006;144:554–62.
26. Tanaka H, Kaneda F, Suguro R, Baba H. Current system for regulation of health foods in Japan. LMAJ. 2004;47(9):436–450.
27. Weil A. Dr. Andrew Weil's Integrative Medicine Clinic. Healing Cancer [Internet]. Available from: http://www.healingcancer.info/ebook/andrew-weil [Accessed: 2016–09-13].

28. Gahche J, Bailey R, Burt V, Hughes J, Yetley E, Dwyer J, Picciano MF, McDowell M, Sempos C. Dietary supplements use among U.S. adults has increased since NHANES III (1988-1994). NCHS Data Brief. 2011;Apr(61):1–8
29. Situation of dietary supplements in Japan. (in Japanese) Holistic Health Academy [Internet]. 2013. Available from: http://h-h-a.com/archives/417/ [Accessed: 2016-09-04].
30. Kaimoto T, Shibuya M, Maeda H. Study on dietary supplement use of university students: Comparative study between educational students and nutritional sciences students. (in Japanese) Bull. Shikoku Univ. 2012;35(B):23–27

CHAPTER 18

Romanian Aromatic and Medicinal Plants: From Tradition to Science

Radu Claudiu Fierascu[1], Irina Fierascu[1], Alina Ortan[2]*, Sorin Marius Avramescu[3], Cristina Elena Dinu-Pirvu[4] and Daniela Ionescu[5]

[1] *Specified Nonprofit Corporation of Horticulture Therapy and Zen, Hamamatsu, Japan*

[2] *GREEN FLASK Inc., Tokyo, Japan*

[3] *University of Massachusetts, Amherst, MA, USA*

ABSTRACT

From ancient times, plants have been used by humans for food, fodder, fibre and medicinal purposes. Several plants were empirically considered as treatments for a large array of illness and medical conditions. Each community had specific natural remedies, based on the geographical area, environmental conditions and other factors. Thus, the use of plants can be considered as part of the intangible cultural heritage of each community. In the geographical area of today's Romania, the ancient inhabitants, Dacians, had very good knowledge regarding the use of plants for medicinal purposes, as presented by several historical sources. The present work describes protocols for the extraction and purification of natural extracts, analytical characterisation, *in vitro* and *in vivo* evaluation of their potential applications as well as some practical examples of their application on selected Romanian native medicinal and aromatic plants. The presented results offer scientific support to their traditional use, suggesting in the same time some modern applications, for example in the nanotechnology field.

Keywords: aromatic and medicinal plants, Romanian, traditional remedy, scientific evidence

1. INTRODUCTION

1.1. Traditional Use of Medicinal and Aromatic Plants

The first human's tries to treat diseases were aiming for the environmental plants, so natural products could be considered the main mean of diseases treatment across the globe until the advent of scientific medicine. During their evolution on Earth, plants have developed the ability to synthesize certain chemical compounds for protection in the fight against world's predators, such as insects, fungi, herbivorous mammals, etc.

Although some of these compounds are toxic to predators, they proved to have beneficial effects in the treatment of human diseases. In the past centuries, the practice of empirical use of plants for therapeutic purposes was passed either in writing or orally, from generation to generation. Thus, the oldest way of treatment, phytotherapy, whose beginnings lie in Palaeolithic and whose traces are preserved in folk medicine up to our days, was developed on an empirical base, but also in the context of a magic vision of the world, where the symbol, the analogy and the correspondences principle have played and are playing an important role in the choice of useful plants [1–3].

People always appealed to nature, mainly to plants, for treating and curing various diseases; harvesting plants with everything they prove useful and their complete exploitation gave humans the opportunity to familiarize with their curative properties [4]. The first more precise data about the use of plants for healing in what is today Romania were given by the Greek doctor Discorides (doctor in Nero's army) in the five-volume treatise on plants "De Materia Medica", the precursor of all modern Pharmacopoeia and one of the most important botanical atlases in history. He pointed out that in Dacia numerous plant species were used on a large scale, making a vague and incomplete description of them. Published in 77, the book was describing 600 species of medicinal plants, of which 40 species were specific to Dacia's territory. Among these species, 27 plants have Daco-Thracian names, 8-Latin and 5-Greek, which is a confirmation of the age of phytotherapy in Romania [2, 3, 5, 6].

Over the time, the inhabitants of Romania's lands kept the continuity of rich traditions in the use of plants from spontaneous flora, which proved to be effective in curing physical or psychological suffering (injuries, fractures, bleeding, poisoning, animal bites, sunstroke, frostbite, infectious diseases, etc.).

Due to its geographic position, with a varied landscape and a climate favourable for rich vegetation, Romania is the meeting place of the Eurasian and Mediterranean flora. Here, grow more than 3600 species of higher plants, of which over 700 have become medicines, thanks to the long experience of the Romanian people in their use to cure diseases. In the Romanian plant heritage, there are numerous wild and cultivated plants that have different uses [1–9].

1.2. Short Presentation of Plants Selected for The Study

Among the Romanian traditional plants, *Heracleum sphondylium* L. (hogweed) occupies a special role due to its extremely intense vitalizing effects, therefore being called "Romanian Ginseng". The plant is known since antiquity, from the time of Plinius the Elder, under the name of *Heracleum*, derived from Heros, Heracles, Hercules, who had discovered its medical effect. It is an herbaceous plant, perennial or biannual that belongs to the family of *Apiaceae*, frequently found in meadows, scrublands, rarely in forests, from the plain region up to the mountains. The flowers are grouped in rich clusters, white or pink pal. The Romanian name of the plant comes from the leaves' shape which resembles a bear's paw. The herbal medicine uses the whole plant (especially leaves, buds, roots and seeds). The leaves and the buds are harvested from April to October; the whole plant is harvested during flowering, while the roots are harvested in spring and autumn. It is dryed in the shadow, in a thin layer. Heracleum's root and seeds quickly refresh and rejuvenate the body, increase exercise capacity and psychological tone, eliminate sterility, impotence and frigidity. It is a remedy indicated to the third age, relieving the discomforts of the premature physical decline, restoring tone and appetite for life [2, 3, 5, 6].

Anethum graveolens L. (dill) is an herbaceous, annual, terofite, vegetable, spicy, aromatic, cultivated and sub-spontaneous plant, which belongs to the *Apiaceae* family. It is known and cultivated since antiquity by Egyptians, Greeks, Romans and Gauls. In Romania, it is grown throughout the country, especially in assorted cultures. The stem, leaves and fruits have therapeutic uses in the traditional human medicine and in cult or traditional veterinary medicine. The harvesting of the stems and leaves is made when inflorescence is formed, while the harvesting of fruits is made when they are mature. From ancient times, culinary and medicinal uses were attributed to the dill. Decoction of the dill's seeds is used to treat stomach aches; the tea obtained by boiling the floriferous stems and dried inflorescences of dill is used to treat kidney diseases, hearts diseases and atherosclerosis and to increase lactation at women. Also, crushed dill's seed are used to treat intestinal worms at children. The dill is recommended in anorexia, indigestions, difficult digestion, heart diseases, urinary diseases, haemorrhoids, flatulence, various internal and external inflammations and menstrual pains [2, 5].

Taraxacum officinalis (L.) Weber ex F.H. Wigg (dandelion) is an herbaceous and perennial plant, which belongs to the *Asteraceae* family, common throughout Romania, in sunny places or in semidarkness, forests, meadows, pastures, on the roadside, from plains until subalpine area. The whole plant is harvested in spring, before and after the flowering, aerial parts or just the leaves are harvested in spring and the root is harvested in autumn or in early spring. Drying is made in the shadow, in a thin layer, in very airy places. The root and aerial parts of the plant have therapeutic uses in human and veterinary medicine, cult and traditional. Since Dacian times, the flowers and the juice resulting from squeezing dandelion's leaves were used in various skin disorders, the root was used in rheumatic diseases, the decoction of the leaves was used to treat liver diseases, while the decoction from the root was used to treat kidney diseases and against bleeding. The dandelion is recommended in liver diseases, urinary disorders, circulatory disorders, obesity, constipation, haemorrhoids, anaemia, cataracts, acne and endocrine diseases. In addition, the traditional medicine uses the plant to stimulate the pancreatic secretion, in the treatment of cancer and as an easy purgative in preventing constipations [2, 6].

Arctium lappa L. (burdock) is an herbaceous and biannual plant, which belongs to the *Asteraceae* family, common throughout Romania, from plain to mountain areas, wild lands, roadsides, along fences and flood groves. Known since antiquity by Geto-Dacians, the burdock was one of the most important and the oldest medicine practiced by our human folk medicine. The uses of the plant passed under the name of empirically remedies. Raw leaves were put on wounds and swellings; they were used to relieve the fever, to treat corns and to remove back pains. Tackling the lost hair and the stimulation of their growth was made by washing the head with broth resulting from boiling the burdock's leaves and stems. The harvesting of burdock's roots is made in spring for the 2 years plants and in autumn for the 1 year plants. After washing, the aerial plants are cleaned and dried at the sun, in thin layer. The leaves are harvested without petiole, in May–June, before flowering and dry in the shadow in one place. The roots show the therapeutically importance for the human and veterinary medicine. They have stronger and laster detoxifying effects, with an impressive range of uses, from food poisoning, skin diseases or rheumatism to glandular or metabolism disorders. In the long term, they have beneficial effects on the activity of the liver and gallbladder and they prevent diabetes and tumour diseases. In case of respiratory diseases and flu, the burdock is a valuable remedy, helping to stop the growth of the microorganisms, reducing fever and preventing complications. It is used the decoction, infusion, hot steeping, tincture and powder of burdock's root [6].

Anthriscus sylvestris (L.) Hoffm. (wild chervil) and *Anthriscus cerefolium* (L.) Hoffm. (the garden species) are herbaceous, biannual (respectively annual), terofite plant, belonging to the *Apiaceae* family, with pleasant smelling, white flowers arranged in small umbrellas, related with parsley and carrot. It is common in temperate regions, on the roadsides and forests, in pastures and hay fields. In the Romanian folk medicine, the aerial flowered parts of the plant are used, harvested in May-June, dried in a thin layer in well ventilated spaces. It has been used traditionally as a stimulant, diuretic, anti-haemorrhoidal, antipyretic, anti-inflammatory and analgesic. In the human folk medicine, it is also used in digestive disorders, having slightly laxative action, in bronchitis and chronic lung diseases, as an antitussive. Externally, it is used in inflammations of the eyelid, dermatitis and eczemas [4].

2. OBTAINING AND CHARACTERISATION OF NATURAL EXTRACTS WITH BIOMEDICAL APPLICATIONS

2.1. Extraction and Separation of Active Compounds

The concentration of biologically active compounds of natural extracts directly depends on a series of factors, such as genomic composition, biological value of the cultivar, maturity stage of the plant, climatic zone, environmental conditions, post-harvest and storage conditions, as well as *the extraction methods applied* [10]. Processing of medicinal plants can be divided into two stages: *primary processing*, which consists of plants drying, conditioning and packaging, and *advanced processing* that consists of transforming raw materials obtained from primary processing into the desired final product. Extracts may be categorized according to several parameters, as follows [10]: considering the *nature of the solvent* extracts may be: aqueous extracts (infusion or maceration-type), hydroalcoholic extracts (tincture-type), oily extracts, medicinal vinegars (macerated in vinegar) or medicinal wines (macerated in wine); considering the *obtaining method*: selective extract (targeting the compounds of interest), non-selective extracts (tincture, macerated type, etc.), extracts obtained by pressing (fruit and vegetables juice); considering the *parts of the plant* subject to extraction: partial extract (only certain anatomical parts of the plant) or total extract (whole plant); considering the *water content* of the plant (humidity): extract obtained from dried plant or fresh plant extract; considering the *method of preparation*: simple, successive or multiple extracts.

Specific biologically active compounds are obtained from the vegetal products (different parts of plants or various mixtures of aromatic and medicinal plants) using appropriate solvents. To obtain water-soluble active substance at a pH close to neutral one (such as acids, bases, salts, sugars, phenols and polyphenols, amino acids, glycosides, gums, tannins, enzymes) water is used as solvent. Given that water is not a good solvent for resins, alkaloid, and oils - type compounds, etc., for obtaining them, alkaline or acidified water can be used. By-products as volatile oils, pigments, lecithin, resins, etc., are obtained using alcohol as a solvent (alcoholic or hydroalcoholic extracts). To prepare extractive solutions, different concentrations are used, ensuring the best yield, but also different solvents, depending on the nature of the substance to be extracted or the nature of raw material. When preparing extracts, it must be taken into account the influence of the following factors: *nature of the solvent* (water for salts of alkaloids, glycosides, sugars, proteins, enzymes, tannins, etc.; alcohol 50 or 70% for volatile oils, hydrocarbons, tannins, alkaloids bases and their salts, glycosides, resins, chlorophyll, etc.; ethyl ether for alkaloid bases, resins, volatile oils, etc.; ethanol for polyphenols, reducing compounds, alkaloids, salts, amino acids, polyphenolic glycosides, sterol glycosides), *the shredding degree of plant, the ratio between the amount of plant and solvent, time and temperature*. The extraction techniques can be *batch processes* (maceration, percolation, infusion, decoction, accelerated solvent extraction, microwave-assisted extraction, supercritical fluid extraction) or *continuous processes* (organic solvents continuous extraction, continuous percolation, Soxhlet extraction) [10].

Maceration is the extraction process for slightly soluble and thermosensitive principles, which consists of treating the vegetal product with a solvent, for well-established time periods, followed by separation by filtration or decantation; maceration time depends on the type of used solvent. *Percolation* is the cold extraction process using solvents in counterflow. *Infusion* is used for the extraction of compounds that are not affected by high temperature and consists of wetting the shredded vegetal material with water, followed by the addition of boiling water and leaving them in contact for a specific period of time (usually 30 min). This process is not used for plants containing volatile oils; in this case, the shredded vegetal product is wetted with a diluted alcohol solution, followed by the method described above. *Decoction* is a technique commonly used for roots, rhizomes and bark, and it is similar with infusion process: the plant is soaked in cold water, macerated, heated water is added and then brought near boiling temperature (this method is also not suitable for plants containing high content of volatile oils). *Microwave extraction* is the technique that uses microwave energy to heat the solvent and sample, in order to increase the rate of mass transfer between the substances dissolved in sample matrix and solvent, contributing to their easier passage into the solvent [11]. *Ultrasonic extraction* involves the use of ultrasounds with varying frequencies, increasing the permeability of cell walls, thus favouring the extraction of biologically active compounds with considerably yields [12, 13]. *Supercritical fluid extraction* (alternative to classical solvent extraction) is performed with supercritical fluids (CO_2), at critical pressure (74 bar) and low temperature [14, 15]. *Continuous extraction* in special equipments (Soxhlet type) is based on a large difference between the boiling points of the solvent and the targeted compounds, representing a highly efficient method [16]. *Extraction via alcoholic fermentation* is the extraction technology based on fermentation processes and *in situ* generation of alcohol. *Continuous extraction with organic solvents* is based on the continuous solvent recycling in the mass of vegetal material. *Accelerated solvent extraction* is a method based on the use of high temperature and pressure in order to accelerate the kinetics of dissolution and to break the bonds of analyte/matrix interaction. It has the advantage of using smaller amounts of raw materials, resulting in high extraction yields [17]. *Cold extraction* methods is used for plants with thermosensitive biologically active compounds (maceration, percolation and counterflow extraction); for thermostable active principles, other methods (Soxhlet extraction, decoction, microwave or accelerated solvent extraction, etc.) are selected.

At the basis of extraction of *volatile oils* from vegetal products are their physico-chemical properties, especially high vapour pressure and solubility in non-aqueous volatile solvents and fatty substances. Vegetal materials subjected to extraction can be both fresh and dried, as a whole or fragmented (leaves, grass), crushed (groundwater bodies) or as sawdust (wood). Choosing appropriate extraction method is based on the amount of oil in vegetal product, disposing and physico-chemical properties. For example, by pressing may be processed vegetal materials with high content of volatile oil, for the ones with medium content is preferably to use steam distillation and for the ones with low content—volatile solvents or lipophilic substances extraction [18, 19]. The main methods for *obtaining* volatile oils are as follows: *hydrodistillation* has several disadvantages: esters hydrolysis caused by the high temperature, obtaining other compounds (such as coumarins), oxidation of alcohols, aldehydes and ketones; the main advantage is the direct obtaining of volatile oil; *steam distillation*—oils having superior quality are obtained, compared with the hydrodistillation; *organic solvents extraction*—especially used for obtaining thermosensitive natural compounds that cannot be obtained by steam distillation; the active principles are isolated and then the solvent is removed by distillation; *animal fat extraction* is a technique exclusively used in cosmetics, and it is applied for fresh flowers. This process enables the extraction of natural fragrances without altering the composition, and it is suitable even for small amounts of plant material [20]; *extraction with liquefied gases*—applies especially for flowers, in the cosmetics industry, using high pressure and supercritical fluids. Using this method are obtained better quality and less colourful extracts than those obtained with organic solvents, rich in waxes and fatty components [21–23]. *Pressing extraction* is applied to aromatic and medicinal plants with high content of volatile oils. It is a suitable method to obtain essential oils from citrus fruits and consists of mechanical pressing, obtaining volatile oils but also mucilage, pectin, proteins, liposoluble colorants, etc. *Adsorption on an adsorbent material* is applied for the extraction of odorants from flowers by their adsorption on a substrate such as activated charcoal, alumina, etc. *Purification and separation* of biologically active compounds can be achieved by chromatographic techniques, membrane techniques, as well as liquid-liquid extraction. *Chromatographic techniques* are effective methods of separation and purification of organic compounds, based on the components distribution of a mixture between two phases: one fixed and one mobile; *liquid-liquid extraction* is applied for the separation of compounds of interest from impurities and is based on solubility difference of the extracted component in one or more solvents immiscible or partially miscible between them. *Membrane techniques* are categorized according to the size of particles to be filtered as follows: microfiltration—the process used for the separation of particles ranging in size from 0.1 to 10 μm; ultrafiltration—used for the separation at low pressure of colloidal substances and compounds with molecular weight between 500 and 500,000 Da, such as viruses, bacteria, colloidal substances biomolecules; nanofiltration—process at low pressure through which the molecules with the size of approximately 0.001 μm are removed; reverse osmosis (hyperfiltration)—process where an important factor is the osmotic pressure [24].

2.2. Analytical Characterisation of Extracts/Active Compounds

The plants are considered in our days as established sources of pharmaceutical, aromatic and industrial compounds. Various biocompounds gives the colour, odour or therapeutic actions. Used as pure compounds [25], impregnated in different supports [26–28] or used as intermediaries (for example for nanoparticle phytosynthesis) [29], bioactive compounds offers a natural health source. The potential applications of medicinal plants are determined by their compositional profile and possible synergies between those compounds. Nevertheless, as previous stated, the composition of the natural extracts varies with a series of factors. So, a variation in the phytochemical profiles of extract of the same plant, harvested from different areas, in different seasons or using different techniques is inherent [30]. In the following paragraphs, we will present the main methods used for the analytical characterisation of natural extracts.

2.2.1. Phytochemical assays

The phytochemical assays are currently used for the preliminary assessment of the extracts composition, following some major type of compounds, such as sesqui- and monoterpenoids, phenolics, anthocyanins, flavonoids, saponins, oligomeric proanthocyanidins, flavan-3-ols, tannins, o-quinone or other parameters, such as the polyphenol index or the potential browning. *The terpenoids* (especially simpler mono- and sesquiterpenoids) represents the main constituents of the essential oils. The non-volatile di- and tri-terpenoids are usually obtained from plants, tree gums and resins [31]. *Sesquiterpene alcohols* represent substances consisting of 15 carbon atoms and contain an alcohol group. They usually accompany other compounds that are in excess in some oils, having specific antidepressant and sedative action [32]. *Sesquiterpene hydrocarbons* are compounds containing 15 carbon atoms, usually accompanying monoterpenes compounds in the volatile fraction. They possess anti-inflammatory, anti-allergic and emmenagogue action [32]. *Polyphenolic carboxylic acids* belong to the heteroside group, having an acetal structure. To this group belong a series of well-known phenolic heterosides such as rosmarinic, caffeic, gentisic, vanillic, siringic acid [32]. *Triterpenic compounds* are actually triterpenic saponozides with a pentacyclic chemical structure, with immunostimulatory, pharmacodynamic, antimicrobial, anti-inflammatory and hypoglycaemic action [32]. *Saponins* are natural occurring substances with sterol or triterpenic structure that, in colloidal solution with water, foams under stirring [32]. *Flavonoids* are natural occurring phenolic substances containing plant pigments. Their anti-inflammatory, anti-allergic and antioxidant properties are well-established [32]. *Phenolics* represent the naturally occurring compounds with one or more aromatic rings and one or more hydroxyl groups. They are the most abundant secondary metabolites of plants. Their widespread in the plant kingdom makes them also one of the most studied classes of natural occurring compounds. They are of special interest not only due to their antioxidant activity, but also due to their presence in plant foods and beverages [33]. The *anthocyanins* are pigments found in flowers, fruits, leaves, roots that change colour depending on the cell pH. The most known anthocyanins are: peonin, malvin, cyanin, rutin, etc. [32]. *Tannins* are naturally occurring plant compounds with a very complex chemical structure (comprising of several phenolic hydroxyl and carboxyl groups). They possess antidiarrheal, antifungal, antiviral and antiseptic properties [32].

The phytochemical assays usually involve a specific reaction, standard substances and spectrophotometric measurements at specific wavelengths [34]. In the following paragraphs will be presented the most common photochemical assays; it must be mentioned that other particular recipes and standards are also used in the literature. The presented recipes could also be applied for the study of essential oils, with proper dilution in alcohols.

The total sesquiterpenoids are determined through the reaction between 98% acetyl chloride, 70% perchloric acid and the extract. The determinations are performed by scanning the 350–800 nm region and measuring at 608 nm. The standard used for calibration is α-santalol, and the results are presented as μg/mg of extract [34, 35]. *The total monoterpenoids* determination involves the reaction of the extract with 2% vanillin-H_2SO_4 reagent, the heating and cooling of the mixture and reading the absorbance at 608 nm. The most common standard is linalool [34, 36]. The results are presented as μg linalool equivalents. One of the earliest phytochemical assays, the *total phenolics content,* is also the most encountered in the literature. The determination involves the reaction between extract, Folin-Ciocalteu reagent and sodium carbonate. The absorbance is measured at 765 nm with a gallic acid calibration curve [34, 37–39]. The results are presented as gallic acid equivalents (GAE) or as percent [38]. *The total anthocyanins* are determined through the reaction of the extract, n-butanol:HCl (95:5) and 2% $NH_4Fe(SO_4)_2$ solution in 2 M HCl, with the absorbance reading at 550 nm. The standard used is cyanidin chloride [34, 40]. *The total flavonoids* content is determined by the reaction of the extract with ethanol, aluminium chloride (10%), 1 M potassium acetate in bidistilled water [39, 41]. The absorbance is recorded at 415 nm and rutin is used as standard. The results are presented as rutin equivalent. *The total saponins* are determined by the reaction between the extract, 8% vanillin in ethanol and 72% H_2SO_4; the absorbance is recorded at 544 nm, using as standard sapogenin [34]. The results are presented as mg saponins/g dry weight. The determination of *oligomeric proanthocyanidins* involves the reaction of the extract 0.5% vanillin reagent and 4% HCl in methanol. The absorbance is recorded at 500 nm, and catechin is used as standard [34]. The results are presented as catechin equivalents. *Determination of total polyphenols and casein/BSA(bovine serum albumin)/PVPP (polyvinylpolypyrrolidone)-bound tannins*—the extract diluted with water reacts with BSA or casein or PVPP; polyphenols are determined as above for total phenolics; absorbance is recorded at 720 nm on a catechin calibration curve and the *bound tannins* = (total polyphenols) – (unbound polyphenols); the result are presented as mg catechin/g dry plant extract [34]. *Total flavan-3-ols* are determined from the reaction between the extract and p-dimethylaminocinnamaldehyde (DMACA) (0.1% in 1N HCl in methanol); the absorbance is measured at 640 nm; catechin is used as standard and the results are presented as mg catechin/g dry plant extract [34, 42]. *The polyphenol index, potential browning and soluble o-quinone content* is determined following the same recipe as presented for the total polyphenols and casein/BSA/PVPP-bound tannins, reading the absorbance at 280 nm (for polyphenol index), 320 nm (for potential browning) and 437 nm (soluble *o*-quinone) [34].

2.2.2. Chromatographic methods

The main objective of the analytical methods is to determine both quantitatively and qualitatively the target compounds from plant extract. It is a difficult task since generally an extract contain numerous compounds (some of them being highly labile) with a broad range of polarities, volatility, molecular weight and quantities. Therefore, it is unrealistic to believe that a complete evaluation of an extract can be performed using a single method. Several aspects must be taken into consideration for appropriate selection of analytical methods: what type of information we need from the sample, amount of sample available for analysis, relative quantities of different components present in the sample. Due to the extract complexity, the analytical tools fitted for this task fall mainly in the chromatographic methods area: thin-layer chromatography (TLC) [43], gas chromatography (GC) [44], high-performance liquid chromatography and capillary electrophoresis (CE) [45]. These techniques show a good performance in plant extract analysis due to simple treatment required for samples and diversity of detectors corresponding to different molecules properties (**Table 1**).

Table 1. Characteristics of the main chromatographic techniques used for the study of natural extracts.

Methods	Target biomolecules	Sample preparation	Detector
GC	Non-polar, thermostable	Difficult, derivatisation	FID, TCD, NPD, MS, HePD
TLC	Large range	Simple	Colour reagents
HPLC	Polar	Simple	UV-VIS, RI, MS, ELSD, NMR, fluorescence
CE	Ionic	Simple	UV, MS

Chromatographic methods represent a suitable choice for nearly all biomolecules that come across in plant extracts. These methods have the advantages of high specificity and also allow us to determine a large number of compounds in a single analysis and to benefit from a high dynamic range. Impact of gas chromatography in biomolecules analysis is somewhat hindered by low thermostability of some of these species which involve the necessity of time consuming and expensive pretreatment of samples, like derivatisation [46, 47].

TLC has been improved in the last years and has a large potential to determine some group of compounds using different detection methods and colouring reagents. Although high-performance thin-layer chromatography (HPTLC) has a good efficiency, the limiting factors in its use remain the low resolution and reproducibility [48, 49]. In contrast, *HPLC* use increase tremendously in the last years due to high efficiency and flexibility. Some certain advantages over other methods are as follows: different separation techniques, large range of HPLC columns suitable for all type of extracts, optimization of measurement using various mobile phases, gradient and isocratic approach extend the options during analysis, reverse phase techniques allow simultaneous analysis of compounds with large differences in properties. In addition, HPLC can be used as preparative or purification method, most analysis require ambient temperature, in most cases analysis are performed in short time interval, large number of detectors and the possibility to be connected in series are available.

Counter-current chromatography (CCC) is a special form of chromatography using only liquid-liquid partition without solid support [50]. This approach give the possibility to overcome some of the classic techniques drawbacks such us: column deactivation, samples contamination, etc. CCC was constantly improved during the years in terms of efficient separation in a short period of time hence the recent development in this field are focused on high-speed CCC (HSCCC) with two main technology high performance centrifugal partition chromatography (HPCPC), and fast centrifugal partition chromatography (FCPC). The performance of these methods allows the separation of plant extracts with complex matrix such us: separation of non-polar compounds, isolation of vitamin B12, chiral resolution of carboxylic acids, complete resolution of isoleucine, DCCC of anthocyanins, etc.

A further improvement of analytical methods in the biological samples is represented by hyphenation techniques, which are a combination or coupling of two or more analytical techniques using an appropriate interface. Most common link between techniques is represented by separation methods (chromatography) with an online spectroscopic detection technology (mass spectrometry or NMR). Also hyphenation techniques can be extended in both ways: in the separation parts (two separation methods) or in the spectrometry zone (two or more spectrometry methods) such us: SPE-LC-MS, LC-PDA-MS, LC-MS-MS, LC-NMR-MS. Albeit these techniques are more expensive than former methods the advantages overcome the capital costs: fast analysis, better automation, large number of sample processed in a period of time, higher reproducibility, less contamination, etc.

Most common and established hyphenation technique is *GC-MS* with a large number of improvements comparing with the components techniques used separately: direct transfer of components from GC to MS, accurate chemical identification of target compounds, possibility to expand analytical capability using the same equipment (GC-MS-MS) or even GC-MS^n). However, the same restrictions apply for biomolecules as in the case of using GC combined with classic detectors: cannot detect non-volatile, polar and thermally labile compounds, requires time consuming sample preparation (hydrolysis/derivatisation).

LC-MS is another relatively frequently used method that became more and more used in biological samples analysis. LC-MS advantages compared to GC-MS are evident: higher sensitivity and specificity, easy sample preparation (aqueous matrix is frequent but forbidden in GC or GC-MS), co-eluting compounds can be more easily separated, can be applied to detect non-volatile, polar and thermally labile compounds, several mass analysers can be used: quadrupole ion traps, time of flight (TOF), time of flight reflection (TOFR) and ion cyclotron resonance (ICR). Among major disadvantages there are lack of mass spectral libraries, hindering of the analyses by matrix effect, high capital costs and the need for qualified operators [51].

A novel hyphenation technique is represented by *LC-HMRS* (liquid chromatography-high resolution mass spectrometry) with sensitivity sub pictogram level that distinguishes compounds with the same nominal mass but different exact mass, avoids development of time-consuming protocols like in LC-MS/MS and requires operators with less experience. The main advantage of *continuous-flow HPLC-NMR* [52] is its simplicity, although the technique is less sensitive than other hyphenated methods previously described. This is mostly due to the limited residence time of the analytes in the flow cell, but also because the volume of most HPLC peaks exceeds the volume of the NMR flow-cell, thereby placing much of the analyte outside the cell during data acquisition. But still these methods remain an important direction in the development of methods for biological samples characterization.

2.2.3. Other methods

Besides the above presented analytical methods, other techniques are currently applied for the characterisation of natural extracts.

The *UV-Vis absorption spectroscopy* is one of the oldest methods of molecular spectroscopy very adequate in characterizing chemical extracts. This technique can be used both to obtain a general spectra of the extracts, but also for other types of measurements such phytochemicals assays. In the UV-Vis spectrophotometric measurements, the biologically active compounds present specific wavelengths (for phenolic acids adsorption shows between 220 and 280 nm, flavonoids, quinones and furanocoumarins between 290 and 420 nm, chlorophylls 600–660 nm, carotenoids between 400 and 500 nm, sesquiterpenoids at 608 nm and so on) [39, 41, 53]. *Fourier transform infrared spectroscopy* (*FTIR*) is based on the detection of molecular vibrations and is a technique particularly suitable for samples that may contain many pigments, being able to detect both organic functional groups and inorganics [54]. With a spectra library, FTIR can provide considerable information useful to the analysis, due to the presence of peaks in specific regions [29, 39, 55–57]. Very often used complementary with FTIR, *Raman spectroscopy* offers information on the molecular vibrations of a system and can be used for qualitative and quantitative determinations [58, 59]. *Nuclear magnetic resonance spectroscopy* (*NMR*) is used to determine the properties of atoms or molecules by exploiting the magnetic properties of the atomic nuclei. It is currently used for qualitative and quantitative determination on natural extracts, either by itself or combined with other techniques [60, 61]. *Matrix-assisted laser desorption/ionisation* (*MALDI*) represents a technique used in mass spectrometry that allows the analysis of thermally labile and high molecular weight compounds (synthetic polymers, biopolymers, macromolecules, etc.). Most often, for biochemistry applications, MALDI is associated with TOF (time-of-flight mass spectrometer) [62]. Less encountered, the *X-ray diffraction (XRD)* can be used for the identification and quantification of natural crystalline compounds, either from dried extracts or in liquid matrixes [63, 64].

By determining the concentrations of carbon, hydrogen, nitrogen, sulphur or oxygen, the *elemental analysis* technique can be an important step in characterizing chemical extracts, given that it can provide information that can be corroborated with other analyse results to get a complex image of the sample [65]. Very often, the elemental composition of an extract is needed, as the mineral components can enhance the therapeutic applications of plants, while other elements can be hazardous for human health [66]; also, the trace element profile can very well be used to characterize a specific soil-plant system [67]. The mineral composition of the extracts is most often quantified using *inductively coupled plasma* (*ICP*) techniques [68] (with variants *atomic/optic emission spectrometry—ICP-AES/OES* and *mass spectrometry—ICP-MS*) or *atomic absorption spectroscopy* (*AAS*) (also with several variants). Lately, with the development of the technique, *X-ray fluorescence* is also used for the elemental characterisation of extracts, having the advantages of rapid and multi-element analysis, lower analysis costs and the possibility of portable instruments [29].

2.3. *In vitro* protocols for the evaluation of natural extracts

The *in vitro* characterisation methods cover a very large area of applications. The most common application is the evaluation of the *antioxidant activity*. Among the methods for evaluating the antioxidant potential, the most encountered in the literature data are as follows: *DPPH assay*—a method based on the reduction of DPPH-free radical, quantified by absorbance reading at 517 nm [39]. The *Trolox equivalent antioxidant capacity* (*TEAC*) measures the antioxidant capacity of a given substance compared with the standard Trolox. Commonly, the antioxidant capacity is determined using the ABTS decolourisation assay [69]. *The ferric-reducing ability of plasma* (FRAP) assay measures reduction of the ferric iron induced by antioxidants. The method measures the absorption changes at 593 nm generated by the reduction of it is based on the reduction of the complex of ferric iron and TPTZ to the ferrous form at low pH [70]. *The ferrous oxide-xylenol orange* (FOX) assay directly measures the peroxide-dependent oxidation of Fe(II) to Fe(III) in acidic environment. The absorbance is measured at 560 nm [71]. The *ORAC assay* (oxygen radical absorbance capacity) measures the oxidative degradation of a fluorescent molecule (β-phycoerythrin or fluorescein) by a radical source (AAPH). The antioxidants protect the fluorescent molecule by quenching, thus inhibiting the fluorescence decay [72]. The *HORAC assay* (hydroxyl radical antioxidant capacity) is based on the oxidation of fluorescein by hydroxyl radicals, free radicals being generated by hydrogen peroxide. The hydroxyl radical mediated oxidation of fluorescein is blocked by the antioxidants present in the samples [70]. *TRAP method* (total radical antioxidant potential) is based on the influence of antioxidants on the fluorescence decay of R-phycoerythrin (R-PE) during a controlled peroxidation reaction, using as radical generator ABAP [70]. The total oxyradical scavenging capacity (TOSC) assay measures the decrease in ethylene production caused by antioxidants. In the presence of antioxidants, KMBA inhibits the thermal hydrolysis of ABAP. The determinations are carried out through gas-chromatography. In the last decades, new electrochemical methods emerged having several advantages when compared with the classical methods: sensitivity, fastness, simple and inexpensive instrumentation, small volumes of samples. The electrochemical methods cover a large area of electrochemical techniques [73].

Another very important group of *in vitro* methods is represented by the *antimicrobial assays.* All the antimicrobial techniques involve the determination of the effect of the sample against relevant strains, compared with a negative control (the solvent used for extraction) and a positive control (an appropriate antibiotic). *The agar diffusion test* (or the Kirby-Bauer antibiotic testing) represents a highly standardized assay [74]. The media used is Mueller-Hinton, at a pH between 7.2 and 7.4. The inoculation is made with a 0.5 McFarland broth culture. *The plate-hole diffusion assay* is commonly used to evaluate the antibacterial potential of natural products. The test material diffuses from pre-cut wells through agar seeded with bacteria. The antibacterial effect is evaluated by the clear zone surrounding the well [75]. *The well diffusion assay* is performed in order to establish the antimicrobial potential of extracts both on bacteria and fungi. Few drops of extract are placed in the appropriate wells and, after incubation, the zones of inhibition in millimetres are measured [75]. *The agar dilution method* involves the incorporation of extracts with varying concentrations into an agar medium (using serial two-fold dilutions), followed by inoculation with the test microbial onto the surface of the plate. The minimum inhibitory concentration is determined as the lowest at which the sample completely inhibits the microbial growth [75]. *The broth micro- or macro-dilution* represents one of the basic techniques for establishing the antimicrobial potential of natural extracts. The procedure uses two-fold dilutions of the sample in a liquid growth medium dispensed in tubes of minimum 2 mL (macrodilution) or in 96-well microtitration plate (microdilution). After inoculation, the tubes or plates are incubated under suitable conditions. The MIC determinations for this method are more difficult, so often viewing devices are used for reading microdilution tests. *The liquid-dilution method* can determine whether a sample has a cidal or static action at a certain concentration. The minimal bactericidal or fungicidal concentration (MBC or MFC) can be determined using completely inhibited dilution cultures and assessing growth (static) or no-growth (cidal) after incubation [76]. *The disc diffusion technique* represents a modified version of the Kirby-Bauer assay. It involves the use of discs impregnated with the test substance for the determination of inhibition area, by comparison with a standard substance [75]. *The agar slant scheme* is particularly useful for the storage of bacteria over extended periods. Agar slants also aids at the identification of bacteria by characteristic patterns of movement [77]. In *the cup plate method*, the test samples diffuse from a cup through an agar layer in a Petri dish; after incubation, around the cup appears zones of inhibition. Similarly, *the agar well procedure* involves the formation of wells in the agar media and the addition of samples and standard in the respective wells [77].

The *antiviral potential* of extracts can be evaluated using some cell-based assays. The most encountered techniques are plaque inhibition assay, plaque reduction assay, inhibition of virus-induced cytopathic effect (CPE), virus yield reduction assay and endpoint titration technique (EPTT). *The plaque inhibition assay* is based on the infection of a monolayer of host cells (pre-incubated with a solution containing the sample) with the targeted virus at various concentrations (expressed as plaque forming units—*pfu*); the surface of the layer is also covered with a sample-containing solution. After incubation, the plaques are stained (for example, with neutral red) and counted; the percent of inhibition is calculated by reference to negative control (without sample) [76]. *The plaque reduction assay* implies the mixing of the sample (or controls) with the viral suspension, incubation and subsequently addition of the mixture to host cells. The overlay used in this case is composed of agar or carboxymethyl cellulose. After specific periods of time, the *pfu* are measured by microscopy, fluorescent antibodies or specific dyes [78]. *The virus-induced cytopathic effect (CPE) assay* is particularly useful for the evaluation of the effect on viruses that induce CPE, but not form plaques. It is based on the detection of structural changes in host cells that are caused by viral invasion, with and without treatment with a non-toxic dose of tested substance [76]. *The virus yield reduction assay* is a two-step procedure. The cells (incubated with the tested materials) are infected with specific viruses. After incubation, the supernatants are collected and the virus titers are determined. *The endpoint titration technique* involves the determination of virus titer reduction, using 2-fold dilutions of the tested substance [76]. For the quantification of viruses, modern methods developed can complete the presented assays: tuneable resistive pulse sensing, flow cytometry, quantitative polymerase chain reaction (*qPCR*), enzyme-linked immunosorbent assay (ELISA), etc. [76].

Unlike the previous presented assays, the *antiparasitic assays* are highly species-specific. Several studies were published, especially covering the tropical infectious diseases, such as leishmaniasis, malaria, African sleeping sickness or Chagas disease [76]. In order to perform the *in vitro* assays, the models used for each illness (*Leishmania donovani*, *Plasmodium falciparum*, *Trypanosoma brucei*,respectively, *Trypanosoma cruzi*) are incubated in specific growth medium (with or without host cells) with the tested materials. By evaluating the parasite multiplication and total parasite burdens (using methods specific for each species), the results can be expressed as % reduction (compared to control), IC_{50} and IC_{90} (50% and 90% inhibitory concentration).

The *cytotoxicity assays* are commonly used in *in vitro* studies. The assays study the effect of tested substances on specific cellular lines. The cytotoxic effect can be manifested in a series of cell fates: the cells may undergo *necrosis* (premature death of cells by lysis) can stop growing and dividing (decrease in cell viability) or they can undergo *apoptosis* (controlled cell death). As the cytotoxic compounds often compromise the cell membrane integrity, *the cellular membrane integrity assays* are the most common ways to determine the cytotoxicity of specific compounds. The cellular integrity can be determined either by determining the effect produced inside the cells of specific vital dyes (for example, *trypan blue* or *propidium iodide*, dyes that are excluded from the healthy cells) [79] or by monitoring the passing to the outside of substances normally sequestered inside the cells (such as *lactate dehydrogenase—LDH*). Other methods of monitoring the cytotoxicity are the colorimetric methods (MTT assay, XTT assay, MTS assay, sulforhodamine B (SRB) assay, other water-soluble tetrazolium salts—WSTs assay, etc.) [80]. *The neutral red assay* can be also used to measure cell viability, as living cells take it up and concentrate it in the lysosomes. *The protein assay* represented an indirect measurement of cell viability, measuring the protein content of viable cells after washing of the treated plates [80].

The genotoxicity assays represents a very important parameter when proposing new drugs (either of natural origin or synthetic ones), determining aberrations in the affected cells (chromatid and chromosome gaps, chromosome breaks, chromatid deletions, fragmentation, translocation, complex rearrangements, etc.). The most used assays are the *Ames test, in vitro chromosome aberration assay, in vitro micronucleus test* and *comet assay.* The *Ames test* represents an *in vitro* assay for bacterial gene mutations using strains of *Salmonella typhimurium* developed by Ames et al. [81]. The general procedure involves the spreading of bacteria on agar plate containing small histidine amounts, allowing it to develop for a short time and have the opportunity to mutate. The surviving bacteria after histidine depletion are those that gain the histidine-producing ability trough mutation. The mutagenicity potential of the tested substance will be proportional to the number of colonies observed. In the *in vitro chromosome aberration assay*, cell cultures are exposed to the test substance, with and without an exogenous source of metabolic activation. At specific intervals, the cells are treated with a metaphase-arresting substance, harvested, stained and analysed in order to determine the presence of chromosome aberrations [82]. For the *in vitro micronucleus test*, cell cultures are exposed to the test substance, with and without an exogenous source of metabolic activation. The cells are grown to allow the formation of micronuclei in interphase cells, harvested and analysed for the presence of micronuclei. For the *comet assay* or single cell gel electrophoresis, cells are embedded in agarose on a microscope slide and are lysed, followed by electrophoresis staining and observation by fluorescence microscopy. A particularly interesting application of the cytotoxicity assays is the evaluation of *anticancer potential* of natural products. In order to establish the anticancer effect, the previously presented assays for cell viability are applied on a series of specific cancerous cells, like bladder cancer (i.e. UM-UC-3), breast cancer (MCF-7), colon cancer (HCT116), cervical cancer (HeLa), gastric cancer (BGC823), kidney cancer (M-1), leukaemia (Jurkat), lung cancer (SHP-77), melanoma (A375), pancreatic adenocarcinoma (SW-1990), ovarian cancer (A2780) and others. Many studies present the anti-cancer effects of various natural extracts [83].

2.4. *In Vivo* Protocols for The Evaluation of Natural Extracts

When exploring crude natural extracts, the use of *in vivo* assays can prove to be problematic, as many of those bioassays can be expensive, time-consuming (turnaround time approximately 2 months) and require high-level expertise. Also, the screening of crude extracts, consisting of multiple components, can be questionable. Thus, many scientists uses the *in vitro* assays for screening the natural extracts and reserve the *in vivo* assays for purified compounds [84]. The *in vivo* methods currently used can be divided by the biological entities subjected to test into assays using plant systems and assays using animals. Several types of *plant-using assays* are available, i.e. specific-locus tests (using maize, corn, etc.) [85] or *Arabidopsis* multilocus assay [86], appropriate for genetic risk assessment. Cytogenetic and chromosomal aberration tests were developed using *Tradescantia* [87] and, respectively, onions or beans [29]. Also, DNA adducts analysis is applicable to somatic and germinal plant cell systems [88] and cucumber seeds can be used for the study of the germination inhibitory effect [39]. What all those tests have in common is the recognized quality of plants to be very goof indicators of genotoxic and cytotoxic effects.

The *animal-usingassays*, in contrast, are much more regulated by the national and international legislation. In Europe, studies involving animals for experimental and other scientific purposes have to be carried out in compliance with EU legislation (Directive 2010/63/EU) [89]. For all *in vivo* methods, the tested samples are administered to the test animals (fishes, mice, rats, etc.) at a definite dosage, described by the specific method. After a period of time, the animals are usually sacrificed and blood or tissues are used for the assay [70]. Practically, for most of the already presented *in vitro* assays, there is an *in vivo* correspondent. However, due to the legislation and moral issues raised by the use of animal *in vivo* models, the current focus is on the development of viable *in vitro* alternatives. Before any *in vivo* study can be performed, *the acute toxicity* of any tested material should be established. This is expressed as the concentration that is lethal to 50% of the test organisms in a specific time interval. The test organism could be vertebrate organisms, but for obvious reasons, very often testing on invertebrate is preferred (for example, the *Daphnia magna assay*) [41]. The animal models can be used for the evaluation of the *antioxidant potential*, in assays such as ferric reducing ability of plasma, reduced glutathione (*GSH*) estimation, glutathione peroxidase (*GSHPx*) estimation, glutathione S-transferase (*GSt*), superoxide dismutase (*SOD*) method, γ-glutamyl transpeptidase activity (*GGT*) assay, glutathione reductase (*GR*) assay, lipid peroxidation (*LPO*) assay, *LDL* assay or catalase assay [70]. For the evaluation of *anti-diabetic potential* of natural extracts, the test animals (usually rodents) are induced diabetes (i.e. with *streptozotocin*) and the variation of peak blood glucose is recorded [90]. The *anti-inflammatory potential* can be evaluated using paw-oedema assay (induced by carrageenin, dextran, kaolin, etc.) and observation of the effect of the tested material on the evolution of the oedema [91]. The *gastroprotective effect* can be determined using animals to which gastric ulcer is induced (usually using absolute ethanol). After treatment, the animals are sacrificed, and gastric tissues are collected to evaluate the ulcers and to measure enzymatic activity [92]. For the evaluation of *anti-microbial* and *anti-viral potential*, the microbial strain or virus is induced to the test animals and the effectiveness of the natural extract is subsequently evaluated [93, 94]. The *antitumor activity* can be determined on tumour bearing animals (i.e. Ehrlich ascites carcinoma-induced or Dalton's ascites lymphoma induced animals). The effect is determined using survival time, haematological studies, lipid peroxidation, solid tumour mass, etc. [95]. The *antipyretic activity* can be evaluated using animals with induced pyrexia (i.e. by Brewer's yeast). The animals' temperatures are measured at specific time intervals after treatment [96]. The *analgesic effect* can be determined using either the hot plate test (pain reflex in response to an external thermal stimulus is observed after treatment) or the acetic acid induced writhing test (after treatment, 0.7% acetic acid is administered and the constriction of abdomen, turning of trunk and extension of hind legs are observed) [96]. Other several assays can be performed using *in vivo* models, such as *subacute/subchronic oral/dermal toxicity*, *eye/dermal irritation, reproduction toxicity, chromosome aberration, micronucleus assay, carcinogenicity, aquatic toxicology, bioaccumulation*, very often provided by specific companies [97].

3. EVALUATION OF SOME TRADITIONALLY USED PLANTS

The following paragraphs will shortly present selected results published by the authors, regarding the obtaining, characterisation and application of natural extracts from selected medicinal and aromatic plants, presented in Section 1.2.

Recently, our group presented the preliminary evaluation of the crude hydroalcoholic extract (50% ethanol) obtained from the upper aerial part of *Heracleum sphondylium* L. subsp. *Sphondylium* [39]. The evaluation consisted of chemical evaluation, as well as antifungal, antioxidant and germination inhibitory properties. The chemical evaluation consisted of UV-Vis, FTIR and GC-MS analyses, as well as phytochemical assays—total phenolics, total terpenoids and total flavonoids. The tested extract revealed good antioxidant potential (determined by the DPPH assay) and good antifungal properties (evaluated by *in vitro* disc diffusion testing, against *Aspergillus niger* and *Penicillium hirsutum*, using miconazole nitrate as positive control). These findings support the traditional use of hogweed as a natural antifungal and antioxidant. The *germination inhibitory effect*, determined on seeds of *Cucumis sativus* Cornichon Wisconsin, recommends its use for the development of environmentally friendly and safe bio-herbicide. The described antifungal and germination inhibitory effects are most probably due to the content in furanocoumarins and total phenolics, as found by analytic determinations.

The essential oils extracted from various parts of *Anethum graveolens* L. (fruits, flowers and leaves), characterised by gas chromatography [98], TLC or phytochemical assays [99], were the subject of several pharamacognostic and applications studies previously published, revealing not only their composition [98, 99], consisting of flavonoids (aglycones and glycosides), hydroxycinnamic acid derivatives, coumarins, sterols, terpenoids (e.g. carvone) and mucilages, but also the antifungal activity (against *Candida albicans* [98]) determined by disk diffusion method, using as positive standard miconazole impregnated disc) and the possibilities of incorporating the essential oils in liposomal delivery vehicles [100–102]. The entrapment of *A. graveolens* essential oils in such delivery vehicles represents a very important step in developing topical pharmaceuticals capitalizing their good antimicrobial properties.

The hydroalcoholic extracts (40% alcohol) obtained from the leaves of *Taraxacum officinalis* (L.) Weber ex F.H. Wigg and roots of *Arctium lappa* L. were characterised using HPLC, phytochemical assays and absorption emission atomic spectrometry (to determine the mineral content) [7]. The content in total polyphenols and total flavone derivatives was found to be higher in the burdock extract compared with the dandelion extract. Several polyphenol carboxylic acids (caffeic, chlorogenic, ferulic, cichoric, cinaric) and flavone derivatives (rutin, quercetrin, quercetin, luteolin, apigenin) were identified in the two extracts by means of HPLC. Also, several minerals (Ca, Mg, Na, K, Mn, Fe, Zn, Cu) were quantified, minerals that could offer the extracts antioxidant properties [7]. The extracts exhibited bactericidal effects against *Escherichia coli* and *Salmonella abony enterica* [8].

Anthriscus with two variations (wild chervil—*A. sylvestris* (L.) Hoffm. and garden chervil—*A. cerefolium* (L.) Hoffm.) represented the subject of an ethnomedicinal and a phytochemical and pharmacological study [4] and of a paper dealing with the obtaining, characterisation and applications of hydroalcoholic extracts (50% ethanol) [41]. The previously mentioned review paper [4] contained the botanical and ethnomedicinal presentation of *A. sylvestris*, the phytochemical profile of its roots, aerial parts and fruits, and its possible applications, such as antitumor, anti-microbial, anti-inflammatory, antioxidant and biotechnology applications. *A. cerefolium* extracts were obtained from the aerial parts, via classical hydroalcoholic method and microwave extraction [41]. The extracts were characterised using UV-Vis, GC-MS and phytochemical assays. The sample obtained by microwave extraction was not only richer in terpenoids, flavonoids and phenolic compounds, but also had a better antifungal (against *A. niger* and *P. hirsutum* fungal lines) and antioxidant activity (as determined by the DPPH assay and a chemiluminescence method). Finally, the paper described the applications of the extracts in nanotechnology, for the phytosynthesis of silver nanoparticles. The nanoparticles obtained using the microwaves-extract had smaller dimensions (evaluated by UV-Vis, SEM and TEM) and enhanced antifungal and antioxidant properties. The phytosynthesised nanoparticles presented a relatively good stability in time; also, no evidence of toxicity exhibited by the silver nanoparticles was shown by an *in vivo* toxicity assay (*Daphnia magna* bioassay) [41].

The present work aims not to exhaustively present the obtaining and characterisation methods of natural extracts, but to bring its contribution to the field of phytochemistry, by addressing the most common methods for obtaining/characterisation of natural products, supported by some examples from our previously published works.

4. ACKNOWLEDGEMENTS

This work was partially supported by the Romanian UEFISCDI—"Partnerships in priority areas" program, project number 176/01/07/2014

(PN-IIPT-PCCA-2013-4-0953). All the authors had an equal contribution to the present work.

REFERENCES

1. Parvu C., Enciclopedia plantelor, Plante din flora Romanie (*Plants encyclopedia, Plants from Romanian flora*). Bucharest: Ed. Tehnica; 2002.
2. Scarlat M.A., Tohanescu M. Mic tratat de fitomedicina (*Small treaty of phytomedicine*). Ploiesti: World Galaxy; 2003.
3. Tudor I., Minoiu M. Plantele medicinale miraculoase din flora Romaniei (*Miraculous medicinal plants from Romanian flora*). Bucharest: Artmed; 2004.
4. Olaru O.T., Nitulescu G.M., Ortan A., Dinu-Parvu C.E. Ethnomedicinal, phytochemical and pharmacological profile of *Anthriscus sylvestris* as an alternative source for anticancer lignans. Molecules. 2005;20:15003–15022.
5. Bojor O., Popescu O. Fitoterapie traditionala si moderna (*Traditional and modern phytotherapy*). Bucharest: Fiat lux; 2001.
6. Milica C. Medicina naturista. Tainele tineretii, sanatatii si frumusetii (*Naturist medicine. Secrets of youth, health and beauty*) v.1 and 2. Iasi: Doxologia; 2013.
7. Ionescu D., Popescu M., Rizea G.D., Mihele D., Bulearca G., Ivopol M., et al. Polyphenols and minerals, antioxidants in the plants used in the natural treatment of hepatobilliary disorders. Rev. Chim. 2014;65(5):507–511.
8. Ionescu D., Predan G., Rizea G.D., Mihele D., Ivopol G., Ionita C. Antimicrobial activity of some hydroalcoholic extracts of artichoke (*Cynara scolymus*), burdock (*Actium lappa*) and dandelion (*Taraxacum officinale*). Bull. Transilvania Univ. Basov. 2013;6(55):113–120.
9. Ionescu D. Studii privind influenta unor noi preparate de natura vegetala cu actiune hepato-colecistica asupra parametrilor biochimici la animale (*Studies on the influence of new preparations of plant acting on the liver and gallbladder biochemical parameters in animals*) [PhD thesis]. Bucharest:2014.
10. Handa S.S., An overview of extraction techniques for medicinal and aromatic plants. In: Handa S.S., Khanuja S.P.S., Longo G., Rakesh D.D., editors. Extraction technologies for medicinal and aromatic plants. Trieste: Int. Centre Sci. High Technol; 2008.
11. Deng C.H., Xu X.Q., Yao N., Li N., Zhang X.M. Rapid determination of essential oil compounds in *Artemisia selengensis* Turcz by gas chromatography–mass spectrometry with microwave distillation and simultaneous solid-phase microextraction. Anal. Chim. Acta. 2006;556:289–294.
12. Chen L., Jin H., Ding L., Zhang H., Li J., Qu C., et al. Dynamic microwave assisted extraction of flavonoids from *Herba epimedii*. Sep. Purif. Technol. 2007;59(1):50–57.
13. Schinor E.C., Salvador M.J., Turatti I.C.C., Zucchi O.L.A.D., Dias D.A. Comparison of classical and ultrasound-assisted extractions of steroids and triterpenoids from three *Chresta* spp. Ultrason. Sonochem. 2004;11(6):415–421.
14. McHugh M., Krukonis V., Brenner H. Supercritical Fluid Extraction, Principles and Practice. 2nd ed. Amsterdam: Elsevier B.V.; 1994.
15. Paun-Roman G., Radu G.L. Obtinerea unor principii biologic active din plante in forma purificata si concentrate prin tehnologii avansate (*Obtaining biologic active principles from plants in a concentrated and purified form, using advanced technologies*). In: Cascaval D., Galaction A.I., editors. Biotehnologia, intre stiinta si arta. Iasi: Ed. Venus; 2007. p. 61–94.
16. Butler M.S. The role of natural product chemistry in drug discovery. J. Nat. Prod. 2004;67(12):2141–2153.

17. Fines-Neuschild S., Boucher E., de Vernal A., Gelinas Y., Leclerc P. Accelerated solvent extraction—An efficient tool to remove extractives from tree-rings. Dendrochronol. 2015;36:45–48.
18. Lawrence B.M. Commercial essential oils: Truths and consequences. In: Swift K.A.D., editor. Advances in flavours and fragrances: From the sensation to the synthesis. Cambridge: Royal Society of Chemistry; 2002. pp. 57–83.
19. Ortan A. Nanostructuri lipidice – vectori de distributie pentru principii active volatile (*Lipid nanostructures - distribution vectors for active volatile principles*). Bucharest, Romania: Ed. Printech; 2013.
20. Hui Y. H., editor. Handbook of food products manufacturing, Vol.1, Hoboken, New Jersey: Wiley; 2007.
21. Bhat S.V., Nagasampagi B.A., Sivakumar M. Chemistry of natural products. India: Springer Science & Business Media; 2005. 840 p.
22. Keville K., Green M. Aromatherapy: a complete guide to the healing art. 2nd ed. USA: Crossing Press; 2008. 256 p.
23. Garland S. The complete book of herb and spices. UK: Frances Lincoln; 2004. 288 p.
24. Paun G., Gheorghe O., Diaconu M. Curs procesare avansata a plantelor medicinale (*Course Advanced processing of medicinal plants*). 2011. Available from: http://medplanet.dbioro.eu/doc/Curs%20procesare%20avansata%20RO.pdf
25. Bucar F., Wube A., Schmid M. Natural product isolation – how to get from biological material to pure compounds. Nat. Prod. Rep. 2013;30:525–545.
26. Mustapa A.N., Martin A., Sanz-Moral L.M., Rueda M., Cocero M.J. Impregnation of medicinal plant phytochemical compounds into silica and alginate aerogels. J. Supercrit. Fluids. 2016;116:251–263.
27. Stoica R., Pop S.F., Ion R.M. Evaluation of natural polyphenols entrapped in calcium alginate beads prepared by the ionotropic gelation method. J. Optoel. Adv. Mat. 2013;15(7–8):893–898.
28. Lima S.S.P., Lucchese A.M., Araujo-Filho H.G., Menezes P.P., Araujo A.A.S., Quintans-Junior L.J., et al. Inclusion of terpenes in cyclodextrins: preparation, characterization and pharmacological approaches. Carbohydr. Polym. 2016;151:965–987.
29. Sutan N.A., Fierascu I., Fierascu R.C., Manolescu D.S., Soare L.C. Comparative analytical characterization and in vitro citogenotoxic activity evaluation of *Asplenium scolopendrium* L. leaves and rhizome extracts prior to and after Ag nanoparticles phytosynthesis. Ind. Crops Prod. 2016;83C:379–386.
30. Dhami N., Mishra A.D. Phytochemical variation: How to resolve the quality controversies of herbal medicinal products? J. Herb. Med. 2015;5(2):118–127.
31. Gutensohn M., Dudareva N. Tomato Fruits-A Platform for Metabolic Engineering of Terpenes. In: O'Connor S.E., editor. Methods in Enzymology. Synthetic Biology and Metabolic Engineering in Plants and Microbes Part B: Metabolism in Plants. 576th ed. Amsterdam: Elsevier B.V.; 2016. pp. 333–359.
32. Istudor V. Farmacognozie, Fitochimie, Fitoterapie (*Pharmacognosy, Phytochemistry, Phytotherapy*). vol. 2 ed. Bucharest: Editura Medicala; 2001.
33. Dai J., Mumper R.J. Plant phenolics: extraction, analysis and their antioxidant
34. Misra B.B., Dey D. Phytochemical analyses and evaluation of antioxidant efficacy of *in vitro* Callus extract of East Indian sandalwood tree (*Santalum album* L.). J. Pharmacog. Phytochem. 2012;1:7–16.
35. Jones C.G., Ghisalberti E.L., Plummer J.A., Barbour E.L. Quantitative co-occurrence of sesquiterpenes; a tool for elucidating their biosynthesis in Indian sandalwood, *Santalum album*. Phytochemistry. 2006;67(22):2463–2468.
36. Doneva-Sapceska D., Dimitrovski A., Bojadziev T., Milanov G., Vojnovski B. Free and potentially volatile monoterpenes in grape varieties from the Republic of Macedonia. Bull. Chem. Technol. Macedonia. 2006;25(1):51–56 .
37. Singleton V., Rossi J. Colorimetry of total phenolics with phosphomolybdic phosphotungstic acid reagents. Am. J. Enol. Viticult. 1965;16:144–158.

38. ISO 14502-1:2005, Determination of substances characteristic of green and black tea-1: Content of total polyphenols in tea-Colorimetric method using Folin-Ciocalteu reagent.
39. Fierascu R.C., Padure I.M., Avramescu S.M., Ungureanu C., Bunghez R.I., Ortan A., et al. Preliminary assessment of the antioxidant, antifungal and germination inhibitory potential of *Heracleum sphondylium* L. (*Apiaceae*). Farmacia. 2016;64(3):403–408.
40. Porter L.J., Hrstich L.N., Chan B.G. The conversion of procyanidins and prodelphinidins to cyanidin and delphinidin. Phytochemistry. 1986;25:223–230.
41. Ortan A., Fierascu I., Ungureanu C., Fierascu R.C., Avramescu S.M., Dumitrescu O., et al. Innovative phytosynthesized silver nanoarchitectures with enhanced antifungal and antioxidant properties. Appl. Surf. Sci. 2015;358:540–548
42. Arnous A., Makris D.P., Kefalas P. Effect of principal polyphenolic components in relation to antioxidant characteristics of aged red wines. J. Agricult. Food Chem. 2001;49(12):5736–5742.
43. Jesionek W., Majer-Dziedzic B., Choma I.M. Separation, identification and investigation of antioxidant ability of plant extract components using TLC, LC-MS and TLC-DPPH. J. Liq. Chromatogr. Relat. Technol. 2015;38(11):1147–1153.
44. Seenivasagan T., Paul A.V.N. Gas-chromatography and electroantennogram analysis of saturated hydrocarbons of cruciferous host plants and host larval body extracts of *plutella xylostella* for behavioural manipulation of cotesia plutellae. Ind. J. Exp. Biol. 2011;49(5):375–386.
45. Rivasseau C., Boisson A.M., Mongelard G., Couram G., Bastien O., Bligny R. Rapid analysis of organic acids in plant extracts by capillary electrophoresis with indirect UV detection: directed metabolic analyses during metal stress. J. Chromatogr. A. 2006;1129(2):283–290.
46. Akhgari A., Laakso I., Seppanen-Laakso T., Yrjonen T., Vuorela H., Oksman-Caldentey K.M., et al. Determination of terpenoid indole alkaloids in hairy roots of *Rhazya stricta* (*Apocynaceae*) by GC-MS. Phytochem. Anal. 2015;29(5):331–338.
47. Qureshi M.N., Stecher G., Sultana T., Abel G., Popp M., Bonn G.K. Determination of carbohydrates in medicinal plants-comparison between TLC, mf-MELDI-MS and GC-MS. Phytochem. Anal. 2011;22(4):296–302.
48. Sripathi S.K., Poongothai G., Lalitha P. Identification of pinitol in plants extracts by HPTLC. J. Chem. Pharmaceut. Res. 2011;3(5):544–549.
49. Turkmen Z., Mercan S., Cengiz S. An HPTLC method for the determination of oleandrin in Nerium plant extracts and its application to forensic toxicology. J. Planar. Chromatogr. - Modern TLC. 2013;26(3):279–283.
50. Wang J., Yu J., Li L., Liu C.M., Zhang Y., Wang Q. Isolation of high purity ginsenosides from plant extract of panax ginseng by high performance centrifugal partition chromatography coupled with evaporative light scattering detection. J. Liq. Chromatogr. Relat. Technol. 2013;36(5):583–590.
51. Fong S.Y., Wong Y.C., Zuo Z. Development of a SPE-LC/MS/MS method for simultaneous quantification of baicalein, wogonin, oroxylin A and their glucuronides baicalin, wogonoside and oroxyloside in rats and its application to brain uptake and plasma pharmacokinetic studies. J. Pharm. Biomed. Anal. 2014;97:9–23.
52. Staerk D., Lambert M., Jaroszewski W.J. HPLC-NMR techniques for plant extract analysis. In: Kayser O., Quax W.J., editors. Medicinal plant biotechnology: from basic research to industrial applications. Weinheim: Wiley Verlag GmbH; 2006. pp. 29–48.
53. Bunghez F., Socaciu C., Zagrean F., Pop R.M., Ranga F., Romanciuc F. Characterisation of an aromatic plant-based formula using UV–Vis spectroscopy, LC–ESI(+)QTOF-MS and HPLC-DAD analysis. Bull. UASVM Food Sci. Technol. 2013;70(1):16–24.
54. Scoog D.A., Holler F.J., Hrouch S.R. Principles of instrumental analysis. 6th ed. USA: Cengage Learning; 2006. 1056 p.
55. Gonzales G.F., Miranda S., Nieto J., Fernandez G., Yucra S., Rubio J., et al. Red maca (*Lepidium meyenii*) reduced prostate size in rats. Reprod. Biol. Endocrinol. 2005;3(6):1–9.

56. Ragavendran P., Sophia D., Raj C.A., Gopalakrishnan V.K. Functional group analysis of various extracts of *Aerva lanata* (L) by FTIR spectrum. Pharmacol. onLine 2011;1:358–364.
57. Chaudhuri C., Pathak A.K., Sevanan M. Phytochemical screening and antimicrobial activity of extracts from leaves and stem of *Ecbolium linneanum*. Bangladesh J. Pharmacol. 2011;6:84–91.
58. Edwards H.G.M., Farwell D.W., de Oliveira L.F.C., Alia J.M., Le Hyaric M., de Ameida M.V. FT-Raman spectroscopic studies of guarana and some extracts. Anal. Chim. Acta. 2005;532(2):177–186.
59. Chylinska M., Szymanska-Chargot M., Zdunek A. FT-IR and FT-Raman characterization of non-cellulosic polysaccharides fractions isolated from plant cell wall. Carbohydr. Polym. 2016;154:45–54.
60. Yan K.J., Chu Y., Huang J.H., Jiang M.M., Li W., Wang Y.F., et al. Qualitative and quantitative analyses of compound danshen extract based on 1H NMR method and its application for quality control. J. Pharm. Biomed. Anal. 2016;131:183-187.
61. Karioti A., Giocaliere E., Guccione C., Pieraccini G., Gallo E., Vannacci A., et al. Combined HPLC-DAD–MS, HPLC–MSn and NMR spectroscopy for quality control of plant extracts: The case of a commercial blend sold as dietary supplement. J. Pharm. Biomed. Anal. 2014;88:7–15.
62. Lopez-Garcia M., Garcia M.S.D., Vilarino J.M.L., Rodriguez M.V.G. MALDI-TOF to compare polysaccharide profiles from commercial health supplements of different mushroom species. Food Chem. 2016;199:597–604.
63. Martins F.T., dos Santos M.H., Coelho C.P., Barbosa L.C., Dias G.C., Fracca M.P., et al. A powder X-ray diffraction method for detection of polyprenylated benzophenones in plant extracts associated with HPLC for quantitative analysis. J. Pharm. Biomed. Anal. 2011;54(3):451–457.
64. Leonell F., Mostarda A., De Angelis L., Lamba D., Demitri N., La Bella A., et al. Proof of the structure of the *Stemodia chilensis* tetracyclic diterpenoid (+)-19-acetoxystemodan-12-ol by synthesis from (+)-podocarpic acid: X-ray structure determination of a key intermediate. J. Nat. Prod. 2016;79(4):1155–1159.
65. Rodrigues D., Freitas A.C., Sousa S., Amorim M., Vasconcelos M.W., da Costa J.P., et al. Chemical and structural characterization of *Pholiota nameko* extracts with biological properties. Food Chem. 2017;216:176–185.
66. Kostic D., Mitic S., Zarubica A., Mitic M., Velickovic J., Randjelovic S. Content of trace metals in medicinal plants and their extracts. Hem. Ind. 2011;65(2):165–170.
67. Kabata-Pendias A. Trace Elements in Soils and Plants. 4th ed. London: CRC Press; 2010.
68. Soare L.C., Visoiu E., Bejan C., Dobrescu C.M., Fierascu I., Iosub I., et al. Research on the in vitro bioaccumulation capacity of lead in some *Pteridophyte* species of the Romanian flora. Rev. Chim. 2015;12:2017–2020.
69. Rice-Evans C.A., Miller N.M. Total antioxidant status in plasma and body fluids. Meth. Enzimol. 1994;234:279–293.
70. Alam M.N., Bristi N.J., M. Rafiquzzaman. Review on *in vivo* and *in vitro* methods evaluation of antioxidant activity. Saudi Pharm. J. 2013;21:143–152.
71. Sochor J., Ruttkay-Nedecky B., Babula P., Adam V., Hubalek J., Kizek R. Automation of methods for determination of lipid peroxidation. In: Catala A., editor. Lipid peroxidation. Rijeka: InTech; 2012. d
72. Tomer D.P., McLeman L.D., Ohmine S., Scherer P.M., Murray B.K., O'Neill K.L. Comparison of the total oxyradical scavenging capacity and oxygen radical absorbance capacity antioxidant assays. J. Med. Food. 2007;10:337–344.
73. Pisoschi A.M., Cimpeanu C., Predoi G. Electrochemical methods for total antioxidant capacity and its main contributors determination: a review. Open Chem. 2015;13:824–56.
74. Bauer A.W., Kirby W.M.M., Sherris J.C., Turck M. Antibiotic susceptibility testing by a standardized single disk method. Am. J. Clin. Pathol. 1966;36:493–496.

75. Balouiri M., Sadiki M., Ibnsouda S.K. Methods for *in vitro* evaluating antimicrobial activity: a review. J. Pharm. Anal. 2016;6(2):71–79.
76. Cos P., Vlietinck A.J., Berghe D.V., Maes L. Anti-infective potential of natural products: how to develop a stronger *in vitro* proof-of-concept. J Ethnopharmacol. 2006;106:290–302
77. Ullah N., Parveen A., Bano R., Zulfiqar I., Maryam M., Jabeen S., et al. *In vitro* and *in vivo* protocols of antimicrobial bioassay of medicinal herbal extracts: a review. Asian Pac. J. Trop. Dis. 2016;6(8):660–667.
78. Schmidt N.J., Dennis J., Lennette E.H. Plaque reduction neutralization test for human cytomegalovirus based upon enhanced uptake of neutral red by virus-infected cells. J. Clin. Microbiol. 1976;4(1):61–66.
79. Riss T.L., Moravec R.A. Use of multiple assay endpoints to investigate the effects of incubation time, dose of toxin, and plating density in cell-based cytotoxicity assays. Assay Drug Dev. Technol. 2(1):51–62.
80. Fan F., Wood K.V. Bioluminescent assays for high-throughput screening. Assay Drug Dev. Technol. 2007;5(1):127–136.
81. Ames B.N., Durston W.E., Yamasaki E., Lee F.D. Carcinogens are mutagens: a simple test system combining liver homogenates for activation and bacteria for detection. Proc. Natl. Acad. Sci. 1973;70(8):2281–2285.
82. Stammberger I., Czich A., Braun K. Drug discovery and evaluation. In: Vogel H.G., Hock F.J., Maas J., Mayer D., editors. Genotoxicity. Heidelberg: Springer; pp. 829–840.
83. Kittakoop P. Anticancer drugs and potential anticancer leads inspired by natural products. Stud. Nat. Prod. Chem. 2015;44:251–307.
84. Attaway D.H., Zaborsky O.R., editors. Marine biotechnology, Vol. 1, Pharmaceutical and bioactive natural products. New York: Springer Science+Business Media; 1993.
85. Griffiths A.J.F., Miller J.H., Suzuki D.T., Lewontin R.C., Gelbart W.M. An introduction to genetic analysis. 7th ed. New York: W.H. Freeman; 2000.
86. Redei G., Acedo G., Andhu S. Sensitivity, specificity and accuracy of the *Arabidopsis* assay in the identification of carcinogens. U.S. Environ. Prot. Ag., Washington, D.C. 2002:EPA/600/D-84/105.
87. Misik M., Ma T.H., Nersesyan A., Monarca S., Kim J.K., Knasmueller S. Micronucleus assays with *Tradescantia pollen* tetrads: an update. Mutagenes. 2011;26(1):215–221.
88. Teaf C.M., Middendorf P.J. Mutagenesis and genetic toxicology. In: Williams P.L., James R.C., Roberts S.M., editors. Principles of toxicology: environmental and industrial applications. 2nd ed. Hoboken: John Wiley & Sons; 2000. pp. 293–264.
89. Directive 2010/63/EU of the European parliament and of the Council on the protection of animals used for scientific purposes. 2010.
90. Ali R.B., Atangwho I.J., Kuar N., Ahmad M., Mahmud R., Asmawi M.Z. *In vitro* and *in vivo* effects of standardized extract and fractions of *Phaleria macrocarpa* fruits pericarp on lead carbohydrate digesting enzymes. BMC Complement. Alt. Med. 2013:201313:39.
91. Lattanzio F., Greco E., Carretta D., Cervellati R., Govonic P., Speroni E. *In vivo* anti-inflammatory effect of *Rosa canina* L. extract. J. Ethnopharmacol. 2011;137(1):880–885.
92. Al Batran R., Al-Bayaty F., Abdulla M.A., Al-Obaidi M.M., Hajrezaei M., Hassandarvish P., et al. Gastroprotective effects of *Corchorus olitorius* leaf extract against ethanol-induced gastric mucosal haemorrhagic lesions in rats. J. Gastroenterol. Hepatol. 2013;28(8):1321–1329.
93. Itelima J.U., Agina S.E. *In vivo* antimicrobial activity of plant species on *Escherichia coli* O157:H7 inoculated into albino rats. World J. Microbiol. 2014;1(1):002–009.
94. Betz U.A., Fischer R., Kleymann G., Hendrix M., Rubsamen-Waigmann H. Potent *in vivo* antiviral activity of the herpes simplex virus primase-helicase inhibitor BAY 57–1293. Antimicrob Agents Chemother. 2002;46(6):1766–1772.

95. Robertson S., Narayanan N., Kapoor B.R. Antitumour activity of *Prosopis cineraria* (L.) Druce against Ehrlich ascites carcinoma-induced mice. Nat Prod Res. 2011;25(8):857–862.

96. Bhowmick R., Sarwar S., Dewan S.M.R., Das A., Das B., Uddin M.M.N., et al. *In vivo* analgesic, antipyretic, and anti-inflammatory potential in Swiss albino mice and *in vitro* thrombolytic activity of hydroalcoholic extract from *Litsea glutinosa* leaves. Biol. Res. 2014;47(1):56.

97. http://www.genpharmtox.de/downloads/Toxicology.pdf

98. Ortan A., Ferdes M., Dinu Pirvu C. Comparative microbiological activity of volatile oils from *Anethum graveolens* species. Stud. Univ. V. Goldis, Life Sci. Ser. 2012;22(4):531–535.

99. Ortan A., Popescu M.L., Gaita A.L., Dinu-Pirvu C., Campeanu G. Contributions to the pharmacognostical study on *Anethum graveolens*, dill (*Apiaceae*). Rom. Biotechnol. Lett. 2009;14(2):4342–4348.

100. Ortan A., Ferdes M., Rodino S., Pirvu C., Draganescu D. Topical delivery system of liposomally encapsulated volatile oil of *Anethumgraveolens*. Farmacia. 2013;61(2):361–370.

101. Ortan A., Dinu Parvu C., Ghica M.V., Popescu L.M., Ionita L. Rheological study of a liposomal hydrogel based on Carbopol. Rom. Biotechnol. Lett. 2011;16(S1):47–54.

102. Ortan A., Campeanu G., Dinu Pirvu C., Popescu L. Studies concerning the entrapment of *Anethum graveolens* essential oil in liposomes. Rom. Biotechnol. Lett. 2009;14(3):4411–4417.

CHAPTER 19

Medicinal Plants to Calm and Treat Psoriasis Disease

Azadeh Izadyari Aghmiuni[1*] and Azim Akbarzadeh Khiavi[2]

[1] *Department of Chemical Engineering, Ayatollah Amoli Branch, Islamic Azad University, Amol, Iran*

[2] *Pilot Biotechnology, Pasteur Institute of Iran, Tehran, Iran*

ABSTRACT

Psoriasis is a chronic and inflammatory multifactorial disease. For psoriasis treatment, topical chemical agents are applied, in spite of inefficient effects or less effectiveness. But medical plants can be one of the alternative treatment methods. In the field, herbal creams are the most used. In fact, they are helping to inhibit leukotriene formation, inflammation and blocks cyclooxygenase enzymes, then heal skin wounds due to the plant's flavonoids and tannins. The aim of this study is the making of new herbal cream for treating psoriasis. In the mentioned cream, synergistic effects of medicinal herbals extracts were evaluate on damaged skin. Some of these extracts include *Santalum album*, *Arctium lappa*, *Matricaria chamomilla*, *Glycyrrhiza glabra*, *Lavandula angustifolia*, *Avena sativa*, *Aloe barbadesis*, *Pinus eldarica* and *Cydonia seed-Mucus*. Cream was prepared by mixing water-in-oil (W/O) and was proposed to five patients who suffer from psoriasis. Results were remarkable. All five patients were satisfied from itching inhibition and skin inflammation in first week. After 2 weeks applying cream, fading skin redness and increasing skin flexibility and repair were noticeable. An important point in this cream was the mixed herbal extract with high effectiveness than each of them alone. In fact, *S. album* and *L. angustifolia* were caused softening of skin corneous *layer. Flavonoids* and tannins in *G. glabra*, *A. lappa*, *P. eldarica* and *A. sativa* are effective for treating skin lesions such as psoriasis. Polysaccharides in *A. barbadensis* and mucilage in *C. seed-Mucus* not only are healing skin wounds but also their malic acid make peeling skin dead cells. Moreover, pectin and provitamins (A) act as antioxidants and prevent damage of skin healthy cells. Herbal β-sitosterols are factor of fading skin redness and anti-itching, a-bisabolol (*M. chamomilla*) as anti-inflammation; blocks cyclooxygenase enzymes and inhibits leukotriene formation to prevent redness. In fact, this treatment cream is effective for collagen synthesis, wound improvement, epidermal-moisture maintenance, inflammation relief and boost immune system and will inhibit psoriasis common symptoms in shortest time and no side effect.

Keywords: psoriasis, herbal extract, therapeutic cream, skin inflammation

1. INTRODUCTION

In most societies, especially the third world, the importance of skin diseases is often overlooked. This is on the terms that they are usually not life threatening, but they are beauty threatening and a significant problem all over the world.

Some skin diseases are as follows: eczema, fungal/yeast infections, bacterial infections, viral infections, parasitic infections, autoimmune disease and miscellaneous skin disease [1]. Among the types of the mentioned skin diseases, psoriasis and eczema (such as seborrhoeic eczema) can be one of the varieties of autoimmune diseases that environmental factors intensify their symptoms [2–7]. Also, these diseases, especially psoriasis, are so important, throughout the world due to the direct impact on quality of life. So, 1 day of the year is allocated to it—World Psoriasis Day Consortium [8, 9].

According to NORD (National Organization for Rare Disorders) definition, psoriasis can be classified as a rare inflammatory chronic recurrent skin disease [7, 8]. It can say that 125 million people worldwide (about 2–3% of the total population) have psoriasis. Moreover, statistics of the national psoriasis foundation in 2014 showed that psoriasis prevalence in African Americans was 1.3% compared to 2.5% of Caucasians, while in Bulgaria, it was 0.2–3% [8, 10].

Psoriasis is the diseases that has great impact on patient's life (physically, psychologically and socially) and studies have shown that the impact of this disease on quality of life is similar to diseases such as diabetes, hypertension, heart diseases, the type of cancers, arthritis and depression [11, 12]. So, talking about this disease, its symptoms and treatment of common types and novel such as herbal extracts (alternative of chemical drugs), is important. But, it is firstly necessary to explain the structure and function of skin. In fact, knowing the skin structure helps better treatment disease.

2. STRUCTURE AND FUNCTIONS OF THE SKIN

2.1. Skin Definition

The definition of skin in English and Medical dictionary describes as the external layer of the body. But this description is inadequate. In fact, the skin is external covering that holds organs and protection of human tissues against environmental factors [13, 14]. In a normal state, the immune system within the skin is quiet and relatively inactive. If the skin is challenged in any way (bacteria, fungus, etc.), a range of immunological cascades are set in motion. In some conditions such as psoriasis disease, these immunological changes are not within a normal range and cause severe inflammation and hyper-proliferation of skin cells [15]. The skin plays an important role in all fields, especially in the esthetics. Also, the skin color represents the racial disparity and it is symbolic of different cultures and ethnic differences.

2.2. Biology of The Skin

The skin is the largest organs in the body, with surface area of 1.8 m^2 and making up about 16% of body weight (between 2 and 6 kg) [13, 14]. It has many functions such as a barrier to protect the body from noxious external factors and to keep the internal systems intact, protection of tissues against microorganisms, ultraviolet radiation and mechanical pressers [13, 16]. The skin is also responsible of many body functions such as sense, regulation of body temperature and elimination of waste products by sweating, producing vitamin D, etc. Moreover, skin can prevent the entry of harmful substances in the body and control vital substances of body. Therefore, skin is important to perform the mentioned essential functions.

2.3. Skin Anatomy

Skin is the dynamic organ in a constant state of change, as cells of the outer layers are continuously shed and replaced by inner cells moving up to the surface [17]. The skin contains the number of accessory organs which assist in its protective role and there are three structural layers to the skin:

- Epidermis
- Dermis
- Subcutis (hypodermis)

Also, hair, nails, sebaceous, sweat, apocrine glands, etc. are regarded as derivatives of skin (**Figure 1**). The thickness of the skin varies depending on the site, with thicker skin being present on areas of the body such as the soles of the feet and palms of the hand. In fact, this thickness depends on epidermis layers such as stratum granulosum [14, 18].

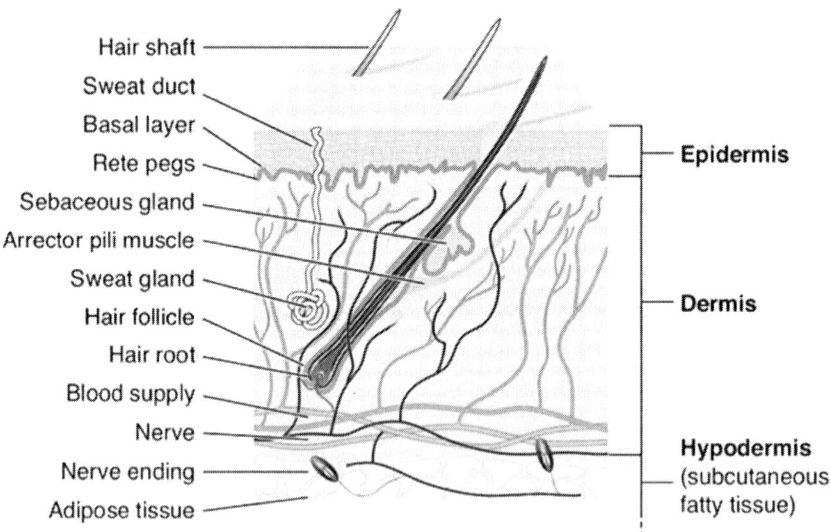

Figure 1. Cross section of the skin.

2.3.1. Epidermis

The epidermis is the top layer of skin with about 0.1–1.4 mm thick (on areas of the body). The main cells of the epidermis are keratinocytes, which produce the protein keratin. This layer is also containing other cells such as Langerhans, Melanocyte and Merkel cells. This layer is constantly proliferating and shedding millions of dead cells [14, 17]. It is estimated that normal skin sheds at the rate of a million cells (every 40 min), which equates to around 18 kg over a lifetime. This process of skin cell shedding is known as desquamation. It could be noted that desquamation (from stratum basale to stratum corneum) in epidermis is between 28 and 30 days [13, 19]. **Figure 2** shows different layers of the epidermis and **Table 1** describes them.

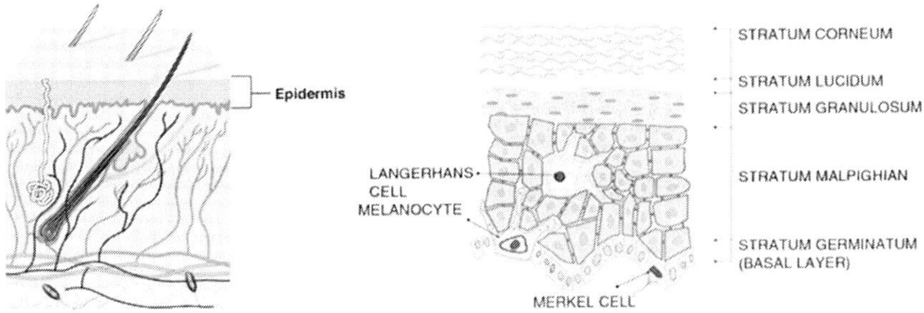

Figure 2. Epidermal layers.

Table 1. Description of epidermis layers.

The name of epidermis layers	Description
Stratum basale	The basal cell layer of the epidermis is composed keratinocytes, which are either dividing or nondividing. These cells contain keratin tonofibrils and melanocytes (to synthesize melanin). Also, Merkel cells are also found in the basal cell layer (to create a sensation in the skin such as sense of touch)
Stratum spinosum (Malpighian)	Daughter basal cells migrate upwards to form this layer of polyhedral cells, which are interconnected by desmosomes. Langerhans cells (immunologically active cells) are mostly found in this layer
Stratum granulosum	In these layers, cells become flattened and lose their nuclei. In fact, there is high lysosomal activity (to digest the cell contents and to disintegrate cells) and the keratohyalin granules become more prominent within the cell and the lipid-filled membrane coating vesicles. These lipids include 40% ceramides, fatty acids, cholesterol and cholesterol sulfate
Stratum lucidum	This layer, in which their cells are nucleated, lies between the stratum granulosum and stratum corneum and can be found in the palms and soles, where the skin is very thick
Stratum corneum	The end result of keratinocyte maturation found in the horny layer. In this layer, sheets are overlapping, which are polyhedral cornified cells and no nuclei (corneocytes)

2.3.2. Dermis

The dermis is defined as a tough supportive connective tissue matrix, with physical support and nutrients to the epidermis. Connective tissue consists of a ground substance with protein fibers that contains water and a mixture of large organic molecules (combination of polysaccharides as complex carbohydrates and proteins). The most common type of polysaccharides in this tissue is glycosaminoglycans (GAGS) that include hyaluronic acid [13, 16]. In fact, this tissue matrix is called the extracellular matrix that its structure has been shown in **Figure 3** [20].

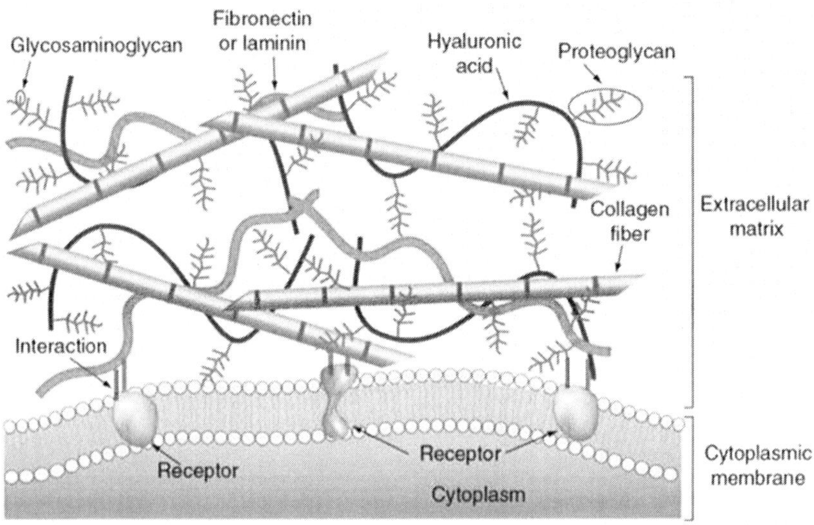

Figure 3. Model of complex 3D structure of the natural extracellular matrix and the interactions between cells and the extracellular matrix components.

Dermis found immediately below the epidermis and it varies in thickness (about 0.6 mm on the eyelids and 3 mm or more on the back, palms and soles). The two layers identified within the dermis are the papillary layer and the reticular layer (**Table 2**). Also, dermis includes elastin, fibrillin, collagen and laminin (**Figure 3**). This layer also contains nerve endings, sweat glands, sebaceous glands, hair follicles and blood vessels [17, 19]. **Figure 4** shows a cross section of dermis layer.

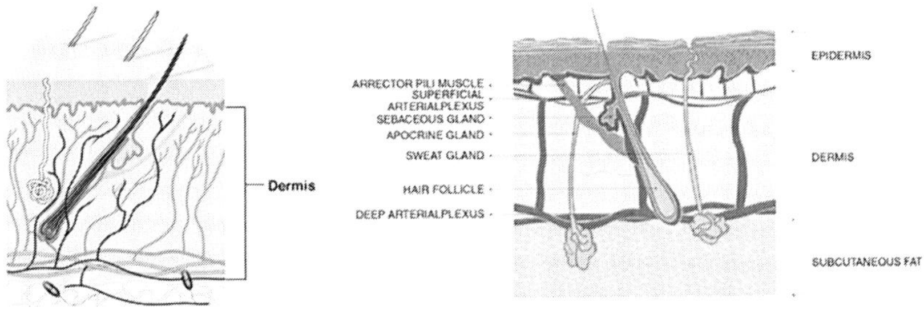

Figure 4. Dermal annexes.

Table 2. Cell types in the reticular layer of dermis.

The types of cells	Definitions
Fibroblasts cells	An important cell that is involved to repair damaged tissues
Mast cells	These cells are playing an important role to fight infection
Lymphatic vessels	The lymphatic system as body's defense system has an important role to fight infection
Epidermal appendages (rete pegs)	epidermal appendages are way to link epidermis and dermis for preventing skin damages
Ground substance	This substance as a gel-like component is supporting the dermis cells and helping to provide its structure

2.3.3. Subcutaneous tissue or the hypodermis layer

This layer lies below the dermis and provides support for it. Hypodermis is made up largely of fat cells and connective tissue for protection of internal structures. Moreover, this layer acts as a heat insulator and can be useful against trauma [13, 16, 19]. Hypodermis schematic has been shown in **Figure 5**.

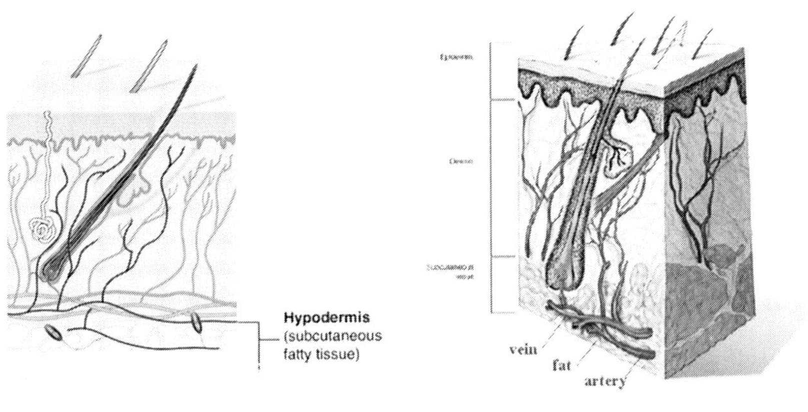

Figure 5. Hypodermis layer.

3. PSORIASIS DISEASE

Psoriasis was first distinguished and diagnosed as a chronic inflammatory disease in nineteenth century; prior to, this disease was often mistaken for other disease such as leprosy. Nowadays, psoriasis patients are estimated to be about 2–3% at all world [13, 21]. Severity ranges this disease from a fingernail pit to skin lesions (on a part of the skin to a total body that can be associated with arthritis) [13]. **Tables 3** and **4** show the types of psoriasis disease. Psoriasis plays an important role on a patient's quality of life and its negative effect can be seen on the beauty of patient and their mental and physical functioning. The cause of this vexing disease is still not known and clinical studies show hyper-proliferation and abnormal differentiation of the epidermis, although this disease can be hereditary [22, 23].

3.1. Biology of Psoriasis Disease

Psoriasis is characterized as a disease with over proliferation of keratinocytes such as skin cells (around 3–4 days) and cell development with abnormal keratinocyte differentiation. Proliferation of cells is in the basal layer doubles and the normal cell cycle (which is around 28 days) and hyperkeratosis (the hyperkeratosis leads to induration or skin thickening) and parakeratosis develop (the granular layer is either absent or reduced) [33]. The color of the psoriatic plaque may be masked by a covering of silvery, white skin scales. Also, high level of vascular activity within the psoriatic plaques, there is the change in color of the skin and inflammation, itching and redness of skin [13].

In fact, psoriasis is a hyper-proliferative disorder with significant inflammatory components and belies a complex cascade of immune reactions, finally, stimulation of the skin. Indeed, psoriasis is recognized as the most prevalent T-cell-mediated[1] - inflammatory condition in humans [13, 15, 34]. Moreover, TNF plays an important role in inflammatory processes of psoriasis (**Figure 6**) [35].

3.2. Psoriasis Treatment

Psoriasis treatments are divided into two groups [13].

- Topical

Topical therapies are generally used to manage mild-to-moderate psoriasis (**Table 5**). This type of therapy may be also used to treat more severe psoriasis in combination with systemic regimes, for

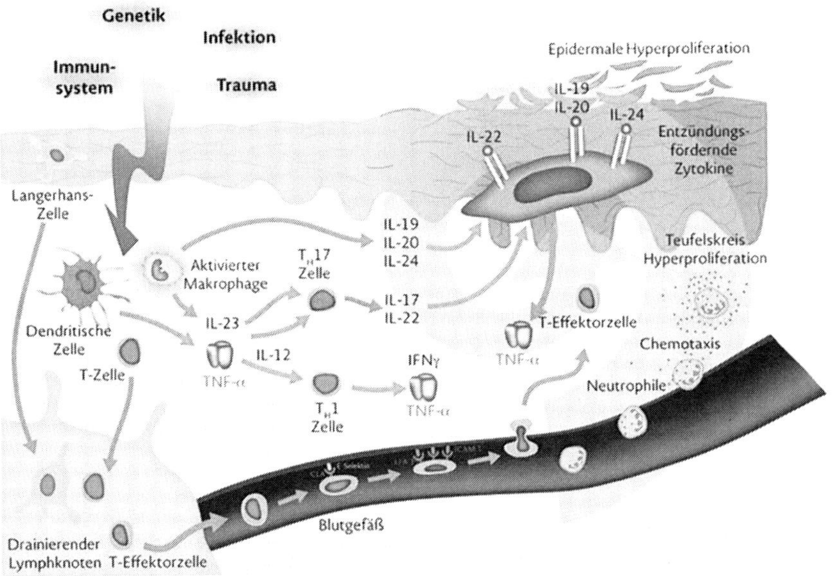

Figure 6. TNF stimulates keratinocyte proliferation and the expression of other inflammatory mediators such as IL-1 and IL-6. Also, TNF stimulates growth factor (vascular endothelial growth factor), VEGF (for angiogenesis) by the induction of adhesion molecules on the vascular endothelium and the immigration of other immune cells.

example, coal tar or short-contact dithranol in combination with UV light.

- Systemic

Systemic therapies used to the more severe spectrum of the disease (**Table 6**).

3.3. Psoriasis and Skin's Barrier Function

Although the psoriasis is a multifactorial disease, the studies show that disruption the homeostasis in skin's barrier can be one of the important factors.

Table 3. Types of psoriasis disease.

Types of psoriasis	Description of the disease	An example of disease picture	References
Skin psoriasis	This disease is one of the types of psoriasis. Skin psoriasis known as a papulosquamous disease that includes the types of vulgaris (common type of skin psoriasis), guttate (spots of drop like), inverse (in the underarms, navel, buttocks, etc.), pustular with symptoms of yellowish pus on the skin and small blisters and erythrodermic psoriasis. **Table 4.** shows completely them	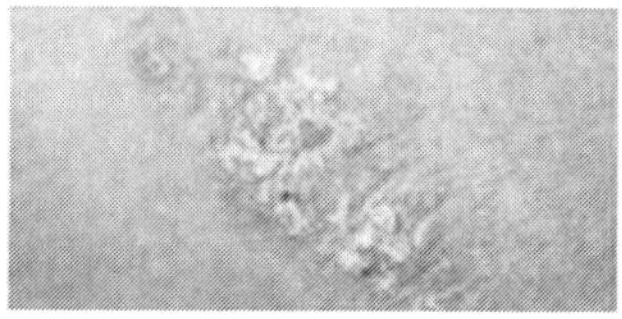	[24–27]
Nail psoriasis	The common signs of nail psoriasis are pitting and distal onycholysis. Also, pitting, yellowish discoloration and paronychia, hyperkeratosis, onycholysis and severe onychodystrophy are other signs of the disease. Approximately 50% of all patients with psoriasis develop characteristic nail changes as a clinical correlate of psoriatic inflammation of the nail matrix and/or nail bed	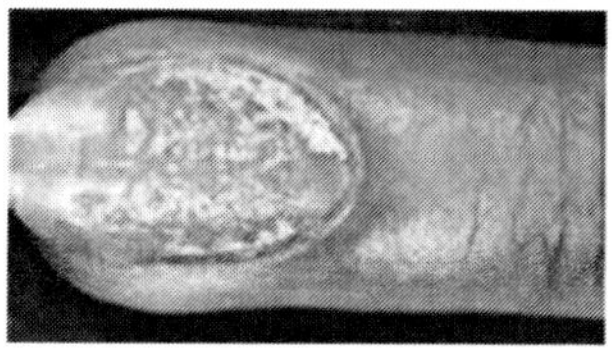	[28, 29]
Psoriatic arthritis	Psoriatic arthritis (PsA) is a chronic inflammatory joint disease occurring in 6–39% of patients with psoriasis (about 0.1–0.25% of world population). This type of arthritis can develop slowly or rapidly. Psoriasis arthritis (PsA) can be also as severe arthritis similar to rheumatoid arthritis (RA). Moreover, this disease specifications are focal bone erosions at the bones junction	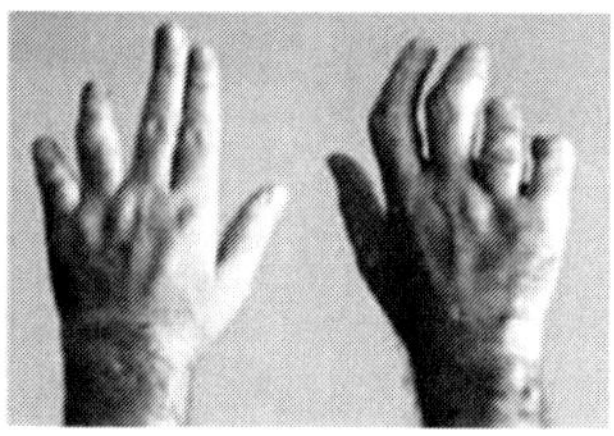	[30–32]

Table 4. Different types of skin psoriasis.

Skin psoriasis		Description of the disease	Disease pictures
Vulgaris psoriasis	1 Chronic stationary psoriasis 2 Plaque-like psoriasis 3 Scalp psoriasis 4 Seborrhoeic psoriasis	This type of psoriasis disease is known by characteristics such as inflamed and red lesions that covered by silvery white scales. These lesions typically found on the elbows, knees, scalp and lower back. Also, it is the most common type of psoriasis (about 80% of those who have psoriasis, suffer from this type)	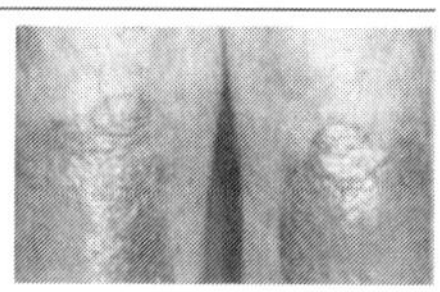 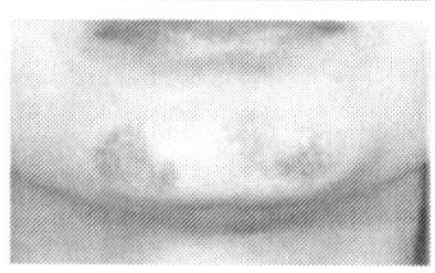
Guttate psoriasis		This type of psoriasis often starts in childhood or young adulthood. Their lesions seen as small, pink, individual spots on the skin of the torso, arms and legs. These spots are not usually as thick as plaque lesions	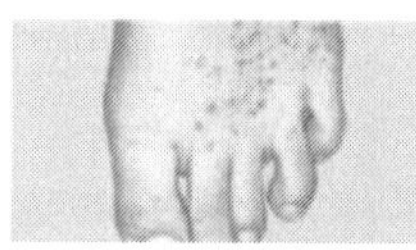 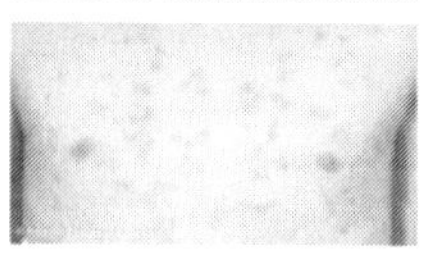

Inverse psoriasis	This type of disease found in the armpits, in the groin, under the breasts and in other areas such as skin folds of the genitals and the buttocks. In this type of psoriasis, lesions appear as bright red that are smooth and shiny	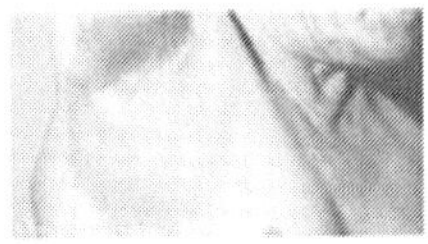 
Pustular psoriasis	This type of psoriasis disease more seen in adult patients. Pustular psoriasis has common symptoms such as rubefaction, white blisters and pus without infection that seen on the hands and feet, or most areas	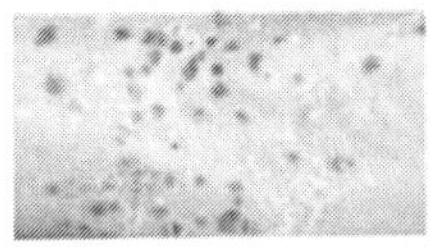 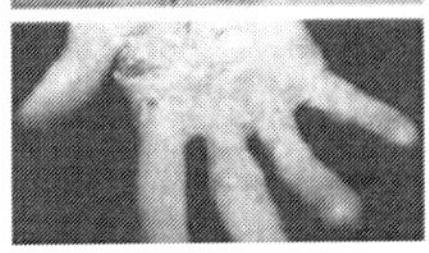
Erythrodermic psoriasis	A particularly form of psoriasis is erythrodermic psoriasis, which affects on the body as an inflammatory disease. The emergence of erythrodermic is to periodically and its symptoms include widespread red lesions	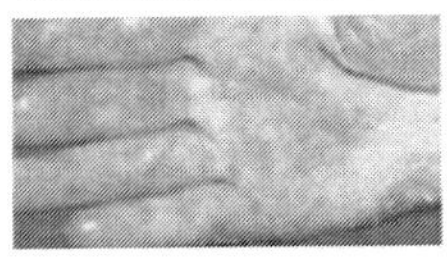 

Table 5. The Types Of Topical Therapies.

Topical therapies	Definition
Vitamin D analogs	Vitamin D3 is naturally synthesized in the epidermis. The mechanism involves natural UVB falling on the skin and converting 7-dehydrocholesterol into vitamin D, which then binds with vitamin D binding protein. The synthetic vitamin D analogs (calcipotriol, tacalcitol and calcitriol) inhibit cell proliferation and encourage the skin cells to mature normally
Vitamin A	Vitamin A also known as a topical retinoid and suitable for mild-to-moderate plaque psoriasis. It seems that vitamin A can modulate cell proliferation and increase differentiation. Its main side effect is skin irritation
Emollients	As psoriasis is a dry skin condition, a key of treatment for all types of psoriasis (except perhaps inverse psoriasis) is emollient therapy. Emollients will not stop hyper-proliferation; however, they reduce the signs of scaling and may be applied topically and/or used for washing
Coal tar	Coal tar is one of the oldest treatments for psoriasis. The studies have shown that mechanism of coal tar may be by reducing epidermal thickness and by suppressing epidermal DNA synthesis. The main potential side effects associated with coal tar include skin irritation and folliculitis. (In the some of the studies, coal tar is as mutation factor)
Dithranol	Dithranol affects the synthesis of cell DNA and has a pronounced anti-proliferative effect. It also appears to have a rapid effect on the normalization of epidermal proliferation. Its side effects include irritation and staining

Table 6. The types of systemic therapies.

Systemic therapies		Definition
Phototherapy	UVB PUVA	In psoriasis disease, exposure to ultraviolet radiation can be improved lesions. But, around 10% of patients, who have light sensitive, exposure to UV radiation can cause their psoriasis to worsen or develop. UVA and UVB are an example of this type of treatment. In fact, they suppress the immune response within the skin and thus dampen down the cascade of immunological changes, which occur to trigger psoriasisPossible side effects of PUVA and UVB include nausea, pruritus, dizziness, headache, erythema (burning), skin cancer, malignant melanoma, squamous cell carcinoma (SCC) and basal cell carcinoma
Methotrexate		Methotrexate is a relatively old treatment (since the late 1960s) and its efficacy is relatively few. In fact, it is an anti-proliferative drug and it affects cell DNA, so stops proliferation skin cells. Side effects of methotrexate include leukopenia and thrombocytopenia, oral and plaque ulceration, lower levels of folic acid, teratogenicity, hepatotoxicity and pulmonary toxicity
Cyclosporin		Cyclosporin is as a relatively modern treatment for psoriasis. Mechanism of Cyclosporin is through effect on the T-cells and helping to inhibit the stimulation of cytokines and cell proliferation. But, like methotrexate, it does have potent systemic side effects that most concern is those related to nephrotoxicity and the resultant kidney function damage and hypertension
Retinoids		Retinoid which is known as acitretin, by altering keratinization and epidermal differentiation, is used as anti-proliferative, anti-inflammatory and anti-keratinizing effects on. Retinoids are complex drugs with many possible side effects. They can affect hepatic function and lead to hyperlipidemia. Also highly teratogenic and therefore not suitable for pregnant women or women who are considering becoming pregnant. Other side effects include dry mucous membranes including eyes, lips and throat

In fact, several factors are effective on the hemostatic [35]:

1. Structure of heterogeneous from lipid/protein can be one of the main causes in hemostasis. The lipid/protein structure is renewed continuously. So, if this barrier is damaged, the lipid/protein structure causes its rapid regeneration.
2. Several key proteases for desquamation.
3. Epidermal junction (EJ) is important in the formation and maintain of barriers such as epithelial and endothelial and desmosomes bands. The studies show that raising calcium concentration stimulates keratinocytes to form strong cell-cell adhesions.
4. In epidermal keratinocytes, both extracellular and intracellular Ca^{2+} can be important to cell differentiation and proliferation.

According to the above-mentioned descriptions, the choice of drug for treating disease plays an important role. In the meantime, there are certain groups of drugs will in some cases trigger an onset or aggravate psoriasis such as lithium, chloroquine-based antimalarials, beta-blockers, etc.

For example, one of the effective therapeutics for people with severe bipolar disease is lithium, which makes sometimes discontinuing it because of worsening psoriasis. The studies show that supplement inositol (found in plants) can be used as alternative therapy to decrease severity of both diseases compared to lithium. So, therapeutic methods and new medicines such as stem cell therapy, use of herbal extracts, etc. can be effective in reducing or eliminating this disorder [36].

4. ALTERNATIVE THERAPEUTICAL METHOD

Plants have been used by men from prehistoric times to get rid of suffering and curing ailments such as skin disease.

Nowadays, herbal resources play a very important role in the management of the skin and inflammatory diseases and herbal medicine is promoted as one of the alternative therapeutical methods for healing skin diseases such as psoriasis [37].

4.1. Therapeutical Plants for Skin

4.1.1. Aloe Vera

Aloe vera (L.) Burm. *f.* syn. *Aloe barbadensis* Miller is most biologically active among 400 species [38]. According to report of World Health Organization, this medicinal herb is the best source to prepare natural drugs (**Figure 7**). The plant is native to southern and eastern Africa. Then, it was introduced into area of northern Africa and other countries.

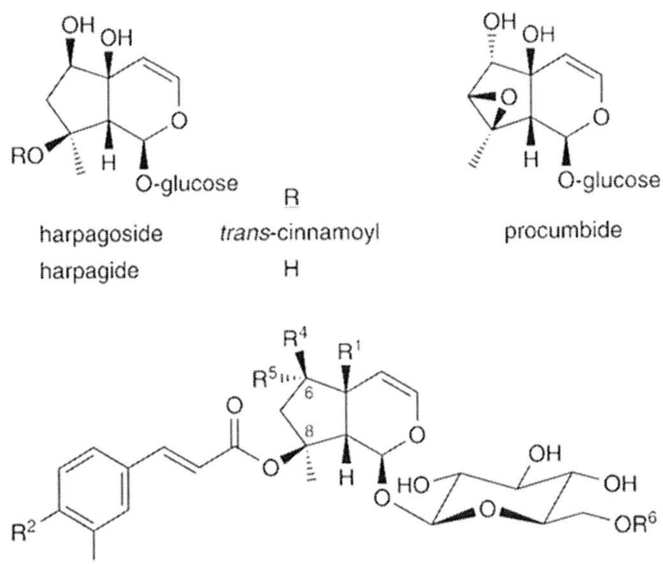

Figure 7. The main constituents of *devil's claw* root.

Aloe vera is from the lily family (*Liliaceae*). Different parts of the plant, especially its gel, are effective on the body.

Gel from the *Aloe vera* plant can help to reduce redness and scaling associated with psoriasis. *Aloe vera* contains anthraquinones, steroids, saponins, mucopolysaccharides and salicylic acid. Active ingredients *Aloe vera* include anthraquinone and acemannan that have antibacterial activity and therapeutic of psoriasis disease. Moreover, salicylic acid as a component in this plant has keratolytic effect to remove psoriatic plaques [39–41].

4.1.2. Silybum marianum

Silybum marianum (L.) Gaertn., common name milk thistle, of the family Asteraceae is a herbaceous annual species of considerable medicinal importance. Having origin in Mediterranian regions of North Africa, middle East and Europe, the species is now found distributed as a weed worldwide. All parts of the plants are edible. Germinated seed raised as sprouts and young fleshy stems and leaves have been in use as antioxidant-rich salad. Seeds of the plant are known to be used in traditional medicine for more than 2000 years for the treatment of liver and gallbladder ailments and to protect live from poisoning by toxins, such as from alcohol, toxic mushrooms, insect stings and snake bites [42].

This plant is very well known for its hepatoprotective activity. Since the numerous changes have been detected in the liver of patients with psoriasis, including steatosis, periportal inflammation, fibrosis, necrosis and cirrhosis, *Silybum marianum* can be affect for skin disorders such as psoriasis [38]. Since the abnormally high levels of cyclic adenosine monophosphate (cAMP) and leukotrienes seen in psoriatic patients, one of the therapies methods can be regulate these levels. The role of silymarin in treating psoriasis is improvement of endotoxin removal by the liver, inhibition of cAMP phosphodiesterase and leukotriene synthesis [43–46]. This plant oil has been also suggested a vitamin E, α-linolenic acid and linoleic acid reach source [47, 48].

4.1.3. Burdock

Medicinal species of Burdock are Arctium lappa, Arctium minus, Lappa major or Bardanae radix. This plant is derived from the Greek with common names of akujitsu, arctii, bardana, beggar's buttons, burdock root, great burdock, burr, burr seed, chin, clot-burr, hardock, hare burr, hurrburr, kletterwurzel, lampaza, lappa and lappola. Burdock is in the family of compositae/Asteraceae (daisy). Burdock has been also used in numerous countries throughout history to treat problems arthritis and skin disorders.

This plant is one of the key herbal ingredients in the twentieth-century cancer remedies [49].

The Chinese used burdock to treat upper respiratory infections. In fourteenth-century Europe, a combination of burdock and wine was used to treat leprosy. Also, Burdock is used for fevers, a variety of dermatologic conditions (eczema, psoriasis, scrapes and burns), syphilis, etc.

Active ingredients in Arctium lappa include:

- Sulfur about of 00.1–0.002%.
- Polysaccharides and mucilages such as xyloglucan.
- Lignans such as arctigenin.
- Other components such as organic acids that includes acetic, butyric, caffeic, chlorogenic, isovaleric, linoleic, linolenic, myristic, oleic, palmitic, proprionic, stearic, tiglic, aldehydes, carbohydrates, sesquiterpene lactones and phytosterols.

This plant acts as anti-inflammatory, antimicrobial (antibacterial, antiviral), antineoplastic (antimutagenic, antitumor) and antioxidant.

4.1.4. Devil's Claw

Harpagophytum procumbens (Pedaliaceae) or *Devil's claw* found in southern Africa (South Africa, Namibia, Botswana). Preparations of the secondary roots have gained a reputation as an anti-inflammatory and antirheumatic agent to relieve pain and inflammation in people with arthritis and similar or skin disorders.

The main constituents of *devil's claw* root are a group of decarboxylated iridoid glycosides (about 3%), including harpagoside (at least 1.2%) as the main component and smaller amounts of procumbide, harpagide and 8-(4-coumaroyl) harpagide (**Figure 7**).

4.1.5. Feverfew

Feverfew is one of the traditional plants to treat arthritis, contact dermatitis, skin difficulties, etc. *Feverfew* is an aromatic herb of the *Compositae/Asteraceae* family. Studies have been shown its effectiveness for treating migraine. This herb inhibits blood platelet aggregation and the release of 5-hydroxytryptamine from platelets. The active ingredients in the herb dried leaves are germacrene and guianane. Parthenolide is also known to be capable of causing some allergic effects, for example, contact dermatitis.

4.1.6. Liquorice

Liquorice (*licorice*; *glycyrrhiza*) is the dried unpeeled rhizome and its root. *Liquorice* is cultivated in Spain, Italy and France. The liquorice extract is usually prepared by extraction with alcohol or masculation in water. The roots of this plant contain about 20% of water-soluble active ingredients and Glycyrrhizic acid, which is comprised 3–5% of the root [46].

The bright yellow color of liquorice root is due to 1–1.5% of flavonoids such as liquiritigenin and isoliquiritigenin. The roots of this plant also include 5–15% of sugars (glucose and sucrose) (**Figure 8**). Its anti-pain properties have also been caused that was used as demulcent natural drugs [46].

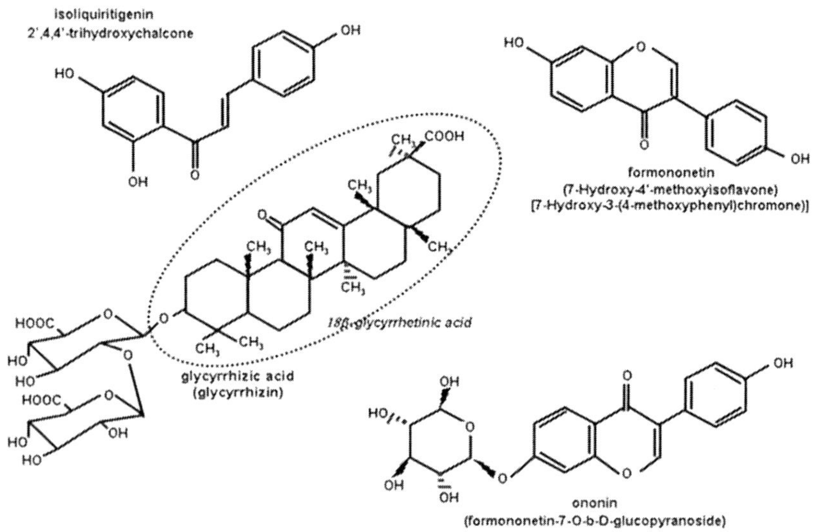

Figure 8. The main constituents of *Liquorice*.

The recent researches indicate the corticosteroid-like and anti-inflammatory activities in liquorice extracts. Glycyrrhetic acid of liquorice extracts is playing a key role for inhibiting enzymes that is important to convert prostaglandins and glucocorticoids into inactive metabolites and increase levels of prostaglandins such as PGE2 and PGF2α [46].

Table 7 shows various herbs that used in the treatment on skin disorders such as inflammation, eczema, psoriasis, scrapes and burns.

Table 7. Various herbs to treat skin disorder.

Common name	Family	Parts used	Useful forms	Chemical constituents	Application
Licorice	Fabaceae	Root, Rhizome	Extract, oil	Glycyrrhizin	Antibacterial
Calendula	Asteraceae	Flower	Infusion	Triterpenoids, Flavanoids	Softener
Rosewood	Arecaceae	Nut	Oil	Linalool	Refreshing
Tea tree	Myrtaceae	Leaves	Oil	Terpinens, EGCG	Antioxidant
Wintergreen	Ericaceae	Fruit, leaves	Oil	Methyl Salicylate, a-pinene, Limonene	Anti-inflammation
Slippery elm	Ulmaceae	Leaves	Extract, seed	Flavanoids	Regeneration
Chamomile	Compositae	Flower, Stalk	Extract, oil, Compress	Flavanoids	Anti-inflammation

Therefore, various herbs are used in pharmaceutical, food, cosmetic, etc. In this regard, there are a lot of researches to produce natural therapeutical products such as creams, lotions, etc. One of these researches was the making of new herbal cream for treating psoriasis by Izadyari et al. [50]. The herbal extract and oil in the mentioned cream were combined with vitamins. Then, the cream was given to five patients who suffer from psoriasis disease. An important point in this study was combined herbal extracts and vitamins. **Table 8** shows the used herbal extracts and oils in this study. Also, Section 5 has explained its method.

4.1.7. Vitamin D

Colecalciferol or cholecalciferol, which is known as vitamin D_3, is a vitamin the fat soluble. This vitamin is found in animals and its derivatives such as vitamin D_2 or ergocalciferol are found in plants [46]. Vitamin D deficiency leads to rickets, an inability to calcify the collagen matrix of growing bone and is characterized by a lack of rigidity in the bones, particularly in children. In adults, osteoporosis may occur. Moreover, based on studies in 1982, increased pigmentation of the skin may reduce and mediate factor (ultraviolet radiation) for synthesis of vitamin D [51].

In fact, when light energy or UV is absorbed by a precursor 7-dehydrocholesterol, vitamin D3 is synthesized. But, this form of vitamin D3 is not active. So, it becomes firstly to 25-hydroxyvitamin D3 (calcidiol) by the enzyme 25-hydroxylase in the liver. Then, within the kidney, 25-hydroxycholecalciferol acts as a substrate for 1-alpha-hydroxylase and producing 1,25-dihydroxycholecalciferol, which is the biologically active form of vitamin D3 (**Figure 9**). Calcitriol is then transported to the bones, intestine and other organs [46, 52, 53].

Table 8. The used various herbs.

The scientific name of the plant	Important properties (in dermatology)	Active ingredients	Plant Pictures
Santalum album	To treat skin disorders (such as psoriasis, etc.), relaxing and calming	Contains high amounts of alpha- and beta-santalol	
Avena sativa	High Repairing properties and collagen synthesis	Linoleic acid, oleic acid, Saponins (steroid, terpenes), alkaloids, vitamins D, E & A	
Cydonia seed-Mucus.	Moisturising and filling skin	Mucilage, various vitamins	
Pinus eldarica	Improving the structure of damaged skin, antioxidant & prevents normal cell death	Containing vitamin A, Magnesium, protein and essential amino acids to repair skin	
Ziziphus spina christi	Improves the structure of the epidermis and skin freshness, antimicrobial	Alpha-pinene, essential oil, flavonoids, tannin	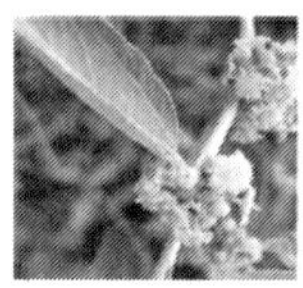
Aloe barbadensis	Restoration properties and moisturizing effects, removing psoriatic plaques	Contains anthroquinones, steroids, saponins, mucopoly-saccharides and salicylic acid	
• *Silybum marianum* • *Matricaria chamomilla* • *Foeniculum vulgare*	Anti-inflammatory and antioxidant properties	• Silybin, flavonolignans • Apigenin, quercetin, patuletin, luteolin, coumarin, essential oil & terpene bisabolol • Mucilage, limonene, flavonoids, alpha-pinene, trans-anethole	 

Rosa damascena	Healthy skin, antioxidant, antiallergy	Flavonoids, anthocyanins, quercetin, tannin, fatty oil, organic acid, vitamin C	
• *Achillea millefolium* • *Lavandula* • *Angustifolia*	Soothing, relaxing and healing of damaged skin, anti-inflammatory	• Alkaloids, coumarins, flavonoids (apigenin,luteolin,quercitin), salicylic acid, volatile oil, tannins, sterols and plant acids • Geraniol, tannin, volatile oil, borneol, linalyl acetate	 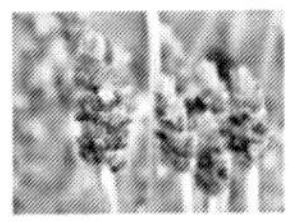
• *Arctium lappa* • *Glycyrrhiza glabra*	Regenerative properties, antibacterial and moisture supply to the skin	• Mucilage, polyacetylenes, guaianolide-type & contain vitamins C, E, B_3 & K • Glycyrrhizinic acid, isoflavones, coumarins, triterpene sterols	 

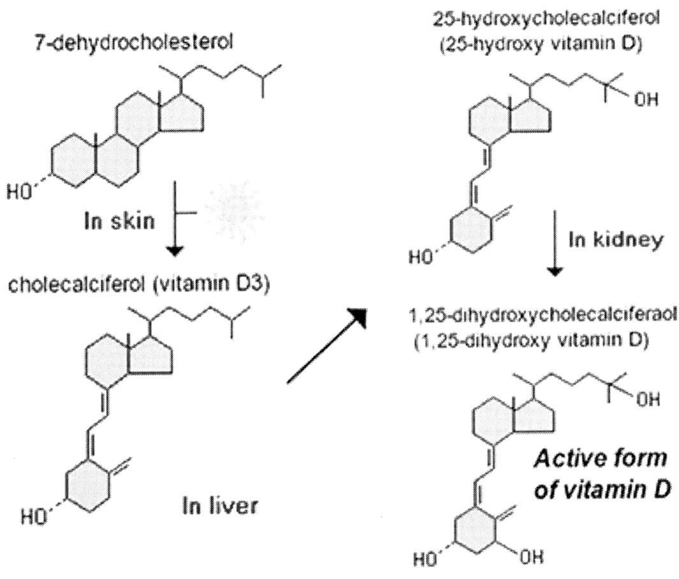

Figure 9. Structure and synthesis of vitamin D.

This form of vitamin D binds to intracellular receptors and function as one of the transcription factors to modulate gene expression in autoimmune disease such as psoriasis, because vitamin D acts as suppressor of immune system. Therefore, it can use in the topical treatment of psoriasis, for inhibiting the cell proliferation [46, 52].

4.1.8. Vitamin A

Vitamin A_1 (retinol) and vitamin A_2 (dehydroretinol) are fat-soluble vitamins found only in animal products, particularly eggs, dairy products and animal livers and kidneys. Dehydroretinol has almost 40% of retinol activity. In fact, there is carotenoid as one of the provitamin in the plants and vegetables, which are changed into retinol in the liver [46].

Vitamin A was applied in the control of psoriasis disease. Derivatives of this vitamin influence on the proliferation rate and the differentiation of epithelia keratin and regulate its disturbances in autoimmune disease such as psoriasis [54].

Vitamin A or retinoic acid is relatively unstable and sensitive to oxidation and light. Antioxidant stabilizers such as vitamin E and vitamin C are sometimes added.

4.1.9. Vitamin C

Ascorbic acid is known as vitamin C (**Figure 10**). However, this vitamin is an outstanding antioxidant in human blood plasma [55], but it is synthesized in the body of most animals except human. Also, it is in fresh fruit and vegetables. Vitamin C is a water soluble and can very fast degrade during cooking or in the presence of air [46, 55].

Deficiency of this vitamin leads to scurvy disease, muscular pain, skin lesions, etc. Vitamin C is known as essential vitamin for the formation of collagen in skin, bone, tendons and ligaments [46]. Skin lesions characteristic in scurvy disease arises from low levels of hydroxylation in the collagen structure due to the lack of vitamin A [56].

Various doses of vitamin C used for skin burns and help to promote healing wound. This vitamin is very important for the prevention of cancer and its therapy. Finally, should be added, vitamin C works as an antioxidant and helps to provide renewal vitamin E [46].

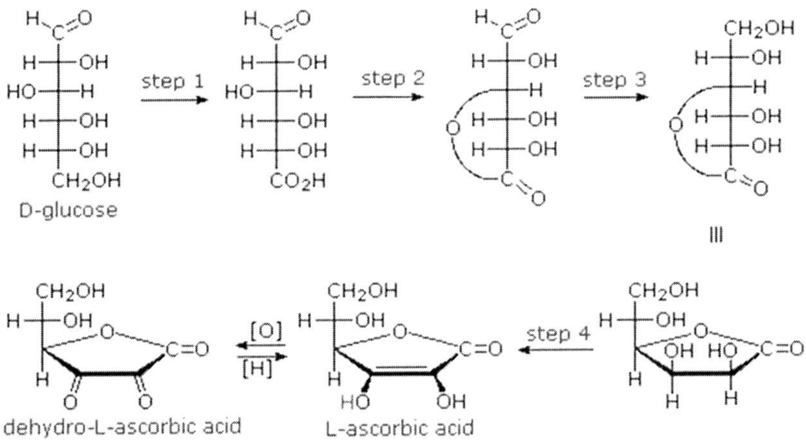

Figure 10. Structure of vitamin C.

4.1.10. Vitamin E

Tocopherol is known as vitamin E. It is one of the fat-soluble vitamins, which are found in more plants. Tocopherols exist in seed oils such as wheat, corn, safflower and soybean [46]. The vitamin has antioxidant properties and it can prevent the tissue destruction by radical. So, it can be effective for treating psoriasis (**Figure 11**). Vitamin E and its derivatives reduce the effects of aging and help to prevent heart disease [46].

4.1.11. Vitamin B_5

Pantothenic acid is known as vitamin B5 (**Figure 12**). This vitamin is water-soluble vitamin. Pantothenic acid, as part of the structure of coenzyme A, is important in metabolisms of carbohydrate, fat and protein. Also, this vitamin can help to treat wound and psoriasis lesions [46].

5. MATERIAL AND METHOD

In the mentioned cream, oily and aqueous extracts of medicinal herbal were formulated with vitamins (E, D_3, B_5, C, F) to apply on damaged skin. Herbal extracts include oily extracts of *Santalum album, Rosa damascena, Arctium lappa, Matricaria chamomilla, Achillea millefolium, Glycyrrhiza glabra, Silybum marianum, Lavandula angustifolia, Foeniculum vulgare* and aqueous extracts of *Ziziphus spina christi, Avena sativa, Matricaria chamomilla, Aloe barbadensis, Pinus eldarica* and *Cydonia seed-Mucus*. The cream was prepared according to the water-in-oil (W/O) method. Then, cream was proposed to patients and was suggested to use two times a day (**Figure 13**).

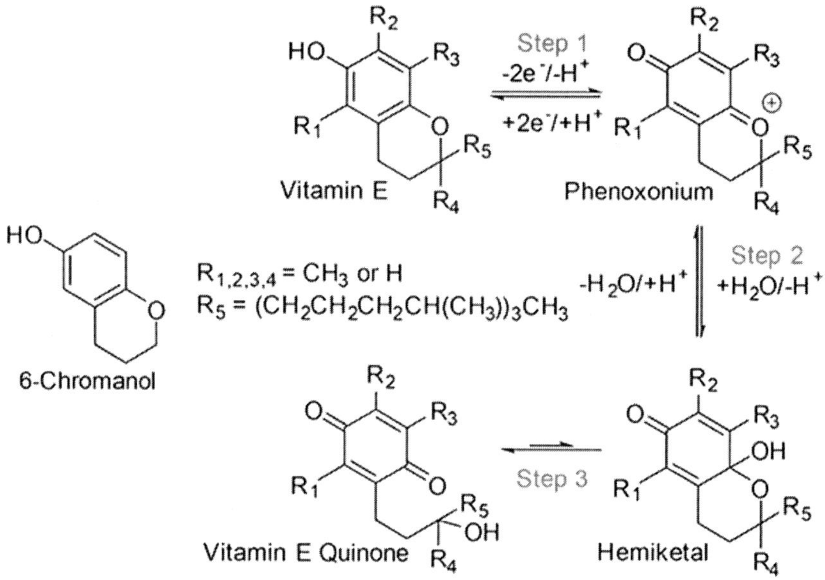

Figure 11. Antioxidant effect of vitamin E.

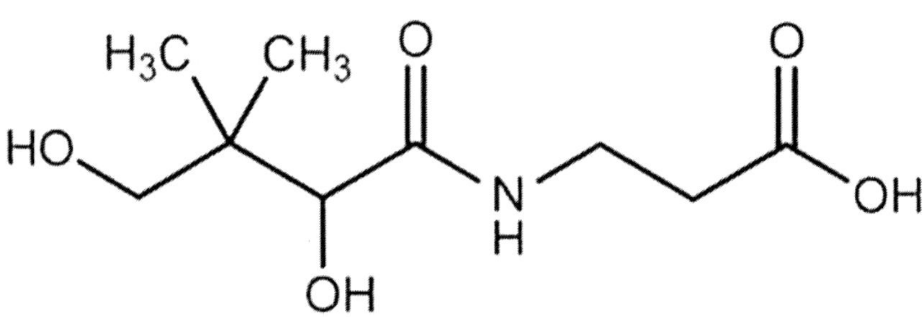

Figure 12. Structure of vitamin B_5.

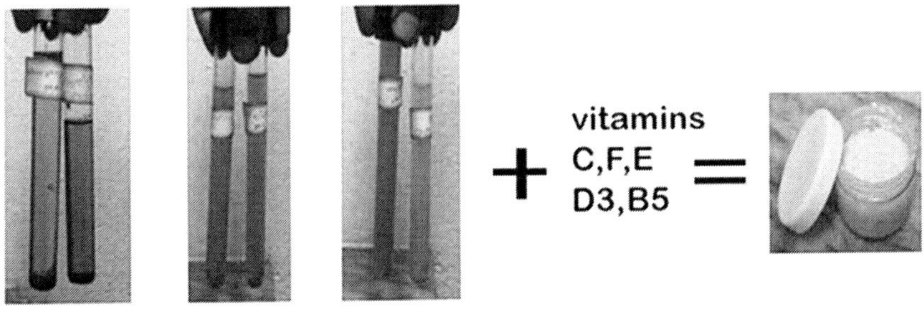

Figure 13. Making therapeutical cream includes oily extracts, aqueous extracts and vitamins.

6. RESULTS

Results were remarkable. All five patients were satisfied from inhibition of itching and inflammation of skin in first week. After 2 weeks applying cream, fading skin redness and increasing skin flexibility and repair were noticeable. An important point of this cream no side effects and combining herbal extracts and vitamins that each alone has little effectiveness. These results are shown in **Charts 1**, **2**, **3** and **4**.

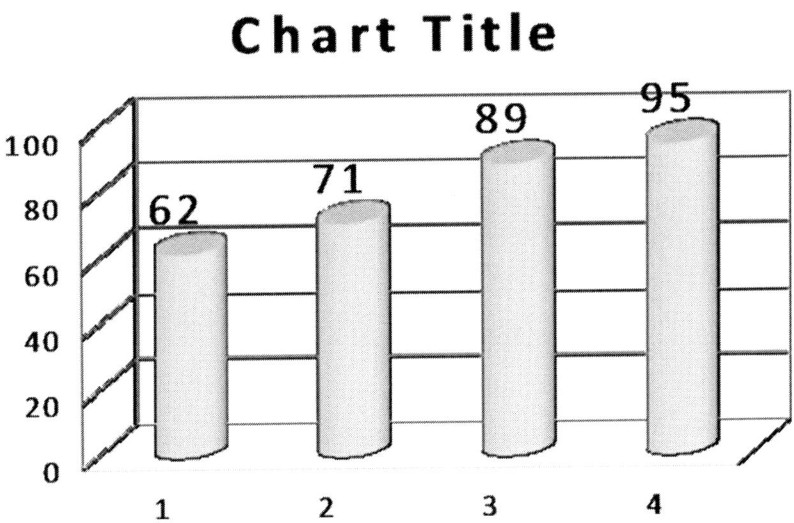

CHART 1. Inhibition of itching and inflammation for all patients, after 4 weeks (about 95%).

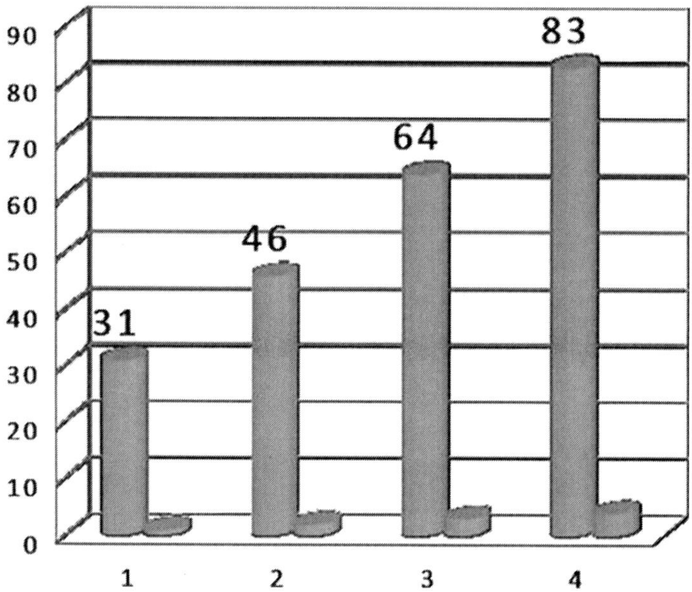

CHART 2. Improvement of psoriasis lesions (in about 83%) for every five patients at the end of 4 weeks.

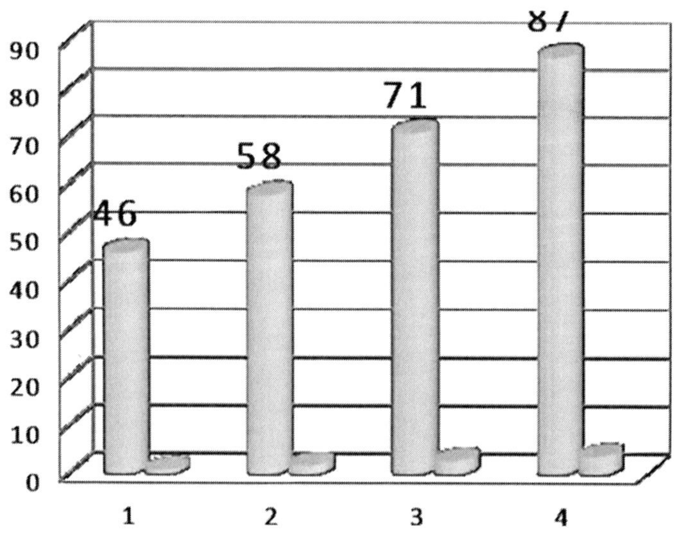

CHART 3. Estimated skin flexibility to be 87% which shows a significant increase after 4 weeks.

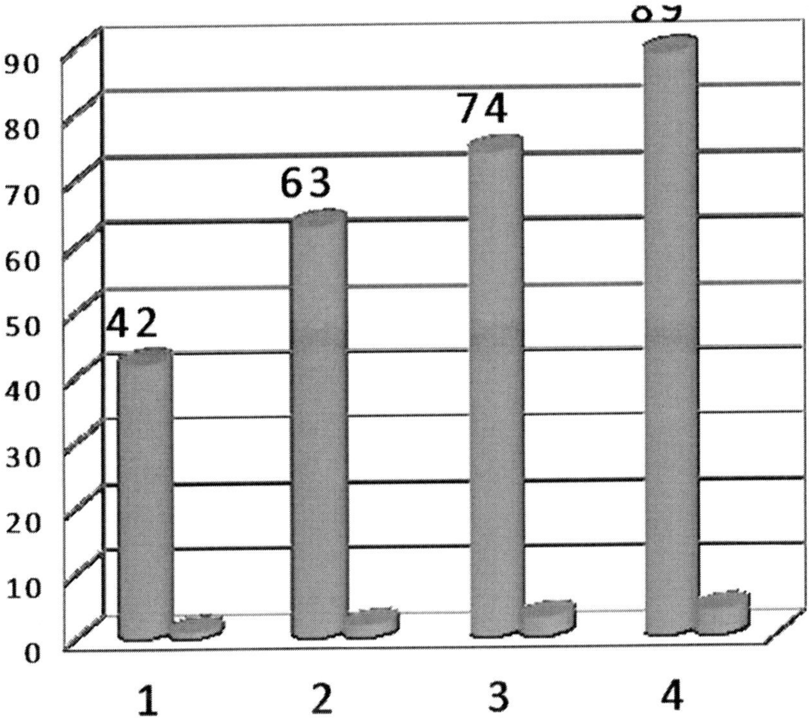

CHART 4. Fading skin redness with cream estimated about 80% in the end of 4 weeks.

According to the results, it can say that the combination of herbal extracts (oily and aqueous) and vitamins is effective in improving the symptoms psoriasis disease.

For example, flavonoids and *tannins* in *Glycyrrhiza glabra*, *Arctium lappa*, *Pinus eldarica* and *Avena sativa* are effective to treat skin lesions such as psoriasis disease and *polysaccharides* in *Aloe barbadensis and mucilage* in *Cydonia seed-Mucus* not only are healing the skin wounds but also their *malic acid* make peeling skin dead cells. Moreover, provitamins such as vitamin A act as antioxidants and prevent damage of skin healthy cells [50]. Other herbal extracts were significant as collagen synthesis and wound improvement, antibacterial and epidermal-moisture maintenance. Furthermore, vitamins E, B_5, F, C and D_3 are used for skin recovery, inflammation relief and boost immune system. It could be noted that this herbal cream will probably inhibit common symptoms of psoriasis in the shortest time.

Nowadays, the using plants developed in many industries. In this regard, the cosmetic industry is not without interest. **Tables 9** and **10** show, respectively, the common plants for skin and mostly used formulas for skin disorders.

Table 9. List of plants to apply in cosmetic [57].

No.	Scientific Name	English name	Used part	Application
1	*Acarus calamus*	Sweet flag	Rhizome	Aromatic, dusting powders, treating skin lotions
2	*Alhagi camelorum*	Jawasa	Leaves	Treating skin disorders
3	*Allium sativum*	Garlic	Bulb	Promotes wound healing, antibacterial properties
4	*Aloe vera*	Aloe vera	Leaf	Moisturizer, Sun screen, Emollient
5	*Alpinia galangal (Zingiberaceae)*	Galanga	Rhizome	Aromatic, dusting powders
6	*Avena sativa (Gramineae)*	Oat	Fruit	Moisturizer, skin tonic
7	*Azadiracta indica(Meliaceae)*	Neem	Leaf	Antiseptic, reduce dark spots, antibacterial
8	*Bauhinia racemosa (Leguminosae)*	Kanchivala	Bark and leaves	Skin disorders
9	*Calendula ofjicinalis (Compositae)*	Marigold	Flower	Skin care, anti-inflammatory, antiseptic creams.
10	*Centella asiatica (Apiaceae)*	Brahmi	Plant	Wound healing, reduce stretch marks
11	*Mesua ferrea (Guttiferae)*	Cobras saffron	Flower	Astringent
12	*Panax ginseng (Araliaceae)*	Ginseng	Root	Stimulate blood flow to skin
13	*Zizyphus jujube (Rhamnaceae)*	Zizyphus	Fruit	Skin care
14	*Zingiber zerumbet (Zingiberaceae)*	Zamabad	Rhizomes	Skin care

Table 10. Used formulas for skin disorders.

Skin disorder	Herbal extract/oil	Quantity (ml)	% Ingredients in cream
Eczema	*Chickweed* succus	20	10
	Calendula tinct	10	5
	Comfrey leaf tinct	10	5
Seborrheic dermatitis	*Calendula* fresh tinct	20	10
	Marshmallow root tinct	10	5
	Comfrey leaf tinct	10	5
Dry skin	*Lemon* oil	4	10
	Aloe vera	2	5
	Rose water	1	2.5
	Orange oil	2	5
	Almond oil	1	2.5

NOTES

[1] - A T cell or T lymphocyte is a type of lymphocyte (a subtype of white blood cell) that plays a central role in cell-mediated immunity.

REFERENCES

1. SEMON H C G 2013 PSORIASIS: (PLATE LXXXIX, XCI) *An Atlas of the Commoner Skin Diseases (Fifth Edition)* ed H C G SEMON (Butterworth-Heinemann) pp 218–21.
2. Wu J J, Nguyen T U, Poon K Y T and Herrinton L J. 2012. The association of psoriasis with autoimmune diseases. *J. Am. Acad. Dermatol.* 67: 924–30.
3. Swindell W R, Sarkar M K, Liang Y, Xing X and Gudjonsson J E. 2016. Cross-disease transcriptomics: unique IL-17A signaling in psoriasis lesions and an autoimmune PBMC signature. *J. Invest. Dermatol.*
4. Ohata C, Ishii N, Koga H, Fukuda S, Tateishi C, Tsuruta D, Furumura M and Hashimoto T. 2015. Coexistence of autoimmune bullous diseases (AIBDs) and psoriasis: a series of 145 cases *J. Am. Acad. Dermatol.* 73: 50–5.
5. Drago L and Toscano M. 2015. *Probiotics: immunomodulatory properties in allergy and eczema.* Elsevier Inc.
6. Yang J, Sundrud M S, Skepner J and Yamagata T. 2014. Targeting Th17 cells in autoimmune diseases. *Trends Pharmacol. Sci.* 35: 493–500.
7. Regehr C H. 1995. *The Cure for all Diseases.* (United States by New Century Press)
8. Petkova V B, Dimitrov M V, Nikolova I N, Voycheva C C, Valchanova V G and Andreevska K G. 2014. Psoriasis influence on the patients'quality of life. *World J. Pharm. Pharm. Sci.* 8: 1942–8.
9. Shenefelt P D 2010 Psychological interventions in the management of common skin conditions. *Psychol. Res. Behav. Manag* 3: 51–63.
10. Anon. 2014. The p is silent National Psoriasis foundation. Statistics prevalence, but we are not. National Psoriasis Foundation.
11. Akay A, Pekcanlar A, Bozdag K E, Altintas L and Karaman A. 2002. Assessment of depression in subjects with psoriasis vulgaris and lichen planus *J. Eur. Acad. Dermatol. Venereol.* 16: 347–52.
12. Sampogna F, Tabolli S, Mastroeni S, Di Pietro C, Fortes C, Abeni D, Alotto M, Antonelli G, Bolli S, Chinni M L, Fazio M, Girolomoni G, Luchetti E, Mazzotti E, Salcedo N M, Moscatelli P, Pasquini P, Piazza P, Picconi O, Pilla M A, Primavera G, Puddu P, Ruatti P, Ruggiero G, Salvatori V, Sera F, Simoni R, Sordi D and Tiago A. 2007. Quality of life impairment and psychological distress in elderly patients with psoriasis. *Dermatology* 215: 341–7.
13. Penzer R and Ersser S. 2010. *Principles of Skin Care: A Guide for Nurses and Health Care Practitioners.* (Wiley-Blackwell).
14. Gawkrodger D and Ardern-Jones M R. 2012. *Dermatology: an illustrated colour text.* Elsevier Health Sciences.
15. Terui T, Ozawa M and Tagami H. 2000. Role of neutrophils in induction of acute inflammation in T-cell-mediated immune dermatosis, psoriasis: a neutrophil-associated inflammation-boosting loop. *Exp. Dermatol.* 9: 1–10.
16. Lanigan S . and Zaidi Z 2010. *Dermatology in Clinical Practice.* (London: Springer London)
17. Bensouilah J and Buck P. 2006. *Aromadermatology: aromatherapy in the treatment and care of common skin conditions.* Radcliffe Publishing.

18. Arens E A and Zhang H. 2006. The skin's role in human thermoregulation and comfort. *Center for the Built Environment.* ed N. Pan and P. Gibsonedited by N. Pan and P. Gibson (Cambridge, England). pp. 560–602.
19. Montagna W. 2012. *The Structure and Function of Skin 3E.* Elsevier (Academic Press).
20. Zhu J and Marchant R E. 2011. Design properties of hydrogel tissue-engineering scaffolds. *Expert Rev. Med. Devices.* 8: 607–26.
21. Camisa C. 2008. *Handbook of Psoriasis,* (Blackwell).
22. Rosbotham J L, Trembath R C, Glover M, Leigh J and Barker J N W N. 1994. An association between psoriasis and hereditary multiple exostoses. A clue for the mapping of a psoriasis susceptibility gene? *Br. J. Dermatol.* 130: 671–4.
23. Cargill M, Schrodi S J, Chang M, Garcia V E, Brandon R, Callis K P, Matsunami N, Ardlie K G, Civello D, Catanese J J, Leong D U, Panko J M, McAllister L B, Hansen C B, Papenfuss J, Prescott S M, White T J, Leppert M F, Krueger G G and Begovich A B. 2007. A large-scale genetic association study confirms {IL12B} and leads to the identification of {IL23R} as psoriasis-risk genes. *Am. J. Hum. Genet.* 80: 273–90.
24. Griffiths C E M, Christophers E, Barker J, Chalmers R J G, Chimenti S, Krueger G G, Leonardi C, Menter A, Ortonne J and Fry L. 2007. A classification of psoriasis vulgaris according to phenotype. *Br. J. Dermatol.* 156: 258–62.
25. Henseler T, Ph D and Christophers E. 1985. Psoriasis of early and late onset: characterization of two types of psoriasis vulgaris. *J. Am. Acad. Dermatol.* 13: 450–6.
26. Zanchi M, Favot F, Bizzarini M, Piai M, Donini M and Sedona P. 2008. Botulinum toxin type A for the treatment of inverse psoriasis. *J. Eur. Acad. Dermatol. Venereol.* 22: 431–6.
27. Asumalahti K, Ameen M, Suomela S, Hagforsen E, Evans J, Munro M, Veal C, Allen M, Leman J, Burden A D, Kirby B, Connolly M, Gri E M, Trembath R C, Kere J, Saarialho-kere U, Jonathan N W and Barker N. 2003. Genetic analysis of PSORS1 distinguishes guttate psoriasis and palmoplantar pustulosis. *J. Invest. Dermatol.* 120: 627–32.
28. Zangeneh F Z and Shooshtary F S. 2013. Psoriasis—types, causes and medication. *Psoriasis Types, Causes and Medication.* InTech, p. 1.
29. Kahl C, Hansen B and Reich K. 2012. Nail psoriasis—an ignored disorder. Pathogenesis, diagnosis and therapy. *Der Hautarzt* 63: 184–91.
30. Moll J M H and Wright V 1973. *Psoriatic arthritis Semin. Arthritis Rheum.* 33: 55–78.
31. Wilson F C, Icen M, Crowson C S, McEvoy M T, Gabriel S E and Kremers H M. 2009. Incidence and clinical predictors of psoriatic arthritis in patients with psoriasis: a population-based study. *Arthritis Rheum.* 61: 233–9.
32. Takayanagi H. 2009. Osteoimmunology and the effects of the immune system on bone. *Nat. Rev. Rheumatol.* 5: 667–76.
33. Farber E M and Cox A J. 1967. The biology of psoriasis. *J. Invest. Dermatol.* 49: 348–57.
34. Coates L C, FitzGerald O, Helliwell P S and Paul C. 2016. Psoriasis, psoriatic arthritis and rheumatoid arthritis: Is all inflammation the same? *Semin. Arthritis Rheum.*
35. Arican O, Aral M, Sasmaz S and Ciragil P. 2005. Serum levels of TNF-α, IFN-γ, IL-6, IL-8, IL-12, IL-17 and IL-18 in patients with active psoriasis and correlation with disease severity. *Mediators Inflamm.* 2005: 273–9.
36. Alonso L. 2003. Stem cells in the skin: waste not, Wnt not. *Genes Dev.* 17: 1189–200.
37. Brown A C, Hairfield M, Richards D G, McMillin D L, Mein E A and Nelson C D. 2004. Medical nutrition therapy as a potential complementary treatment for psoriasis—five case reports. *Altern. Med. Rev.* 9: 297–307.
38. Singh K K and Tripathy S. 2014. Natural treatment alternative for psoriasis: a review on herbal resources. *J. Appl. Pharm. Sci.* 4: 114–21.
39. Syed T A, Ahmad S A, Holt A H, Ahmad S A, Ahmad S H and Afzal M. 1996. Management of psoriasis with Aloe vera extract in a hydrophilic cream: a placebo-controlled, double-blind study. *Trop. Med. Int. Health.* 1: 505–9.

40. Robson M C, Heggers J P and Hagstrom W J J. 1982. Myth, magic, witchcraft, or fact? Aloe vera revisited. *J. Burn Care Res.* 3
41. Dhanabal S, Raj B and Muruganantham N. 2012. Screening of *Wrightia tinctoria* leaves for anti psoriatic activity. *Hygeia JD Med* 4: 73–8.
42. Saleeby JP. 2006. *Wonder Herbs: A Guide to Three Adaptogens.* Google Books, USA.
43. Sabir S, Arshad M, Asif S and Chaudhari S K. 2014. An insight into medicinal and therapeutic potential of *Silybum marianum* (L.) Gaertn. *Int. J. Biosci.* 4: 104–15.
44. Fiebrich F and Koch H. 1979. Silymarin, an inhibitor of lipoxygenase. *Experientia.* 35: 1548–50.
45. Koch H P, Bachner J and Loffler E. 1985. Silymarin: potent inhibitor of cyclic AMP phosphodiesterase. *Methods Find. Exp. Clin. Pharmacol.* 7: 409–13.
46. Demirci F. 2009. *Medicinal Natural Products: A Biosynthetic Approach, 3rd edition.* In: P.M. Dewick. John Wiley & Sons, Chichester, UK, vol 124.
47. Malekzadeh M, Mirmazloum S I, Mortazavi S N, Panahi M and Angorani H R. 2011. Physicochemical properties and oil constituents of milk thistle (*Silybum marianum* gaertn. cv. budakalászi) under drought stress. *J. Med. Plants Res.* 5: 2886–9.
48. El-Mallah M H, El-Shami S M and Hassanein M M. 2003. Detailed studies on some lipids of *Silybum marianum* (L.) seed oil. *Grasas y Aceites* 54: 397–402.
49. Lin C-C, Lin J-M, Yang J-J, Chuang S-C and Ujiie T 1996 Anti-inflammatory and Radical Scavenge Effects of Arctium lappa. *Am. J. Chin. Med.* 24: 127–37.
50. Izadyari1 A, Bakhshadi R K, Motallebi1 S, Akbarzadeh A and Sharifzadeh Baei M 2015 Formulation of Herbal Cream For Psoriasis Treatment and Its Symptoms Inhibition. *Invest. Dermatol.* 135: S1–37.
51. Clemens T L, Adams J S, Henderson S L and Holick M F. 1982. Increased skin pigment reduces the capacity of skin to synthesise vitamin D3. *Lancet (London, England)* 1: 74–6.
52. Tryfonidou M A, Stevenhagen J J, Van Den Bemd G, Oosterlaken-Dijksterhuis M A, DeLuca H F, Mol J A, Van Den Brom W E, Van Leeuwen J and Hazewinkel H A W. 2002. Moderate cholecalciferol supplementation depresses intestinal calcium absorption in growing dogs. *J. Nutr.* 132: 2644–50.
53. Robinson C J, Alanis M C, Wagner C L, Hollis B W and Johnson D D. 2010. Plasma 25-hydroxyvitamin D levels in early-onset severe preeclampsia. *Am. J. Obstet. Gynecol.* 203: 366.e1–366.e6.
54. Orfanos C E, Pullmann H, Runne U, Kurka M, Strunk V, Künzig M and Dierlich E. 1979. Treatment of psoriasis using vitamin A, vitamin A acid and oral retinoids. *Hautarzt.* 30: 124–33.
55. Darr D, Combs S, Dunston S, Manning T and Pinnell S. 1992. Topical vitamin C protects porcine skin from ultraviolet radiation-induced damage. *Br. J. Dermatol.* 127: 247–53.
56. Offord E A, Gautier J-C, Avanti O, Scaletta C, Runge F, Krämer K and Applegate L A. 2002. Photoprotective potential of lycopene, β-carotene, vitamin E, vitamin C and carnosic acid in UVA-irradiated human skin fibroblasts. *Free Radic. Biol. Med.* 32 1293–303.
57. Sumit K, Vivek S, Sujata S and Ashish B. 2012. Herbal cosmetics: used for skin and hair. *Inven. J.* 2012: 1–7.

CHAPTER 20

The Protective Effect of Egyptian Withania somnifera Against Alzeheimer's

Rahma SR Mahrous[1*], Doaa A Ghareeb[2], Hoda M Fathy[1], Rasha M Abu EL-Khair[1] and Abdallah[2]

[1]*Department of Pharmacognosy, Faculty of Pharmacy, Alexandria University, Egypt*

[2]*Department of Biochemistry, Faculty of Science, Alexandria University, Egypt*

ABSTRACT

Withania somnifera popularly known as ashwaganda or Indian Ginseng is a well known plant in India. Furthermore, it is a common plant in Egypt. The objective of this study is to evaluate the effectiveness of the extracts of different parts of the plant in the protection against oxidative stress and its therapeutic potential in the treatment of Alzheimer's. All the tested extracts showed significant anticholinesterase activity, with the highest activity produced by the ripe fruit extract, significant nitric oxide scavenging activity, strong haemolysis protection as well as strong antioxidant activity. Unlike the Indian plant that is reputed for the medicinal use of its roots, the results of the present study also suggest that both the fruits and leaves of Egyptian plant have promising activities. Withanolide S isolated from its leaves extract, could be regarded as an interesting drug candidate for treatment of Alzheimer's as it showed strong in vitro acetylcholinesterase (AChE) inhibitory activity with IC50 value of 0.00035 μM, strong haemolysis protection and antiinflammatory activities. Molecular docking studies was carried out in order to explore its exact mechanism of action.

Keywords: *Withania somnifera* ; Ashwaganda; Alzheimer's; Withanolide S; Aticholinesterase

INTRODUCTION

Alzheimer's disease (AD), the most common progressive form of dementia affecting middle-aged and elderly men and women occurs as a result of malfunctioning of different biochemical pathways especially the hyperactivation of acetylcholine esterase (AChE). Currently, some plant based constituents have been clinically used for AD symptomatic management for example Galanthamine and physostigmine [1].

Withania somnifera (Ashwgandha or Indian ginseng), is used in Ayurvedic medicine as an agent which promotes learning acquisition and memory retrieval. Its extract is used alone or in combination with other herbal drugs for the reversal of cognitive deficits associated with old age, chronic illness and behavioral disorders [2-4]. It contains bioactive compounds (withanolides and withanamides) which represent good candidates for alleviating Alzheimer's. Withanolide A, withaferin-A, 2,3-dihydrowithaferin-A, withanoside IV and VI that were isolated from Withania root extract showed inhibitory effect towards AChE enzyme, and attenuated neuronal dysfunction by decreasing β-amyloid protein formation [5,6]. Withanolide A, has neuritic regeneration properties and facilitated the reconstruction of pre- and postsynaptic neurons, when neuron damage had already progressed [7]. Withanamides isolated from the fruit extracts especially withanamide A and C have been proved to be potent lipid peroxidation inhibitors and protect against β-amyloid formation [8,9].

To the best of our knowledge, few studies were carried on Egyptian *W. somnifera* focusing mainly on its cytotoxic activity [10,11].

This study was carried out to evaluate the pharmacological activities of crude extracts of different parts of *W. somnifera* and their major constituents relevant to the treatment of cognitive disorders including *in-vitro* cholinesterase inhibiting, antioxidant and anti-inflammatory effects.

EXPERIMENTAL

Plant Material

Withania somnifera (L.) Dunal plant was collected from Alexandria, Egypt in November 2014. The plant was kindly identified by Professor Atef M. Ibrahim, Department of Pomology, Faculty of Agriculture, Alexandria

University, Egypt. A voucher specimen (WS105) has been deposited in the Pharmacognosy Department, Faculty of Pharmacy, Alexandria University, Egypt.

Isolation of Withanolide S

The dried powdered leaves (600 g) was defatted with light petroleum, the marc left was air dried then extracted with 75% ethanol (2 L). The ethanolic extract was filtered and the solvent was evaporated under reduced pressure to give a dark green semisolid residue (40 g). The residue was re-dissolved in 70% ethanol then extracted successively with light petroleum. The lower layer was concentrated and then partitioned between 60% ethanol and methylene chloride. The methylene chloride fraction (11.2 g) was chromatographed on a silica gel column. Elution was carried with methylene chloride with increasing concentrations of ethyl acetate till 100% ethyl acetate followed by increasing concentration of methanol in ethyl acetate till 100% methanol. The fraction eluted with 5% methanol in ethyl acetate (2 g) was purified on a silica gel column eluted with 25% ethyl acetate in n-hexane and purified using preparative TLC to isolate withanolide S (50 mg) in a pure form. It was used at different concentrations (0.2-1 mg/ml) as test materials for *in vitro* biochemical analysis.

WITHANOLIDE S(1): White needle shape crystals; UV λmax (MeOH) 225 nm. 1H- NMR (400 MHz, DMSO-d6): δ 5.6 (1H, dd, J = 10, 2 Hz, H-2), 6.5 (1H, ddd, J = 10, 5, 2 Hz, H-3), 1.90 (1H, dd, J = 19.1, 4.6 Hz, H-4a), 3.07 (1H, d, J = 19.1Hz, H-4b), 3.47 (1H, dd, J = 8, 4 Hz, H-6), 1.25 (1H, m, H-7a), 1.95 (1H, m, H-7b), 2.34 (1H, m, H-8), 1.98 (1H,td, J = 12, 5 Hz, H-9), 1.28 (1H, d, J = 14 Hz, H-11a), 2.12 (1H, dd, J = 14, 4.6 Hz, H-11b), 1.38 (1H, m, J = 12 Hz, H-12a), 1.57 (1H, td, J = 12 Hz, H-12b), 1.12 (1H, s, H-15a), 2.15 (1H, m, H-15b), 1.49 (1H,br d, J = 6 Hz, H-16a), 2.37 (1H,br d, J = 3 Hz, H-16b), 0.99 (3H, s, 18- Me), 1.12 (3H, s, 19-Me), 1.20 (3H, s, 21-Me), 4.6 (1H, dd, J = 13, 3 Hz, H-22), 2.3 (2H, m, H-23a,b), 1.73 (3H, s, 27-Me), 1.86 (3H, s, 28-Me). 13C -NMR (100 MHz, DMSO-d6): δ 204(C-1), 128.2 (C-2), 142.7 (C-3), 35.6 (C-4), 76.8 (C-5), 73.9 (C-6), 29.4 (C-7), 33.6 (C-8),34.1 (C-9), 51.8 (C-10), 22.7 (C-11), 32.4 (C-12), 54.7 (C-13), 83.1 (C-14), 31.2 (C-15), 36.3(C-16), 88 (C-17), 21.4 (18-Me), 15.6 (19-Me), 78.6 (C-20), 19.9 (21-Me), 81.9 (C-22), 34.9 (C-23), 151.4 (C-24), 120.7 (C-25), 166.6 (C-26), 12.6 (27-Me), 20.8 (28-Me). ESI-MS m/z 503 [M - H]-.

Preparation of Extracts

Dried leaves, roots, ripe and unripe fruits of *W. somnifera* were powdered separately and 10 g of each powder was extracted with ethanol in 1:10 w/v ratio for 24 h. The filtered ethanolic extracts were concentrated under reduced pressure. The dried extracts were dissolved in dimethylsulfoxide (DMSO) and stock solutions of the extracts were appropriately diluted to get different concentrations (0.2-1 mg/ml) for each extract, and were used for biochemical analysis.

Chemicals

Acetylthiocholine iodide (ACTI), 5,5′-dithiobis[2-nitrobenzoic acid] (DTNB), diphenyl-1-picrylhydrazyl (DPPH) and Sulphanilamide were purchased from Sigma-Aldrich, USA. N-(1-Naphthyl)ethylenediamine dihydrochloride (NED) was purchased from Win Lab, UK. Sodium nitrite, Silica gel (230-400 mesh), precoated TLC plates silica gel 60 (GF-254) were purchased from Merck, Germany. All the solvents used were of analytical grade.

The 1D and 2D NMR analysis were obtained using a Bruker avance III (switzerland) 400 MHz. Residual peaks of the deuterated solvents were used to reference the spectrum.

ESI-MS spectra were measured on Waters Xevo G2-S Q-TOF mass spectrometer (Milford, MA, USA).

Preparation of Brain Homogenate

Six Balb/c mice were obtained from animal house of medical research institute, Alexandria University. After anesthesia, six brains were isolated and washed in cold saline, one gram of the brains was homogenized in 9 ml phosphate buffer saline. The homogenate was centrifuged at 3000 rpm and the supernatant was used for further biochemical estimations.

The experimental protocol was approved by the Animal Care & Use Committee (ACUC), Faculty of Pharmacy, Alexandria University, Egypt and was given number (AUC17/12).

Blood Sample

Fresh whole blood (3 ml) was kindly donated and collected in heparinized tubes to protect blood against coagulation.

Preparation of Erythrocyte Suspension

A volume of isotonic buffer solution (7.5 ml; 50 ml of 1M sodium phosphate buffer, pH 7.4 mixed with equal volume of NaCl (170 g/ml) then complete the volume to 1000 ml with distilled water) was added to 100 μl of fresh whole blood and then was centrifuged at 3000 rpm for 10 min. A volume of isotonic buffer solution equivalent to that of the supernatant was used to re-suspend the red blood pellets (the reconstituted red blood cells).

Acetylcholinesterase Activity

Acetylcholinesterase (AChE) activity was assessed according to the method of Ellman et al in which, 130 µl phosphate buffer (0.1 M pH 7.4) were added to a mixture of 20 µl brain homogenate and 20 µl of different concentration of each extract (test) or DMSO (control), then incubated for 45 min at 37°C. Five µl of substrate ACTI (75 mM) were added, mixed well and incubated for 15 min at 37°C. Finally, 60 µl DTNB (0.32 mM) were added and left for 5 min. The absorbance was measured at 405 nm and the specific activity was calculated using the following equations.

$$\frac{[A] \times [Total\ volume\ in\ cuvette\ (\mu)]}{1.36 \times 10^4 \times [Volume\ of\ brain\ homogenate\ (\mu)]}$$

$$\%\ \text{of inhibition} = \frac{\text{AChE activity(control)-AChEactivity(sample)}}{\text{Acontrol}} \times 100$$

Where, 1.36×10^4 is the molar extinction coefficient of DTNB.

Antioxidant Assay

Determination of Dipheynl-α picrylhydrazyl (DPPH) radical scavenging activity

DPPH radical scavenging assay of the total extract was performed by using the previously established and modified methodology by Choi et al. Assays were performed in flat bottom polystyrene 96 well microtiter plates. To 100 µl of each sample (0.2-1 mg/ml) in DMSO, 25 µl DPPH (1 mM) in ethanol was added. The resultant mixture was briefly shaken and maintained at room temperature, in the dark for 30 min. At the end of this period, the absorbance (A) of the mixture was measured at 490 nm, using ELISA. Scavenging ratio of DPPH assay calculated as follows:

$$\%\ \text{DPPH radicalscavengingactivity} = \frac{\text{Acontrol-Asample}}{\text{Acontrol}} \times 100$$

Antiinflammatory Activities

Nitric oxide scavenging activity: Nitric oxide scavenging activity was determined by the method described by Green et al, Sodium nitroprusside

(5 mM) in phosphate buffered saline (pH 7.2), was mixed with different concentrations (0.2-1 mg/ ml) of the extracts and withanolide S then incubated at 25°C for 30 min. Samples from the above were reacted with Griess reagent (1% sulphanilamide, 2% phosphoric acids and 0.1% Naphthylethylenediamine dihydrochloride) for 150 min at room temperature. The absorbance of the chromophore (A) formed during the diazotization of nitrite with sulphanilamide and subsequent coupling with naphthylethylenediamine was read at 546 nm. The scavenged ratio of NO calculated as follows:

$$\%\ \text{NO Inhibition} = \frac{\text{Acontrol-Asample}}{\text{Acontrol}} \times 100$$

In Vitro Human Red Blood Cell Membrane Stabilization Method (HRBC)

The assay was determined by using the method of Shinde et al. The reconstituted red blood cells were used for the estimation of antiinflammatory property. Different concentrations of each extract and withanolide S (test) or DMSO (control) were separately mixed with 125 µl of isotonic buffer or 125 µl of distilled water to simulate hypotonicity-induced-inflammation and 50 µl of erythrocyte suspension. The mixtures were incubated at 37°C for 60 minutes then centrifuged at 1000 rpm for 10 minutes. The supernatant liquid was decanted and the hemoglobin content was estimated by measuring absorbance at 560 nm. The percentage haemolysis was estimated by assuming the haemolysis produced in the control as 100%. Haemolysis prevention and anti- inflammatory properties were calculated as follows:

$$\%\text{Anti-inflammatory} = 100 - \left[\frac{\text{EH2O sample}}{\text{Emax control}}\right] \times 100$$

$$\%\ \text{Anit-Haemolytic} = \left[1 - \frac{\text{EH2O sample-E}}{\text{Emax control-E}}\right] \times 100$$

Where E is the sample absorbance in isotonic buffer, EH2O is the sample absorbance in water, and Emax is the control absorbance in water.

Docking Studies

The protein structure protein data bank (PDB) code 1EVE, which is three-dimensional (3D) structure of the anti-Alzheimer drug, e2020 (Aricept®), complexed with its target AChE with a resolution of 2.50 Å was retrieved from Brookhaven PDB and used for the validation of the docking algorithm.

The three dimensional structure of withanolide S was derived by Avogadro software and the energy minimized compound was prepared. Cdt X-ray crystallographic structure of human BACE (3TPR Crystal structure of BACE1 complexed with an inhibitor) which was obtained from the RCSB Protein Data Bank. The accurate docking was performed using the docking tool iGEMDOCK v2.0. Based on the binding energy in kcal/mol. numbers of runs taken are 70 and max interactions were 2000 with population size 200 and with an energy threshold of 100 also at each step least 'min' torsions/translations/ rotations are tested and the one giving lowest energy is chosen. The hydrophobic preference and electrostatic preference were set to 1.00. The binding site of the target was identified at a distance 8 Å (**FIGURE 1**).

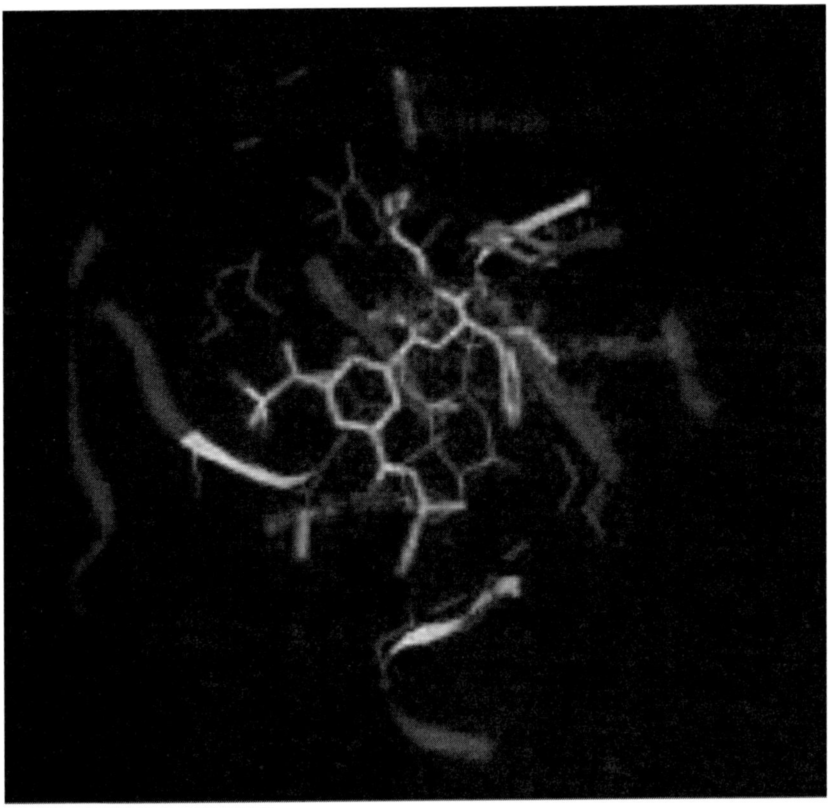

FIGURE 1: Predicted binding model of withanolide S to BACE. The BACE was shown as a ribbon drawing. The side chains of the active site residues and withanolide S are represented as stick model.

Statistical Analysis

All results are presented as means of triplicate experiments. All experiments were conducted in triplicate and values expressed as mean ± standard deviation, correlations were carried out using the correlation and regression in the EXCEL program.

RESULTS AND DISCUSSION

Chemistry

FIGURE 2 shows the structure of compound 1 isolated from leaves of *W. somnifera* . The molecular formula was established as $C^{28}H^{40}O^{8}$ by analysis of the (ESI-MS) spectrum which showed $[M - H]^-$ peak at m/z 503. The ^{13}C-NMR and ^{1}H-NMR spectra showed the characteristic signals of steroidal Δ2-1-one system; a characteristic feature of withanolide skeleton. ^{1}H and ^{13}C-NMR spectra indicate clearly the presence of a 17 α -oriented side-chain where the values of rings C,D and the lactone side chain were similar to those of withanolide E [12], withanolide C [13], and withaperuvin [14].

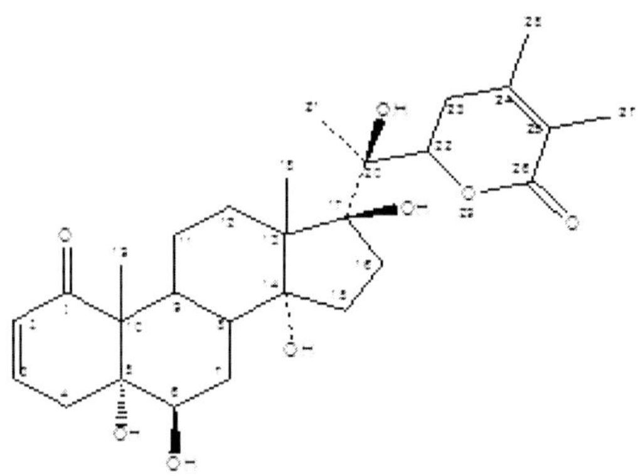

Figure 2: Structure of the Withanolide S isolated from *W. Somnifera* .

The stereochemistry of the two hydroxyl groups at C-5, C-6 were assigned to be 5 α,6 β based on the following data; Resonances of hydrogens/carbons of ring A and B were in complete agreement to those of withanolide compounds, possessing the same configuration. The trans-fusion of the A/B ring system was confirmed by the ^{13}C NMR spectrum [12].

The presence of a β -OH at C-5 causes an upfield shift of C[19] to around δ 8-9.6 ppm because of gauche-interaction and shielding of the C-19 resonance. However, no such upfield shift of C[19] was observed and its signal was observed at δ 15.6 that supports the 5α configuration of OH group. The molecular identity as withanolide S was confirmed by matching its spectral data with previous reports [12,15]. Assignments of all functional groups were achieved by ^{1}H, ^{1}H-COSY, HSQC and HMBC which further ascertained its identity.

Indian chemotypes are characterized mainly by the presence of 6α, 7α-epoxy withanolides with withanone and withanolide A representing the main components along with some 20-hydroxywithanolides like withaferin A. Besides, several sub chemotypes have been detected.

Phytochemical analysis of Egyptian plant suggested that it belongs to the Israeli chemotype III as we have isolated 5-ene withanolides, 17 α-oriented side chain withanolides and 20 hydroxy withanolide with the major withanolide being withanolide S (0.18%) in leaves (results under publication).

Previous study has proved that there is pharmacological difference between Italian chemotypes (equivalent to Israeli chemotype III) and the Indian one regarding the antioxidant activity [16].

To the best of our knowledge, this is the first investigation of this chemotype and of withanolide S concerning the protective effects against alzeheimer's

Recent evidence indicates that the one molecule, multiple targets paradigm is effective for the treatment of complex diseases due to the drug's ability to interact with multiple targets responsible for the pathogenesis of the disease. Given the complex nature of AD and the fact that a single drug acting on a specific target (AChE) may have undesirable clinical effects, the "one molecule, multiple targets" paradigm strategy has also been applied in anti-AD drug design and was recently proven to be a successful strategy for the treatment of this multifaceted disease.

There are several strategies to tackle Alzheimer's disease (AD), primarily, to augment cholinergic neurotransmission, which is fundamental to memory and learning processes [17], and have represented the primary approach in treating AD. Regarding AChE activity, the extracts of different parts of W. somnifera showed significant activity where the ripe fruits extract has strong activity when compared to Ginkgo biloba extract; a well-known cognitive enhancer; which shows IC50 of 0.268 mg/ml as reported in previous studies [18].

Oxidative stress and inflammation are the hallmarks for several diseases that affect mitochondria rich organs such as testicular tissue and brain.

They increase in conditions such as Alzheimer's disease (AD) and male infertility. Therefore, antioxidant and antiinflammatory activities are among pharmacological targets which are relevant to treatment of AD.

Reactive oxygen species play an important role in the AD neurodegeneration and pathogenesis; hence, the antioxidant capacities of plant extracts provide further support for their use in preventing the onset and progression of AD.

Different crude extracts of *W. somnifera* had moderate antioxidant activity when compared to Ginkgo biloba extract with IC50 of 0.04 mg/ml as previously reported [19].

Nitric oxide is a bio-regulator with crucial functions in human body as it is a key mediator in inflammation process. There is a strong relation between NO neurotoxicity and AD development as the reactive species NO- and peroxynitrite have been observed to cause neuronal damage in brains in AD [20,21].

Consequently, Nitric oxide scavenging activity of the extracts as shown in **TABLE 1** has dual effects in management of AD, since the extracts act as antioxidant against reactive nitrogen species (RNS) and as antiinflammatory.

Table 1: IC_{50} for the biological activities of the extracts of different organs and withanolide S isolated from W.somnifera.

Organ used	IC_{50} (mg/ml)				
	% Inhibition of ACHE	DPPH Scavenging	NO Scavenging	Anti-Inflammatory	Anti-Haemolytic
Root	0.6	0.27	0.36	0.048	0.034
Leaves	0.75	0.26	0.17	0.063	0.058
Unripe fruits	0.5	0.23	0.18	0.042	0.04
Ripe fruits	0.17	0.25	0.19	0.108	0.071
Withanolide S	0.18	0.76	0.14	0.03	0.005
Phyostigmine*	2.8× 10-4				
Diclofenac K*				0.27	0.14
Vit C*		0.5	1.65		

The inhibition of red blood cell membrane lysis by injurious substances was taken as a further measure of the mechanism of antiinflammatory activity [22]. The percentage inhibitions of haemolysis shown by all the extracts of *W. somnifera* were significant comparable to that obtained by diclofenac K.

Molecular Docking Analysis of Withanolide S Against Proteins Causing Alzheimer's Disease

The β -amyloid (Aβ) peptide is the major constituent of amyloid plaques in Alzheimer's disease (AD) brain and is likely to play a central role in its pathogenesis. The β -secretase, β -site amyloid precursor protein cleaving enzyme (BACE1), is the enzyme responsible for initiating (Aβ) generation. Thus, BACE is a prime drug target for the therapeutic inhibition of (Aβ) production in AD.

Molecular docking simulation is a powerful and increasingly substantial tool for drug discovery, it can be used to model the interaction between a small molecule and a protein at the atomic level, which allow us to characterize the behaviour of small molecules in the binding site of target proteins as well as to elucidate fundamental biochemical processes.

As Withanolide S showed strong invitro acetylcholinesterase (AChE) inhibitory activity, molecular docking study was undertaken in order to understand its interaction with these target enzymes (**FIGURE 3**).

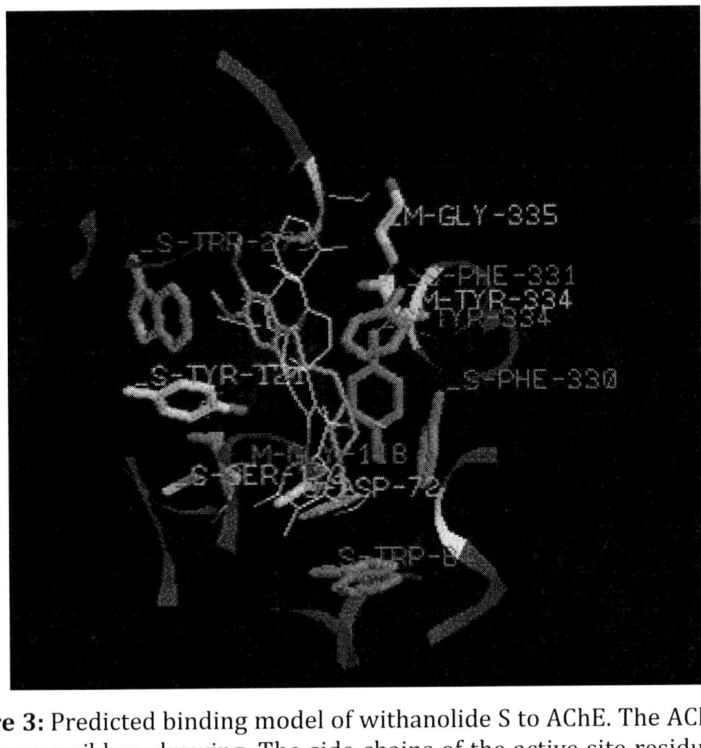

Figure 3: Predicted binding model of withanolide S to AChE. The AChE was shown as a ribbon drawing. The side chains of the active site residues and withanolide S are represented as stick model.

The structure of BACE1 site is generally composed of 11 pockets (also referred to as the "subsite"), i.e., S7, S6, S5, S4, S3, S2, S1, S1′, S2′, S3′ and S4′. The S1 pocket, hydrophobic in nature, is located at one of the entries of the cavity formed by the target protein interface, the S1′ and S2′ pockets are located at the other entry of the BACE1 active cavity. The S1′ pocket which is amphiphilic in nature is surrounded by two hydrophobic residues VAL326 and ILE22 as well as a few hydrophilic residues, such as THR329, LYS224, TYR198, ARG235, ASP228, THR72 and THR231. The S2′ pocket is amphiphilic in nature, formed by both hydrophilic residues such as SER35 and ARG128 and hydrophobic residues such as ILE126 and PRO70 [23].

The energy of binding of Withanolide S towards BACE1 was -95.7. TYR71, TYR198, ASP228, GLY230 and THR231 were identified as the binding sites.

withanolide S docking results showed hydrogen bonds with ASP32, ASP228 and THR232. Van der waal bonds with TYR71, TYR198, ASP228, GLY230 and THR231.

Docking With Ache

The 3D crystallographic structures of AChE revealed a deep and narrow gorge (20 Å) lined by conserved aromatic amino acids with two possible binding sites: the active site located at the bottom of the gorge and the peripheral site at the entrance. The active site comprises two sub-sites: the catalytic triad numbering: SER200, HIS440 and GLU327, which is essential for hydrolyzing the ester bond of ACh, and the 'anionic site' composed of aromatic residues, such as TRP84 and PHE330, that play an important role guiding the substrate towards the active site The second binding site is the peripheral site that lies at the entrance of the gorge and contains two principal amino acids, TRP279 and ASP72 [24].

The energy of binding of withanolide S towards 1eve was −124.5. TRP84, ASP72, TRY121, SER 122, TYR334, GLY118, TYR 121, TRP 279, PHE 330, PHE 331, TYR 334 were identified as interacting residues.

Withanolide S docking results showed hydrogen bonds between ASP72, TRY121, SER 122, TYR334 and TRP 84. Van der waal bonds with TRP84, GLY118, TYR 121, TRP 279, PHE 330, PHE 331, TYR 334.

CONCLUSION

All together the present study shows that the ethanolic extracts of the different parts of Egyptian *W. somnifera*, exhibited high acetylcholinesterase inhibitory effect as well as significant antioxidant activity, nitric oxide scavenging and anti-inflammatory activities [25,26]. Ripe fruit extract showed the best inhibitory activity towards AChE as well as good antioxidant and nitric oxide scavenging activities. Unlike the Indian plant that is reputed for the medicinal use of its roots, the results of the present study suggest that both the fruits and leaves of Egyptian plant have better biological activities than the roots [27,28].

Withanolide S which was isolated previously from the leaves of the Italian and Israelian plant [29,30], was isolated as a major constituent in the Egyptian plant. Withanolide S showed strong acetylcholinesterase (AChE) inhibitory activity so it could be regarded as an interesting drug candidate for treatment of Alzheimer's disease. It also showed strong haemolysis protection and antiinflammatory activities. In addition, it was also observed to elicit significant BACE1 inhibition makes it interesting as potential dual inhibitor.

ACKNOWLEDGEMENTS

This work was supported in part by the University of Alexandria, Alexandria, Egypt.

DISCLOSURE STATEMENT

No potential conflict of interest was reported by the authors.

REFERENCES

1. Ferreira A, Proenca C, Serralheiro ML, Araujo ME (2006) The in vitro screening for acetylcholinesterase inhibition and antioxidant activity of medicinal plants from Portugal. Journal of ethnopharmacology 108:31-37.
2. Ghosal S (1986) In pursuit of standardization of Ayurvedic drugs. Ann Natl Acad Indian Med.
3. Bhattacharya SK, Kumar A, Ghosal S (1995) Effects of glycowithanolides fromWithania somnifera on an animal model of Alzheimer's disease and perturbed central cholinergic markers of cognition in rats. Phytother Res. 9:110-113.
4. Dhuley JN (2001) Retracted: Nootropic-like effect of ashwagandha (Withania somnifera L.) in mice. Phytother Res. 15:524-528.
5. Kuboyama T, Tohda C, Komatsu K (2006) Withanoside IV and its active metabolite, sominone, attenuate Abeta(25-35)-induced neurodegeneration. The European journal of neuroscience. 23:1417-1426.
6. Choudhary MI, Yousuf S, Nawaz SA, Ahmed S (2004) Cholinesterase inhibiting withanolides from Withania somnifera. Chem Pharm Bull 52:1358-1361.

7. Kuboyama T, Tohda C, Komatsu K (2005) Neuritic regeneration and synaptic reconstruction induced by withanolide A. British journal of pharmacology 144:961-971.
8. Jayaprakasam B, Padmanabhan K, Nair MG (2010) Withanamides in Withania somnifera fruit protect PC-12 cells from beta-amyloid responsible for Alzheimer's disease. Phytother res: PTR. 24: 859-863.
9. Jayaprakasam B, Strasburg GA, Nair MG (2004) Potent lipid peroxidation inhibitors from Withania somnifera fruits. Tetrahedron 60:3109-3121.
10. Abdelgawad S, Ma G, Hetta M, Ross S, Badria F (2015) Chemical and biological study of Withania somnifera (L.) dunal leaves growing in Upper Egypt: Beni-Suef region. J Nat Prod. 8:64-74.
11. Ahmed WA, Mohamed A, Nasser E, Doaa E (2015) Potential Toxicity of Egyptian Ashwagandha: Significance for their Therapeutic Bioactivity and Anticancer Properties. Int J Sci Res. 4:2170-2176.
12. Glotter E, Abraham A, Günzberg G, Kirson I (1977) Naturally occurring steroidal lactones with a 17a-oriented side chain. Structure of withanolide E and related compounds. Journal of the Chemical Society, Perkin Transactions 1:341-346.
13. Bessalle R, Lavie D (1992) Biochemistry Withanolide C, A chlorinated withanolide from Withania somnifera. Phytochemistry The International Journal of Plant . 31:3648-3651.
14. Frolow F, Ray AB, Sahai M, Glotter E, Gottlieb HE, et al. (1981) Withaperuvin and 4-deoxyphysalolactone, two new ergostane-type steroids from Physalis peruviana (Solanaceae). Journal of the Chemical Society, Perkin Transactions 1:1029-1032.
15. Gottlied HE, Kirson I (1981) 13C NMR spectroscopy of the withanolides and other highly oxygenated C28 steroids. Organic Magnetic Resonance. 16:20-25.
16. Scartezzini P, Antognoni F, Conte L, Maxia A, Troia A, et al. (2007) Genetic and phytochemical difference between some Indian and Italian plants of Withania somnifera (L.) Dunal. Nat Prod Res. 21: 923-932.
17. Drachman DA, Leavitt J (1974) Human memory and the cholinergic system. A relationship to aging? Archives of neurology. 30:113-121.
18. Das A, Shanker G, Nath C, Pal R, Singh S, et al. (2002) A comparative study in rodents of standardized extracts of Bacopa monniera and Ginkgo biloba: Anticholinesterase and cognitive enhancing activities. Pharmacol Biochem Behav. 11:893-900.
19. Mensor LL, Menezes FS, Leitao GG, Reis AS, dos Santos TC, et al. (2001) Screening of Brazilian plant extracts for antioxidant activity by the use of DPPH free radical method. Phytother Res. 15:127-30.
20. Law A, Gauthier S, Quirion R (2001) Say NO to Alzheimer's disease: the putative links between nitric oxide and dementia of the Alzheimer's type. Brain Res Rev. 35:73-96.
21. Hobbs AJ, Higgs A, Moncada S (1999) Inhibition of nitric oxide synthase as a potential therapeutic target. Annual review of pharmacology and toxicology. 39:191-220.
22. Shenoy S, Shwetha K, Prabhu K, Maradi R, Bairy KL, et al. (2010) Evaluation of antiinflammatory activity of Tephrosia purpurea in rats. Asian Pacific journal of tropical medicine. 3:193-195.
23. Wu Q, Li X, Gao Q, Wang J, Li Y, et al. (2016) Interaction mechanism exploration of HEA derivatives as BACE1 inhibitors by in silico analysis. Molecular bioSystems. 12:1151-1165.
24. Orvig MAT, Brian OP, Cristina RR, Chris (2013) Exploring the multifunctionality of thioflavin- and deferiprone-based molecules as acetylcholinesterase inhibitors for potential application in Alzheimer's disease.
25. Ellman GL, Courtney KD, Andres V, Feather-Stone RM (1961) A new and rapid colorimetric determination of acetylcholinesterase activity. Biochemical pharmacology. 7:88-95.
26. Choi DW, Leininger-Muller B, Kim YC, Leroy P, Siest G, et al. (2009) Differential Role of CYP2E1 Binders and Isoniazid on CYP2E1 Protein Modification in NADPH-

dependent Microsomal Oxidative Reactions: Free Radical Scavenging Ability of Isoniazid. Free Radic Res. 36:893-903.
27. Green LC, Wagner DA, Glogowski J, Skipper PL, Wishnok JS, et al. (1982) Analysis of nitrate, nitrite, and [15N]nitrate in biological fluids. Analytical biochemistry. 126:131-138.
28. Shinde UA, Phadke AS, Nair AM, Mungantiwar AA, Dikshit VJ, et al. (1999) Membrane stabilizing activity - a possible mechanism of action for the anti-inflammatory activity of Cedrus deodara wood oil. Fitoterapia. 70:251-257.
29. Vitali G, Conte L, Nicoletti M (1996) Withanolide composition and in vitro culture of Italian Withania somnifera. Planta Med. 62:287-288.
30. Velde VV, Lavie D (1981) New withanolides of biogenetic interest from Withania somnifera. Phytochemistry 20:1359-1364.

CHAPTER 21

The Use of Aroma-acupoint Therapy and Aromatherapy to Treat Depression in Dementia Patients with Severe Cognitive Impairment

Man-Hua Yang[1*], Li-Chan Lin[1], Shiao-Chi Wu[2], Jen-Hwey Chiu[3]and Jaung-Geng Lin[4]

[1]*Institute of Clinical Nursing, National University of Yang-Ming, Taipei City, Taiwan*

[2]*Institute of Health and Welfare Policy, National University of Yang-Ming, Taipei City, Taiwan*

[3]*Institute of Traditional Medicine, National University of Yang-Ming, Taipei City, Taiwan*

[4]*Institute of Traditional Chinese Medicine, China Medical University, Liaoning Sheng, China*

ABSTRACT

BACKGROUND: Depression, one of the most common behavioral and psychological symptoms of dementia in seniors, adversely affects life quality of both patients and caregivers. Because many antipsychotic drugs can cause severe side effects in patients with dementia, there is a need for effective noninvasive treatments for depression in these patients. This study compared the effect of aroma-acupoint therapy and aromatherapy on depression in DSMIV diagnosed senior dementia patients with Mini-Mental

State Examination (MMSE)-assessed severe cognitive impairment living in long-term care facilities.

METHODS: Participants were randomly assigned to one of three groups: aroma-acupoint therapy (n=56), aromatherapy (n=73), and control (n=57). Outcomes were measured using the Cornell Scale for Depression in Dementia (CSDD) for changes in depressive mood and MMSE for changes in cognitive status in pre-test, post-test, and post-three-week test, and blood pressure and pulse rate, physical indicators of stress, measured daily during the four-week intervention.

RESULTS: The aroma-acupoint therapy and aromatherapy groups were found to have significant improvement in depressive mood in the post-test and post-three-week test, but no significant change in cognitive function. During the four-week intervention period, the two study groups also had significant weekly improvements in blood pressure and pulse rate.

CONCLUSION: Aroma-acupoint therapy and aromatherapy can effectively improve depression as well as the physical indicators of relaxation, blood pressure and pulse rate, but not cognitive status in seniors with dementia and severe cognitive impairment. Future studies with larger populations are needed further develop optimal regimens using aroma-acupoint therapy and aromatherapy to treat depression in this population.

Keywords: Aroma-acupoint therapy; Depression; Dementia; Aromatherapy

INTRODUCTION

Depression is regularly found in senior citizens who have been reported in Taiwan to have a prevalence rate of 35.2% [1]. Senile depression is often accompanied by low mood, sleep disorder, and agitation, and those with severe symptoms may also attempt suicide [2]. It is often neglected clinically and its symptoms like sleep disturbance are commonly considered a natural sign of aging by family members or caregivers. Senile depression is highly related to deterioration in physical functioning sometime caused by cardiac diseases, head injuries, or fractures and also related to psychosocial factors such as living alone and decreased cognitive function [3]. The decreased cognitive function associated with dementia not only easily leads to depression in some seniors but it also makes it very difficult for them to express their symptoms verbally. As a result, depression in senior dementia patients is easily neglected and often requires psychological assessment to identify. Prado-Jean et al. using the Cornell Scale for Depression in Dementia (CSDD) to interview 319 dementia patients and their caregivers, found 42.9% of these patients had the symptoms of depression and 75.9% had *behavioral and psychological symptoms of dementia* (BPSD) [4]. The most common symptom of the BPSD is depression, followed by restlessness, aggressive behavior, and anxiety. Physiologically, the sympathetic nerves are stimulated, and adrenaline will be released from the adrenal medulla, increasing HR, BP, and respiration [5]. Patients with these symptoms are often sent to long-term care facilities. Often, caregivers at these facilities, where the prevalence rate of BPSD is known to be higher, are not well prepared to manage BPSD and find the workload stressful [6-8].

Currently, antidepressants remain the mainstay for treatment of senile depression. The frequent use of Selective Serotonin Reuptake Inhibitor (SSRI) may increase the occurrence rate of side effects. Because serotonin syndrome and serotonin discontinuation syndrome may contribute to agitation, restlessness, or a change in consciousness, the Canadian Consensus Conference on Dementia (3rd CCCDTD) has suggested using antidepressants only after non-pharmacological interventions have not been found to be effective [9].

Non-pharmacological interventions for depression include exercise programs, acupuncture, music therapy, and aromatherapy [10-13]. Aromatherapy has been reported to improve low mood, activate the parasympathetic nervous system, lower heartbeat, inhibit the sympathetic system, and alleviate tension [12]. According to a systematic review by Yim, Ng, Hector, Tsang and Leung, aromatherapy can alleviate depression and improve such physical parameters as blood pressure and heart rate [14]. They reviewed six studies, three of which were not randomized controlled trials and number of subjects ranged from five to eight only. Two of the randomized controlled trials they reviewed performed repeated measures at baseline, at Week 6, and at Week 10. In one of the studies, the depression and anxiety of the experimental group (aromatherapy massage) and the control group (usual care) were significantly improved at Week 6, but there was no difference at Week 10. In the other one, the control group (massage group) was found to have greater improvement in depression than the aromatherapy massage group at post-1-week test. In all six studies, essential oil was applied whenever massage was practiced, making aroma-massage the major technique they studied.

Acupuncture is another well-known non-pharmacological means of treatment. In rats with vascular dementia, acupuncture was found to significantly increase the number of pyramidal neurons in the hippocampus and improve cognitive deficits. It is also reported to activate central nervous system's neurotransmitters, such as serotonin, which reduces oxidative damage, enhances blood flow in the brain, increases quality of sleep, and improves depression [15]. Zuppa et al. conducting a 5-week study investigating the effect on acupuncture on poor sleep in 48 older patients, found improvements in relaxation response, sleep quality, and depression symptoms [13]. Acupressure, the application of pressure to acupuncture points, is thought to function similarly to acupuncture. It might be of interest to investigate the combined use of aromatherapy and this special time of massage and its effect on depression.

In this study, we compare the effects of aroma acupoint therapy and aromatherapy on depression symptoms in senior patients with dementia and severe cognitive impairment. The findings of this study may help determine whether this easy to apply non-pharmacological intervention might be used by caregivers when caring for their patients.

MATERIALS AND METHODS

Participants

For this experimental study, we recruited participants from six dementia care units in different institutions (3 veterans' homes and 3 long-term care facilities) in Taiwan. Residents in each of the two types of facilities were randomly assigned to be one of three groups, either an aroma-acupoint, aromatherapy, and control group, making six groups in total. All recruitment and intervention occurred from February 1, 2012 to May 31, 2013. To be included, each participant had to: (i) be diagnosed with dementia following DSM-IV criteria, (ii) be assessed as having severe cognitive impairment defined as having a Mini-Mental State Examination (MMSE) score under 16, (iii) be residing in one of the six long-term care facilities during the study; and (iv) have no skin tears or infections around the acupoints.

Using G-power 3.0 to calculate the sample size, we determine that we needed 108 subjects. When $\alpha=0.05$ and the effect size was 0.25, we would achieve a statistical power of 80%.

During recruitment, the purpose and the protocol of this study were explained to facility administrators who helped us obtain the consent of the participants or their family members. Each participant or guardian provided informed written consent. The protocol for this study was approved by the IRB of Taipei City Hospital.

Interventions

Five common acupoints were chosen based on those used by previous studies on the use of acupuncture to treat depression. It is thought that treatment of these acupoints can help pacify patients and stabilize their emotions. Chinese medicine experts were requested to review the aroma acupoint protocol, including the selection of acupoints and the duration of the applying pressure to the acupoints. As can be seen in **FIGURES 1-3**, the acupoints we treated were the Baihui (GV 20), Fengchi (GB 20), Shenmen (HT 7), Neiguan (PC 6), and Sanyinjiao (SP 6) [16-19]. Aroma acupoint protocol involved (i) use of a 2.5% lavender oil applied to each acupoint which would be pressed for 2 minutes and (ii) a five-minute warm-up massage. The therapy was performed in patients once a day five days a week for four weeks no longer than 15 minutes per session. In the aromatherapy group, 2.5% lavender oil was applied at the same acupoints and left there for the same length of time as it was for the aroma acupoint therapy group. The control group received their usual daily care without either of the two interventions.

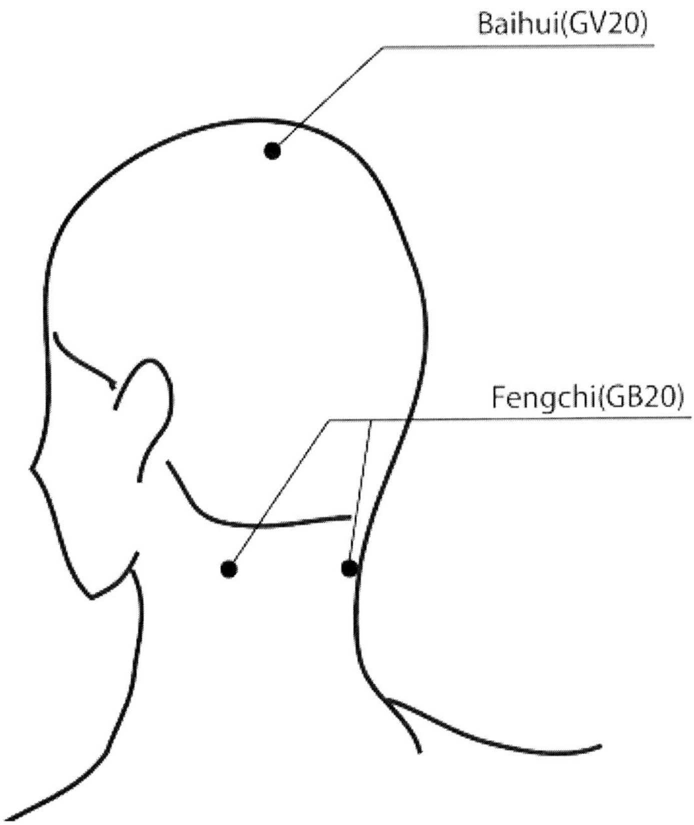

Figure 1: The location of acupoints: Baihui and Fengchi.

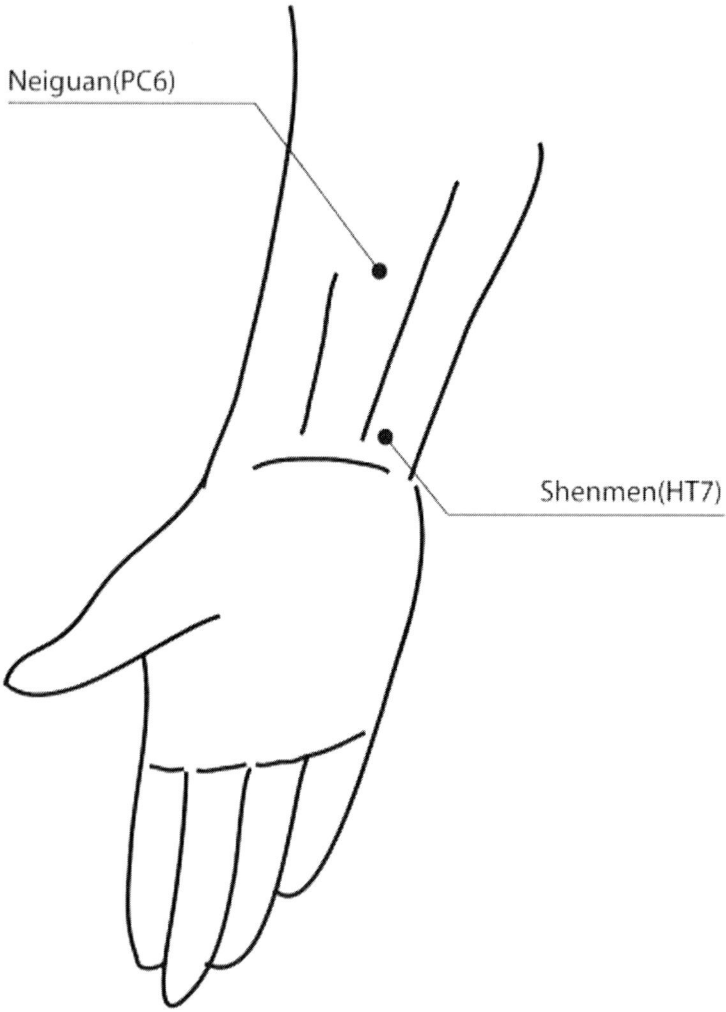

Figure 2: The location of acupoints: Neiguan and Shenmen.

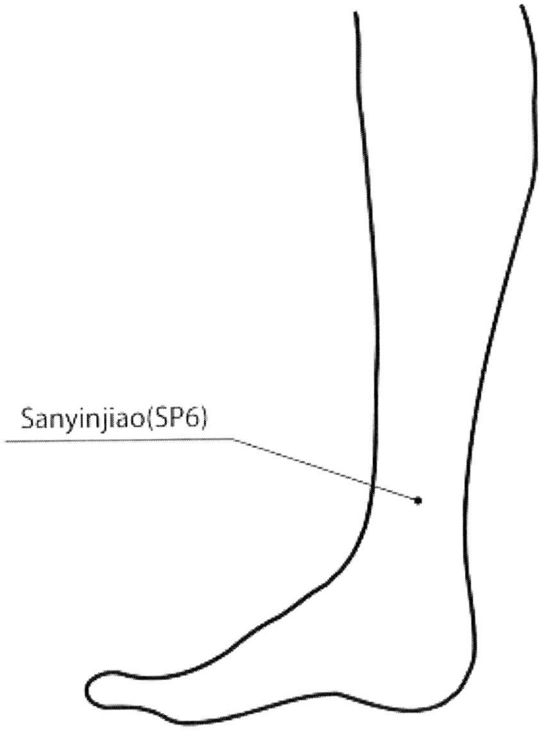

Figure 3: The location of acupoint: Sanyinjiao.

Instruments

To assess depression, we used the Cornell Scale for Depression in Dementia Chinese Version (CSDD-C), originally developed by Alexopoulos, Abrams, Young and Shamoian [20]. The CSDD is used to evaluate depression in seniors with cognitive dysfunction. The tool's 19 item covers five areas: mood-related signs, physical signs, behavioral disturbance, cyclic functioning, and ideational disturbances, 0 for none, 1 for mild, and 2 for severe. A total score less than 6 points is considered normal. The higher the score, the more severe the depression. Lin and Wang established the reliability and validity of CSDD-C, and five experts previously evaluated its content [21]. In that study which included 145 elderly dementia patients in long-term care facilities in southern Taiwan, the Content Validity Index (CVI) was 0.92 and, Kappa showed an inter-rater agreement of 0.43–0.89. The Cronbach's α for internal consistency reliability was 0.84. Thus, the questionnaire has been found to be a stable and reliable tool for the assessment of senior Chinese-speaking patients with dementia patients.

Professor Wang of the previous study authorized the use of CSDD-C for current study.

MMSE, introduced by Folstein et al. in 1975, has been a commonly used tool for assessing cognitive function for clinical and research purposes [22]. It was translated into Chinese by Guo et al. and has been approved by National Health Insurance Administration as a major standard when prescribing medications for Alzheimer's disease [23]. The MMSE covers time and location orientation, immediate memory, short-term memory, attention, arithmetic ability, language ability, and visual graphic ability. The highest score is 30, the higher the score, the better the cognitive function. A total score less than 24 points indicates mild cognitive dysfunction and a score less than 16 severe cognitive dysfunctions.

Each participant's blood pressure and pulse rate were recorded every day before and after the intervention for four weeks to evaluate the changes in these vital signs.

Study Procedure

The CSDD-C and MMSE were administered by a research assistant three times in all three groups: pre-test, post-test, and post-three-week test. Blood pressure and heart rate were taken before and after each intervention session for four weeks.

Statistical Analysis

Patient characteristics were analyzed descriptively. Differences between the three groups were compared using one-way ANOVA and chi-square tests. A generalized estimating equation (GEE) for repeated measurements was used to assess depression and cognition. Based on intention-to-treat analysis, subjects with incomplete data were included in the GEE analysis using the missing-at-random assumption. All statistical operations were performed using Statistical Product and Service Solutions (SPSS) 20.0 (IBM Corporation, Armonk, NY, USA).

RESULTS

A total of 276 qualified participants were identified from the six institutions. Twenty-one participants were hospitalized during the study, one participant died, and 68 participants decided not to participate in the study, leaving 186 participants in six institutions to randomly assign to one of three groups. As can be seen **TABLE 1**, a summary of participant characteristics, the mean age was 85.3 in the aroma-acupoint group, 83.67 in the aromatherapy group, and 81.56 in the control group, with significant differences in age between the groups. There were no significant gender differences among the groups. AD was the most common type of dementia. Two types of restraints were being used: a physical restraint consisting of restrictive strap and a medical restraint involving the use of antipsychotic drugs. There were no significant differences in the type of restraints used.

Table 1: Participant characteristics. Note aone-way ANOVA; bchi-squared test; * p<0.05

	Aroma-acupoint group (n=56)	Aroma Group (n=73)	Control group (n=57)	Fa/χ2b	P
Age, mean ± SD ,n(%)	85.3 ± 5.76	83.67 ± 4.96	81.56 ± 6.79	5.91a/ 20.83b	0.00*/ 0.00*
65-74yrs	4(7.1%)	3(4.1%)	11(19.3%)		
75-84yrs	14(25%)	39(53.4%)	25(43.9%)		
≥85yrs	38(67.9%)	31(42.5%)	21(36.8%)		
Sex, n(%)				5.69b	0.06
Male	46(82.1%)	48(65.8%)	43(75.4%)		
Female	10(17.9%)	25(34.2%)	14(24.6%)		
Diagnosis of Dementia,n(%)				20.40b	0.00*
AD	45(80.3%)	70(95.9%)	55(96.5%)		
Vascular	10(17.9%)	0(0%)	2(3.5%)		
Others	1(1.8%)	3(4.1%)	0(0%)		
Restraint, n(%)				5.67b	0.23
None	29(51.8%)	35(47.9%)	31(54.4%)		
One(chemical/physica)	22(39.2%)	35(47.9%)	25(43.9%)		
Both	5(9%)	3(4.2%)	1(1.7%)		

TABLE 2, a summary of CSDD-C scores, shows both the aromaacupoint group and the aromatherapy group had significantly CSDDC higher average pre- and post-test scores than the control group. Compared to pre-test, the aroma-acupoint group and the aromatherapy group had significantly lower in the post-test and post-3-weeks test, indicating that depression had improved in the two groups. The post-test scores of the two groups were lower than they were in the post-3-week test, suggesting immediate improvement in depression. The post-3-week test, however, revealed reduced improvement.

Table 2: Pre-, post- and post-three-week CSDD-C scores. Note: a. aroma-acupoint therapy group vs. control group; b. aromatherapy group vs. control group; c. post-test vs. pre-test; d. post-3-week vs. pre-test; e. post-test vs. pre-test in aroma-acupoint therapy group; f. post-test vs. pre-test in aromatherapy group; g. post-3-week vs. pre-test in aroma-acupoint therapy group; h. post-3-week vs. pretest in aromatherapy group. * p<.05

	Aroma-Acupoint group	Aroma-group	Control group	p
Group, β(95%CI)	1.98(1.26-2.71)	1.07(.39-1.75)	0(reference)	0.00*a/.00*b
Time				
Pre-test, mean ± SD	9.02 ± 2.94	8.14 ± 1.68	7.19 ± 1.48	
Post-test, mean ± SD	8.20 ± 1.98	7.04 ± 1.16	8.63 ± 2.78	0.00*c
Post 3 weeks, mean ± SD	8.16 ± 2.24	7.84 ± 1.57	8.12 ± 1.66	0.00*d
Time×group				
Post-test*group,β(95%CI)	-2.25(-3.08 to -1.42)	-1.78(-2.62 to -0.93)	0(reference)	0.00*e/0.00*f
Post 3 weeks*group,β(95%CI)	-2.52(-3.30 to -1.74)	-1.19(-1.97 to -0.40)	0(reference)	0.00*g/0.00*h

In **TABLE 3**, the means of MMSE of the three groups in the pre-test, post-test, and post-3-weeks test were shown to be significantly lower than 16 points, suggesting that the cognitive dysfunction of all three groups was extremely severe. There was no difference in cognitive status among the three groups. We also found no significant differences change in cognitive functioning in the three groups over time, when comparing their pre-test, post-test, or post-three-weeks test. As for cognitive function, the comparison of the pre-test and posttest revealed that neither aroma acupoint therapy nor aromatherapy had an effect on cognitive functioning.

Table 3: Pre-, post- and post-three-week MMSE scores. Note: a. aroma-acupoint therapy group vs. control group; b. aromatherapy group vs. control group; c. post-test vs. pre-test; d. post-3-week vs. pre-test; e. post-test vs. pre-test in aroma-acupoint therapy group; f. post-test vs. pre-test in aromatherapy group; g. post-3-week vs. pre-test in aroma-acupoint therapy group; h. post-3-week vs. pretest in aromatherapy group. * p<0.05

	Aroma-Acupoint group	Aroma-group	Control group	p
Group, β(95%CI)	2.02(-0.52-4.56)	0.88(-1.49-3.24)	0(reference)	0.12a/0.47b
Time				
Pre-test, mean ± SD	7.67 ± 6.36	6.54 ± 6.34	5.94 ± 7.03	
Post-test, mean ± SD	8.42 ± 7.09	7.47 ± 6.66	6.39 ± 7.25	0.18c
Post 3 weeks, mean ± SD	7.96 ± 6.39	6.09 ± 6.14	6.37 ± 7.47	0.31d
Time×group				
Post-test*group,β(95%CI)	0.41(-.49-1.30)	-0.13(-1.31-.77)	0(reference)	0.38e/0.77f
Post 3 weeks*group,β(95%CI)	0.33(-.53-1.19)	-0.52(-1.36-0.33)	0(reference)	0.45g/0.23h

TABLE 4 shows average weekly changes in blood pressure and pulse rates in the three groups during the interventions. Based on our one systolic and diastolic, and pulse rates were significantly lower in the aroma acupoint therapy group and the aromatherapy group than in the control group each week.

Table 4: Comparisons of average blood pressure and heart rate during intervention. Note **p<.01

	Mean			SD			F
	Aroma-acupoint	Aroma	Control	Aroma-acupoint	Aroma	Control	
Average Systolic pressure differences							
1st week	-8.10	-0.88	0.28	5.69	7.14	6.60	28.09**
2nd week	-10.08	-3.29	1.51	5.32	8.01	6.31	44.27**
3rd week	-10.91	-3.95	1.14	6.63	7.29	6.64	46.25**
4th week	-4.44	-1.67	0.88	3.29	5.16	5.05	19.17**
Average diastolic blood pressure differences							
1st week	-4.74	-1.61	1.33	4.54	3.96	4.63	28.55**
2nd week	-5.02	-2.17	0.77	3.91	4.63	5.49	22.37**
3rd week	-5.64	-1.97	0.39	3.73	4.41	4.28	31.67**
4th week	-4.44	-1.67	0.88	3.29	5.16	5.05	19.17**
Average pulse rate differences							
1st week	-3.59	-1.36	2.29	2.50	4.27	3.98	37.62**
2nd week	-3.90	-1.89	2.58	2.46	3.49	3.60	61.09**
3rd week	-3.72	-1.27	2.45	2.72	3.90	3.34	51.16**
4th week	-3.49	-2.14	1.78	2.80	4.73	4.04	28.06**

DISCUSSION

This experimental investigation found both aroma-acupoint therapy and aromatherapy to have a significant effect on reducing depression and improving physiologic indicators, but not on cognitive functioning, in the participants seniors with dementia and severe cognitive impairment in this study.

Our finding of significant improvement in depression was consistent with those previous studies of the effect of acupuncture/ aromatherapy on depression [13,14]. The participants with severe dementia in our study had lower CSDD scores in general than other other studies incidence of depression. People in the early and middle stages of dementia are well known to have high rates of depression and anxiety. Depression in early dementia is an emotional reaction to perceived cognitive loss. Severe dementia and depression both commonly produce symptoms such as hypersomnia, fatigue and weight loss, so it is difficult to distinguish which of the two cause these symptoms in one patient. One psychiatrist's study in Taiwan found a lower incidence of depression in his patients with severe dementia [24,25]. Future studies are needed to clarify whether this difference incidence of depression is caused by severe cognitive impairment, speech retardation, or behavioral problems.

Our participants both the aroma acupoint therapy and aromatherapy groups had significant improvement in the physiologic indicators of depression. Blood pressure and pulse rate in both groups significantly declined during the 4-week intervention period, suggesting elevation of parasympathetic activities and physical relaxation. When the sympathetic nervous system was inhibited, blood pressure declines and heart rate slows, proving that both the aroma acupoint therapy and aromatherapy groups experienced had benefited from some adjustment of the autonomic nervous system.

No significant group, time, and group-time interaction differences in cognitive status were found in the three groups, indicating that neither aroma-acupoint therapy nor aromatherapy could improve the cognitive function in our patients with dementia patients, a finding consistent with that of Zuppa et al. [13]. Recently, aromatherapy has been gradually introduced the treatment of dementia patients. In most studies, it is used to reduce agitation and improve life quality not to try to improved depression or cognitive functioning [26,27].

This study has some limitations. One is that epidemiological studies of dementia show there are more women with Alzheimer's disease (AD) than men [1]. Our study recruited more male AD patients because three of the six institutions from which we recruited participants were retirement homes for veterans. Another limitation is that this was a single-blind study, meaning the participants were aware of which group they had been assigned to. In future studies, the participants of the control group can have placebo oil applied or be pressed at fake acupoints in order to create a stricter double-blind experiment.

CONCLUSION

This experimental study assessing the effect of Complementary and Alternative Medicine (CAM) on depression in patients with dementia fond that both aroma-acupoint

therapy and aromatherapy had significant improvement in depression as assessed by CSDD-C and blood pressure and heart rate during the period of intervention. Aroma acupoint therapy was better able to inhibit the sympathetic nervous system and activity of the parasympathetic nervous system compared with aromatherapy alone. Our findings support the efficacy of this non-pharmacological treatment in improving depression. These non-pharmacological protocols should be more fully explored and refined in future studies.

ACKNOWLEDGEMENTS

We thank all participants and staff members of six long-term healthcare facilities (Chih-Shan Senior Citizen Home, Chao-Ju Senior Citizen Home, Pingtung Veterans' Home, Saint Joseph Home, Taipei Veterans' Home and Taoyuan Veterans' Home) for their support.

REFERENCES

1. Lin HW, Yang PO, Yang YS, Chen CC, Chen SC, et al. (2010) Predictive factors of geriatric depression in Taiwan: a ten-year longitudinal study. Taiwan Geriatrics & Gerontology 5: 257-265.
2. Sherrod RA, Collins A, Wynn S, Gragg M (2010). Dissecting dementia, depression, and drug effects in older adults. Journal of Psychosocial Nursing 48: 39-47.
3. Gao S, Jin Y, Unverzagt FW, Liang C, Hall KS, et al. (2009) Correlates of depressive symptoms in rural elderly Chinese. International Journal of Geriatric Psychiatry 24: 1358-1366.
4. Prado-Jean A, Couratier P, Druet-Cabanac M, Nubukpo P, Bernard-Bourzeix L, (2010) Specific psychological and behavioural symptoms of depression in patients with dementia. International Journal of Geriatric Psychiatry 25: 1065-1072.
5. Chan MF, Chan EA, Mok E (2010) Effects of music on depression and sleep quality in elderly people: a randomised controlled trial. Complementary Therapy in Medicine 18: 150-159.
6. Chang CH, Lin YH, Yeh SH, Lin LW (2010) Managing problem behaviors in cognitively impaired older people: the competence and preparedness of healthcare providers in long-term care facilities. Journal of Nursing Research 18: 164-173.
7. Shinan-Altman S, Cohen M (2009) Nursing aides' attitudes to elder abuse in nursing homes: the effect of work stressors and burnout. The Gerontologist 49: 674-684.
8. Testad I, Aasland AM, Aarsland D (2007) Prevalence and correlates of disruptive behavior in patients in Norwegian nursing homes. International Journal of Geriatric Psychiatry 22: 916-921.
9. Sepehry AA, Lee PE, Hsiung GYR, Beattie BL, Jacova C (2012) Effect of selective serotonin reuptake inhibitors in Alzheimer's disease with comorbid depression. Drug Aging 29: 793-806.
10. Raglio A, Bellandi D, Baiardi P, Gianotti M, Ubezio MC, et al. (2015). Effect of active music therapy and individualized listening to music on dementia: a multicenter randomized controlled trial. Journal of the American Geriatrics Society 63: 1534-1539.

11. Shin KR, Kang Y, Park HJ, Heitkemper M (2009) Effects of exercise program on physical fitness, depression, and self-efficacy of low-income elderly women in South Korea. Public Health Nursing, 26: 523-531.
12. Tsou PL (2010) The research on the influence of aroma treatment on senior citizens' self-conscious health-illustrate with the senior citizens in community concerned lots in south districts in Taichung City (unpublished master's thesis). Chaoyang University of Technology, Taichung.
13. Zuppa C, Prado CH, Weick A, Zaparte A, Barbosa A, et al. (2015) Acupuncture for sleep quality, BDNF levels and immunosenescence: a randomized controlled study. Neuroscience Letters 587: 35-40.
14. Yim VWC, Ng AKY, Tsang HWH, Leung AY (2009) A review on the effects of aromatherapy for patients with depressive symptoms. The Journal of Alternative and Complementary Medicine 15: 187-195.
15. Simoncini M, Gatti A, Quirico PE, Balla S, Capellero B, et al. (2015) Acupressure in insomnia and other sleep disorders in elderly institutionalized patients suffering from Alzheimer's disease. Aging-Clinical & Experimental Research 27: 37-42.
16. Wei KW (2008) Clinical study on 30 cases of post stroke depression by acupuncture. Liaoning Journal of Traditional Chinese Medicine, 35: 1563-5164.
17. Lu M, Sun M (2005) Clinical study on 80 cases of post stroke depression by acupuncture. Journal of Emergency in Traditional Chinese Medicine 14: 936-937.
18. Zhao H, Zhao WL (2007) Clinical study on post stroke depression by acupuncturing Baihiu (DU20) point. Chinese Archives of Traditional Chinese Medicine, 25: 275-276.
19. Sun C, Huang DM, Tzeng LY (2007). Clinical study 64 cases of depression by acupuncture. Chinese Journal of Rehabilitation Theory Practice 13: 1186.
20. Alexopoulos GS, Abram RC, Young RC, Shamoian CA (1988) Cornell Scale for Depression in Dementia. Society of Biological Psychiatry, 23: 271-284.
21. Lin JN, Wang JJ (2008) Psychometric evaluation of the Chinese version of the Cornell Scale for Depression in Dementia. Journal of Nursing Research, 16: 202-210.
22. Folstein MF, Folstein SE, McHugh PR (1975) "Mini-mental state". A practical method for grading the cognitive state of patients for the clinician. Journal of Psychiatry Research 12: 189-198.
23. Guo NW, Liu HC, Wong PF, Liao KK, Yan SH, et al. (1988) Chinese Version and Norms of the Mini-Mental State Examination. Journal of Rehabilitation Medicine, 16: 52-59.
24. Bennett S, Thomas A (2014) Depression and dementia: cause, consequence or coincidence? Maturitas 79: 184-190.
25. Hwang JP (2000) Psychiatric Aspects of Dementia. Research in Applied Psychology 8: 19-26.
26. Fu CY, Moyle W, Cooke M (2013) A randomised controlled trial of the use of aromatherapy and hand massage to reduce disruptive behaviour in people with dementia. BMC Complementary and Alternative Medicine 13: 165-174.
27. O'Connor DW, Eppingstall B, Taffe J, Ploeg ES (2013) A randomized, controlled cross-over trial of dermally-applied lavender (Lavandula angustifolia) oil as a treatment of agitated behaviour in dementia. BMC Complementary and Alternative Medicine 13: 315-322.

CHAPTER 22

Medicinal Plants in the Ethno Veterinary Practices of Bensa Woreda, Southern Ethiopia

Yibrah Tekle

Animal Health Researches, Southern Agricultural Research Institute (SARI), Hawassa, Ethiopia

ABSTRACT

Ethno botanical study on traditional medicinal plants were conducted between November, 2013 and September, 2014 in Bensa Woreda, Southern Ethiopia and documented different types of traditional medicinal plants used by the indigenous peoples. The study was focused on identifying medicinal plants, plant habit, disease treated, plant parts used, way of knowledge acquired, methods of preparation, route of administration, importance/indication, and ingredients added. This study was aimed to document the ethno veterinary knowledge and practices used to treat and control of livestock and human diseases in Bensa Woreda, SNNPRs, Ethiopia. A purposive sampling technique study was carried out using a semi-structured questionnaire, field observational and survey to document indigenous knowledge of 33 traditional healers. Descriptive statistics were used to analyze and summarize the ethno-botanical data. Twenty nine plants, which have medicinal value against a total of 16 livestock and animal diseases, were reported and botanically identified as belonging to various 21 plant families. Most of the plant species reported belonged to one of four major families: Asteraceae (13.79%), Lamiaceae (13.79%), Euphorbiaceae (6.9%) and Rosanceae (6.9%). Trees (51.72%) and herbaceous (24.14%) were the major growth plant habit used, whilst leaves

(65.52%) and bark (20.69%) were the major plant parts used in the study areas. The informants mostly practice oral drenching (54%) and topical (22%) of plant technique preparations. Out of the total 29 ethno-veterinary medicinal plant species were identified and documented in the study area 62.1% predominantly used to livestock aliments treatment followed by 37.9% for livestock and human (both) aliments treatment. The distribution of healers indicated that the majority were in the range of 26 - 45 years of age (66.7%, younger) and 33.3% between 46 and 70 years old (elders). Majority of informants accounting for 90.9% were males, and the remaining 9.09% were females. This study indicated that traditional medicine is, playing a significant contribution in obtaining the first aid healthcare needs of the Bensa district community. The persistence and recognition of this indigenous knowledge is due to the limited access to modern healthcare facilities. Documentation of the practices and medicinal plants is a critical issue and essential to safeguard from lose the knowledge and medicinal plants, and can be used to support the country's livestock and human health care system and improve lives and livelihoods.

Keywords: Bensa, Botanical, Healer, Medicinal Plants, livestock, practices, Documentation

1. INTRODUCTION

In most developing African countries like Ethiopia, which has the huge livestock population in Africa with a total contribution of 15% GDP and 33% of the agricultural output, and possess about 41.5 million heads of cattle, 28.2 million sheep and goats, 5.8 million equine (horses, Donkeys and Mules), 1 million camels and over 42 million poultry [1] , livestock production remains crucial and represents a major asset among resource-poor smallholder farmers by providing milk, meat, skin, manure and traction. However, the economic benefits of livestock populations remain low due to prevailing livestock diseases which are among the principal bottle necks of livestock performance and cause of high economic losses of the resource poor farmers [2] [3] . The majority of livestock raisers in Ethiopia are geographically far away from the sites of animal clinic stations; and those that are closer to the sites may not afford the fees for services. The inadequate funding at the national level for the prevention and control of animal diseases adds to the burden, especially among pastoralists who live in the remote arid and semi-arid lowland parts of the country [4] .

Modern veterinary medicines are not well developed in the country, nor are modern drugs available adequately to fight animal diseases. It is estimated that the traditional remedies are sometimes the only source of therapeutics for nearly 80% of human population [5] and 90% of livestock in Ethiopia of which 95% are plant origin [6] [7] . Ethiopian farmers and pastoralists rely on traditional knowledge, practices and locally available materials, plants in particular, to control and manage livestock diseases [7] - [10] and Ethiopians have used traditional medicines for many centuries, the use of which has become an integral part of the different cultures in Ethiopia, due to cultural acceptability, efficacy against certain diseases and economic affordability [11] [12] . The indigenous peoples of different localities in the country have developed their own specific knowledge of plant resource uses, management and conservation [13] [14] . In some parts of the country, livestock diseases such as anthrax, blackleg, anaplasmosis, ascariasis, abscess, leeches, trypanosomiais, lymphangitis, stomatitis, and coccidosis have been treated using various natural plant product combinations [15] [16] . Therefore, due to the inadequate animal health services in the different areas of country traditional ways of treatment appear to be a viable alternative to tackling the problems approach.

The application of traditional medicine to veterinary medicine has been termed as ethno veterinary medicine. It is mainly concerned with folk beliefs, knowledge, skills, methods and practices which are used in the healthcare of animals. The knowledge varies from region to region, and from community to community [17] . In general, ethnoveterinary practices have been developed by trial and error and by actual experimentation [18] . Ethno veterinary medicine comprises of traditional surgical techniques, traditional immunization, magico-religious practices, and the use of herbal medicines to treat livestock diseases [7] [12] . The relationship between the use of medicinal plants in animals and humans is rather complex. However, an overlap in use of plant remedies for the same indications in animals and human beings may occur, pointing to a theory that humans may have tried these remedies in animals before they used them for their own medical problems. Alternatively, humans may have used their overall arsenal of medicinal plants to treat animals, irrespective of whether or not they used the remedies themselves. Some evidence indicates that some animals may develop a natural attraction towards cer- tain healing chemicals in plants [4] . According to some researchers; about 30% of botanical preparations that are used to treat livestock diseases in Africa are probably effective. For example, one research has indicated that among 31 medicinal plants used by the Fulani of Mauritania, 10 have been found to be useful for the treatment of eight types of animal diseases.

In most scenarios, similar to other forms of traditional knowledge, the ethno veterinary medicinal plant knowl- edge, is not compiled, [17] [19] in

Ethiopia is passed verbally from generation to generation and valuable information can be lost whenever a traditional medical practitioner passes without conveying his traditional medicinal plants knowledge, and the younger generation is not interested in living the traditional way of life [20] . This situation is exacerbated by rapid socio economic, technological and environmental changes [21] . In addition, the loss of valuable medicinal plants due to population pressure, agricultural expansion and deforestation is widely reported by different workers [22] . As a result, the need to perform ethno botanical researches and to document the medicinal plants and the associated indigenous knowledge must be an urgent task [23] [24] .

The studies conducted on the traditional remedies used in animal healthcare in Ethiopia are inadequate when compared with the multiethnic cultural diversity and the diverse flora of Ethiopia. Thus, this study, it was necessary and important, was initiated to collect and document the traditional use of medicinal plants available in Bensa Woreda of Sidama Zone, Southern Nations, Nationalities and Peoples Regional State (SNNPR), which suppose that the data could be used as a source for further studies on medicinal plants in Bensa Woreda and for future pharmacological and phytochemical studies. Therefore, this paper presents compiled ethno-veterinary me- dicinal plants used in the study districts, Bensa Woredas of the Southern Ethiopia.

2. MATERIALS AND METHODS

2.1. Study Areas

The study was conducted in Bensa Woreda of Sidama Zone, Southern Nations Nationalities and Peoples Regional State (SNNPR), southern Ethiopia. Bensa has a total of 250,727 population (male = 126,959; Female = 123,768) from both rural 239,139 (Male = 120,582; Female = 118,557) and urban 11,588 (Male = 6377; Female = 5211) [25] Bensa is located 460 km from Addis Ababa. Mixed crop and livestock farming system are the mode of agriculture in the Woreda, and the livestock available this study area are cattle, sheep, poultry, horses, goat, donkey and mule. Cattle, sheep, and poultry, production particularly plays a central role in the farming system.

2.2. Field survey

An ethno-veterinary botanical survey was conducted to congregate information on the traditional usage of plants in animals and human health care system using a semi-structured interview and observation [26] with the traditional healers who were willing to share to their indigenous knowledge. A prior communication was done with the zone and woreda livestock coordinators and veterinarian, kebele administrative and elders, and agricultural developing agents (animal health assistances) up on the objective of study. And the selected traditional healers in the study areas clearly discussed and communicated with kebele administrative and elders, and agricultural developing agents (animal health assistances) on the objective of the study. At this point, the healers raised questions about their payment, safety and how prevent their intellectual from someone copied/stolen while interviewing and collecting plants. Finally, we were mostly arrived to the agreement by avoiding the fear to feed us the genuine information, but no further attempt was made to influence those healers who completely refused to provide information. A total of 33 individuals healers were purposively selected and interviewed based on their knowledge on traditional medicine using semi-structured interviews and field observations.

2.3. Sample Size and Sampling Techniques

In this study three kebeles were selected from the study area using purposive sampling techniques. This is because of the kebeles are typically have an intellectual healers and covered by different plant species and these plants are used for traditional medicinal value to treat different livestock and human diseases. The researcher selected 33 healers using purposive sampling technique to gather the relevant data.

2.4. Data Collection

Specimens of plants that were used by the traditional healers for treatment of livestock and humans ailments were collected. The collection data was based on the information supplied by the healers during the interview. The specimens of plants were collected in the field using standard botanic methods together with the traditional healers, that including the vegetative part, leaves, and floral, fruiting and/or seed parts as it was appropriate for taxonomic identification. During collection information regarding habitat data, general description of the plant and geographical site of collection were recorded. The information collected included local name of the traditional medicinal plant, type (cultivated or wild), diseases treated, parts used, condition of plant used, method of preparation, route of administration, ingredients added, other uses of the plant and existing threats to medicinal species. The collected samples of medicinal plants were coded, pressed, and dried then taken to botanical identi- fication by botany specialists in Science Faculty of Addis Ababa University National Herbarium.

2.5. Data Analysis

The ethno botanical data were analysis using descriptive statistics was used to summarize the collected ethno- veterinary medicinal data.

3. RESULTS AND DISCUSSION

A total of 29 ethno-veterinary medicinal plant species belonging to various 21 families were documented with details on their importance, mode of application, use, ingredient added, traditional preparation, plant part used, habit, family name, scientific name, local name and code (Table 1). Asteraceae (13.79%), Lamiaceae (13.79%) Euphorbiaceae (6.9%) and Rosanceae (6.9%) (Figure 1), and also Clerodendrum myricoides (Hochst.) Vatke (18.19%), Maesa lanceolata Forssk. (15.15%), Bersama abyssinica Fresen. (4.55%), Croton macrostachyus Del. (4.55%) and Ranunculus multifidus Forssk. (4.55%) (Figure 2) were the most frequently used and reported plant families and species for ethno-veterinary practices. Many plants were mentioned against particular diseases, one plant species to one disease, and mixing of two or more different medicinal plants against a single disease was also commonly observed.

The study indicated that the major portion of the farmers in the villages relies on traditional veterinary knowledge, practices and locally available materials [7] primarily medicinal plants used to cure and prevent livestock health problems such as Nasal bot (O. ovis), lumpy Skin Disease, Kidney problem, Cancer (Tumor), Calf diarrhea, Blackleg, Babesiosis, Pasteurollosis, Trypanosomosis, Mastitis, Diarrhea, Equine colic, Internal parasite, Swelling, External parasite, Wound, Constipation, Bloat, and Abdominal pain (Table 2). Although, other local materials are used, the use of plants for the treatment of various ailments prevails.

In this study, trees and herbaceous medicinal plants were the widely used for the treatment of various ailments that constituting the 51.72%, and 24.14% respectively, followed by 13.79% Shrub and 10.34% climbers (Figure 3). On the plant parts basis used for medicinal purposes, different plant parts like leaves, roots, bark and flower were used for treatment. However, leaves were the predominantly (65.52%), used plant part for herbal preparation in the areas, which agrees with the study in other part of Ethiopia [27] ; Leaves, 65.52%, have been used as a remedy more than other parts since leaves seem to contain more active chemicals followed by

20.69% bark, 10.34% root, 3.45% leaf and flower (both) by the ethno-veterinary practitioners (Figure 4). As elucidated in Table 1 the leaf part plant species used for different diseases treatment and prevention were more than the root, bark, and flower, such as Olea europaea L. subsp. cuspidata (WalL. ex G. Don) Cif, Vernonia amygdalina Del, Juniperus procera Hochst. ex Endl, Echinops sp, Acmella caulirhiza Del., Acacia melanoxylon R. Br., Olinia rochetiana A. Juss., Ranunculus multifidus Forssk., Plectranthus glandulosus Hook. f., Clerodendrum myricoides (Hochst.) Vatke., Lepidotrichilia volkensii (Giirke) Leroy, Prunus africana (Hook. f) Kalkm., Vernonia auriculifera Hiern, Solanum benderianum Schimper ex Damme, Lagenaria abyssinica (Hook. f) C. Jeffrey, Ehretia cymosa Thonn., Podocarpus falcatus (Thunb.) R. B. ex. Mirb., Hagenia abyssinica (Brace) JF. Gmel., Phytolacca dodecandra L. (DB.3), and Rhamnus prinoides L. Herit. Such wide harvesting of leaves and flower compared to barks and roots which are important for survival of plants has a less negative impact on the survival and continuity of useful medicinal plants and hence does not affect sustainable utilization of the plants, but in this study area the combine factors like the barks and roots used for ethno-veterinary practice by the traditional healers and the residents also depend on medicinal plants for various purposes such as firewood, construction, agricultural expansion, forage and food they have the negative impact on sustainability of the medicinal plants and climate change. A numerous techniques of preparation were employed before administering the remedies, while the leaves, roots, bark and flower of traditional medicinal plants are crushed/chopped and squeezed and then filtrated to get the plant juice needed for oral drenching (54.7%), topical (22%), ear drop (14%) and nasal (10%) applications (Figure 5). The quantity of plant part used were measured by number of leaves, flower and length of root and bark. The units of measurements used to determine dosage were coffee cup, beer bottle, highland plastic, finger length, number of drops and teaspoon. Some of the plant parts are processed with other ingredients like79.31% water, 3.44% butter and 3.44% coffee while the remaining does not have any ingredients added, 13.79% (Figure 6). Thus, the normality and accuracy dose determination and unit measurements of the medicinal plants were the gaps/problems of the traditional veterinary healers.

3. Results and Discussion

Table 1. List of ethno-veterinary medicinal plants for treatment of livestock ailments in Bensa Woreda.

No	Code	Local name	Botanical/ scientific name	Family	Habit	Preparation	Ingredient use	Use	Mode of application	Importance
1	1.2	Ejerssa	Olea europaea L. subsp. cuspidata (Wal L. ex G. Don) Cif	Oleaceae	Tree	Chopped leaf mixed with water	Water	Pasteurollosis, Calf diarrhea	Oral	Veterinary
2	2.1	Hecho	Vernonia amygdalina Del.	Asteraceae	Tree	Crushed leaf mixed with water and filtered. And topically applied the sediment part	Water	Lumpy Skin Disease. Pasteurollosis. Internal parasite	Oral. Topical	Veterinary
3	2.2	honcho	Juniperus procera Hochst. ex Endl.	Cuperssaceae	Tree	Chopped leaf mixed with water	Water	Internal parasite. Pasteurollosis,	Oral	Veterinary
4	4.1	Bursa	Echinops sp.	Asteraceae	Herb	Chopped leaf mixed with water	Water	Blackleg	Oral, nasal. Ear drop	Veterinary
5	4.2	Bieto	Acmella caulirhiza Del.	Asteraceae	Herb	Chopped leaf mixed with water	Water	Blackleg	Oral, Nasal, Ear drop	Veterinary
6	7.1	Million	Acacia melanoxylon R. Br.	Fabaceae	Tree	Crushed leaf mixed with water	Water	Mastitis. Exttemal Parasite, wound, Cancer* (Tumor)	Oral, Topical	Veterinary , Human
7	8.1	Duwancho	Syzygium guineense (Willd.) DC.	Myrtaceae	Tree	Chopped bark boiled and mixed with water	Water	Calf diarrhea, Pasteurollosis	Oral	Veterinary
8	9.1	Nole	Olinia rochetiana A. Juss.	Oliniaceae	Tree	Chopped leaf mixed with water	Water	Calf diarrhea, wound**	Oral	Veterinary, Human
9	11.1	Qenta/qinta	Ranunculus multifidus Forssk.	Ranunculaceae	Herb	Crushed leaf mixed with water	Water	Mastitis, Internal parasite, Trypanosomosis	Oral	Veterinary
10	11.2	Faficho	Plectranthus glandulosus Hook. f	Lamiaceae	Herb	Chopped leaves mixed with water	Water	Internal parasite, trypanosomosis .	Oral	Veterinary
11	11.3	Gobacho/gowacho	Maesa lanceolata Forssk.	Myrsinaceae	Tree	Chopped bark mixed with water	Water	Babesiosis. Mastitis. Kidney problem*, Internal Parasite, Wound. Cancer (Tumor)	Oral, Topical. Ear drop	Veterinary Human
12	12.3	tewarako	Bersama abyssinica Fresen.	Melianthaceae	Tree	A chopped bark is mixed with water and then applied orally drenching and topically.	Water	Kidney Problems*, wound**, swelling	Oral, Topical	Veterinary, Human
13	16.1	Densicho	Galiniera saxifraga (Hochst.) Bridson	Rubiaceae	Tree	A chopped bark is mixed with water and then applied.	Water	Calf diarrhea. Pasteurollosis	Oral, Nasal. Ear drop	Veterinary
14	21.1	malasincho	Clerodendrum myricoides (Hochst.) Vatke	Lamiaceae	Shrub	A chopped leaf applied orally, topically, nasal and ear drop	Noon	wound**. Blackleg, Nasal Bot (O.ovis). Internal Parasite*. Trypanosomosis, Mastitis, Swelling, Internal parasite, cancer* (Tumor)	Oral, Topical. Ear drop, Nasal	Veterinary Human

15	22.1	Tabecho	Lepidotrichilia volkensii (Girke) Leroy	Meliaceae	Tree	A chopped leaf and fruit is mixed with water and then applied orally and topically.	Water	Wound** Cancer* (Tumor)	Oral, Topical	Veterinary Human
16	23.1	Gebrcho	Prunus africana (Hook. f) Kalkm.	Rosaceae	Tree	A chopped leaves are applied on wound.	noon	Wound, Cancer (Tumor)	Topical	Veterinary
17	24.1	Rejcho	Vernonia auriculifera Hiern	Asteraceae	Shrub	A chopped leaves are applied on wound.	Noon	Lumpy Skin Disease	Topical	Veterinary
18	25.1	Totono/ tontono	Plectranthus garckeanus (Vatke) J. K. Morton	Lamiaceae	Herb	A chopped root mixed with water and filtered then the sediment applied topically (for human mixed with coffee).	Water, Coffee	Mastitis, Abdominal pain*	Oral, Topical	Veterinary
19	26.3	Kote	Salvia nilotica Jacq.	Lamiaceae	Herb	A chopped root is mixed with water and the sediment applied topically.	Water	Mastitis, wound**	Oral, Topical	Veterinary Human
20	27.3	Shehonie/ shishihonie	Rumex abyssinicus Jacq.	Polygonaceae	Herb	A chopped root is boiled and mixed with butter.	Butter	Internal parasite*, wound	Oral	Veterinary Human
21	29.1	tunabaylo	Solanum benderianum Schimper ex Damme	Solanaceae	Climber	A chopped leaves are mixed with water and then applied.	Water	Calf diarrhea	Oral, Nasal, Ear drop	Veterinary
22	35.2	kiendiedicho	Lagenaria abyssinica (Hook. f) C. Jeffrey	Cucurbitaceae	Climber	Crushed leaf mixed with water and filtered. And the sediment applied topically.	Water	Lumpy Skin Disease	Oral Topical	Veterinary
23	37.1	gindicho	Ehretia cymosa Thonn.	Boraginaceae	Tree	Crushed leaf mixed with water and then filtered.	Water	Mastitis, wound**	Oral	Veterinary Human
24	xx.x	Dagucho	Podocarpus falcatus (Thunb.) R. B. ex. Mirb.	Podocarpaceae	Tree	A chopped leaf mixed with water and filtered.	Water	wound**	Oral	Veterinary Human
25	38.1	Semcho	Tragia brevipes Pax	Euphorbiaceae	Climber	A chopped bark mixed with water and filtered and then orally drenching and one droplet use in ear.	Water	Babesiosis	Oral, Ear drop	Veterinary
26	27.3	unknown	Croton macrostachyus Del.	Euphorbiaceae	Tree	A chopped bark filtered and then orally drenching.	Noon	Equine Colic, Abdominal pain, Bloat	Oral	Veterinary
27	3.1	unknown	Hagenia abyssinica (Brace) JF. Gmel.	Rosaceae	Tree	A dried leaves and flowers crushed and mixed with water and filtered.	Water	Internal parasite**, Nasal bot (O. ovis)	Oral	Veterinary Human
28	12.3	Tedo	Rhamnus prinoides L 'Herit.	Rhamnaceae	Shrub	A chopped leaf mixed with water and then filtered	Water	Nasal bot (O. ovis)	Oral	Veterinary
29		unknown	Phytolacca dodecandra L. (DB.3)	Phytolaccaceae	Shrub	A chopped leaf mixed with water and then filtered	Water	Wound**, cancer* (Tumor), calf diarrhea, Constipation, Internal parasite	Oral	Veterinary Human

* = only for animal disorder; ** = for human and animal disorder

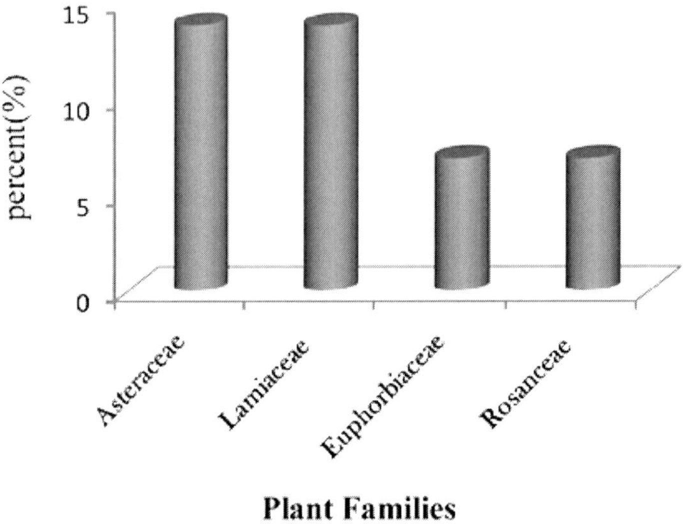

Figure 1. The most frequently used and reported plant families for ethno-veterinary practices in Bensa Woreda, Southern Ethio- pia.

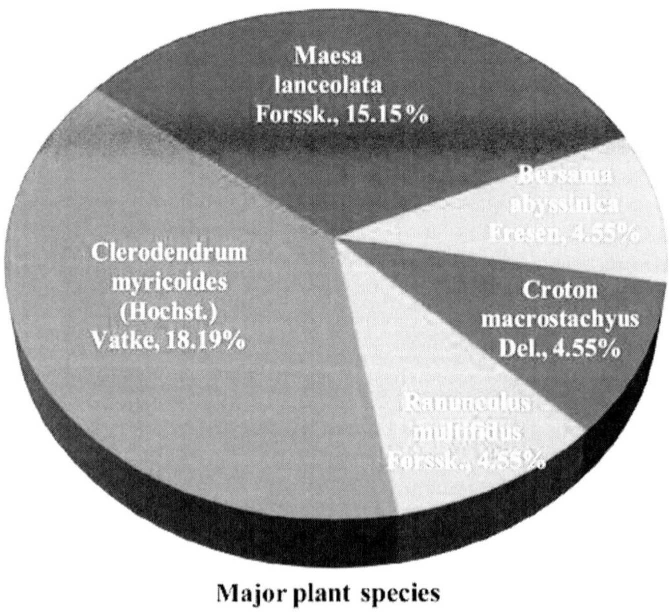

Figure 2. The most frequently used and reported plant families for ethn-veterinary practices in Bensa Woreda, southern Ethio- pia.

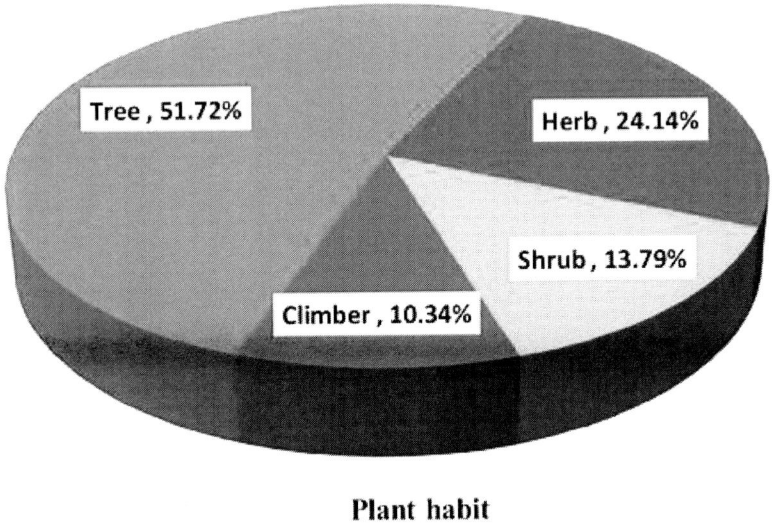

Figure 3. The proportion of growth form of medicinal plant identified for the treatment of different livestock and human disorders in the Bensa Woreda, southern Ethiopia.

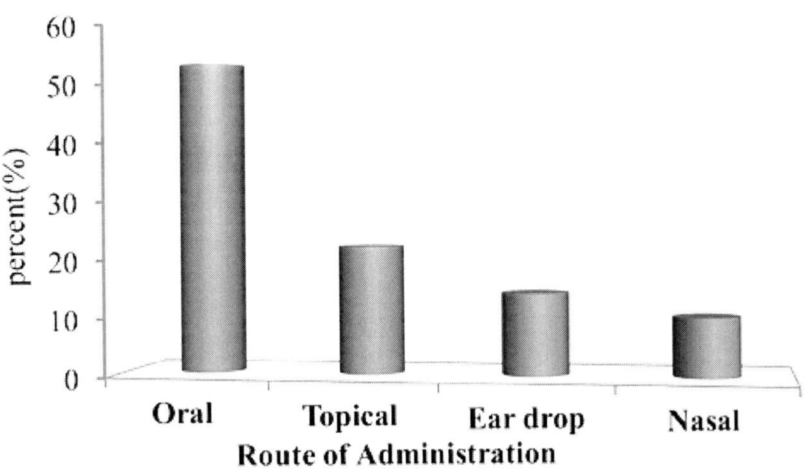

Figure 4. Proportion of administration route of medicinal plants preparation treatment of different livestock and human diseases in Bensa Woreda, Southern Ethiopia.

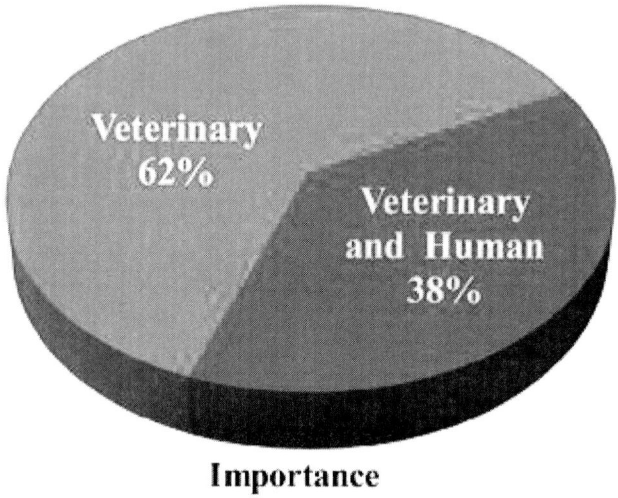

Figure 5. The total proportion of the ethno-veterinary practices used for the treatment of livestock and human diseases by the healers of the Bensa Woreda, Southern Ethiopia.

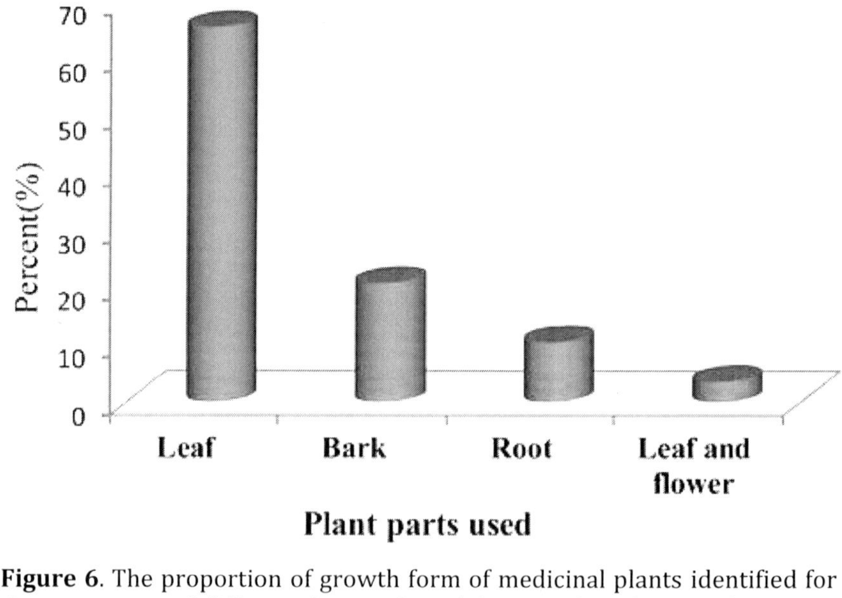

Figure 6. The proportion of growth form of medicinal plants identified for the treatment of different livestock and human disorders in the Bensa Woreda, Southern Ethiopia.

In the study area the practices/knowledge, information on the ethno-veterinary medicinal plants, is rarely found in written form; 9 healers had a written document out of the total healers (33) participated in this study, the traditional healers are acquired from their fathers (21/33, 63.63%), friends (2/33, 6.06%), stranger (2/33, 6.06%), bought from other healers (2/33, 6.06%) and by trial and error approaches discovered (6/33, 18.18%, 4 out of the 6 were females and the remaining 2 were men) (Figure 7) and some of the healers their knowledge transferred to their children, the majority often to the first-born and only some of the traditional healers to their honest and faithful children who can never leak information without regarding the age from generation to gener ation while the others kept the knowledge with them for the sake of secrecy. Derogatory attitudes towards the traditional medicine practitioners had forced healers to keep their knowledge and practices to themselves. Moreover, it is an income generation activities for these healers. This is analogous with the studies done in other part of Ethiopia [19] . Most of the plant species were collected from wild habitats, 72.41%, [10] and few of them obtained from around home garden, 27.59%, and are being exploited to cure and control the livestock and hu- man diseases.

Table 2. The number of ethno-veterinary medicinal plant remedies used to treat ailments/con- ditions in the study area.

Diseases treated	Number of ethno-veterinary medicinal plant remedies used
Mastitis	14
Trypanosomosis	3
Pasteurollosis	5
Nasal bot/O. ovis	3
Babesiosis	2
Blackleg	3
lumpy Skin Disease	3
Calf diarrhea	5
Internal parasite	6
Equine colic	1
Kidney problem	2
Cancer (Tumor)	6
Swelling	3
Wound	45
Diarrhea	1
Bloat	1
External parasite	1
Abdominal pain	2
Constipation	1

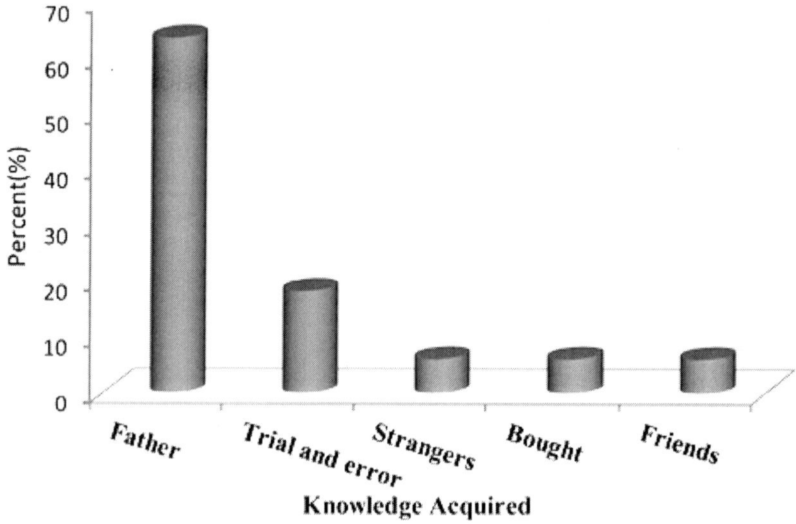

Figure 7. The proportion of knowledge aquired the traditional healers in Bensa Woreda, Southern Ethiopia.

The truth is that the medicinal plants in largely found in the natural habitats faced a major threat to their survival of the mother plants due to the combined effect of factors [28] like continues the mass destruction in their habitats, [19] cultivation of marginal lands/agricultural expansion [29] - [32] grazing, soil erosion, orally transfer of endogenous knowledge from generation to generation, herbal preparation that involves roots and barks [20] draught, urbanization [31] [33] and agricultural expansions. As a result, the stocks of the ethno-veterinary medicinal plants are coming shrink [31] .

Some of the surveyed traditional medicine plant species practiced by informants are recorded in other parts of the country like: Vernonia amygdalina Del., Croton macrostachyus, Rhamnus prinoides L 'Herit., Tragia brevipes Pax., Echinops sp., Clerodendrum myricoides (Hochst.) Vatke Phytolacca dodecandra L 'Herit., Maesa lanceolata Forssk. Hagenia abyssinica (Brace) JF. Gmel. is used to treat internal parasite in livestock in Tigray region, Northern Ethiopia, [34] [35] and in Kofle, Bale, and Debark rural communities of Ethiopi, [36] and also used to treat tapeworm in human in Kofle, Bale, and Debark rural communities of Ethiopia, [36] in Southeast of Ethiopia [4] and by Zay people in Ethiopia [37] . Similar, its use for treatment of livestock ailments has been also documented [2] [3] [38] [39] . And also Croton macrostachyus use for treatment of bloat in Borana Pastoralists, Southern Ethiopia [40] . The result of this study is agreed with the all above citation and documentation.

The distribution of healers indicated that, of the 33 traditional medicine practitioners interviewed, majority were in the range of 26 - 45 years of age. Further analysis showed that 66.7% were over 26 years old, and 33.3% between 46 and 70 years old. Majority of informants accounting for 90.9% were males, and the remaining 9.1% were females. This work indicated that among the informants, nearly 66.7% traditional veterinary healers were Younger's.

Out of the total 29 ethno-veterinary medicinal plant species were identified and documented in the study area 62% predominantly used to livestock aliments treatment followed by 38% for livestock and human (both) aliments treatment (Figure 8).

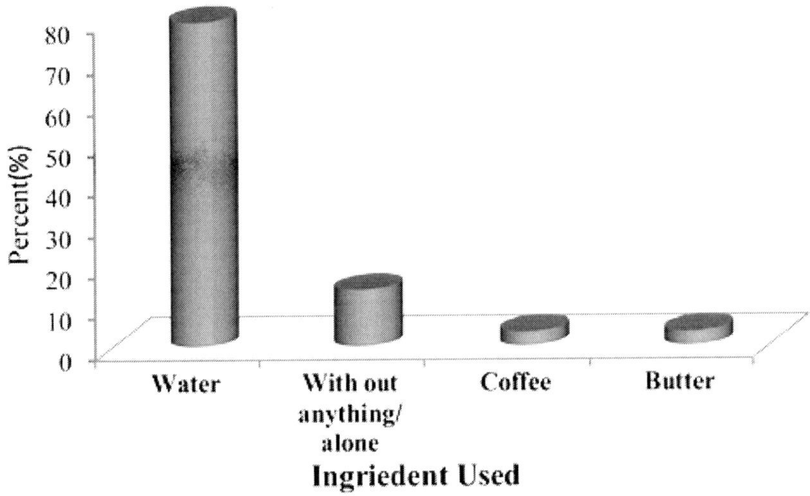

Figure 8. The proportion of knowledge aquired the traditional healers in Bensa Woreda, Southern Ethiopia.

4. CONCLUSIONS

The people of Bensa Woreda were shown in this study to have a wealth of knowledge/practices about traditional veterinary medicines for treating and controlling their livestock and human diseases. For centuries, the local farmers have learnt a great deal about animal and human diseases and their way of management through indigenous knowledge to protect their health and enhance productivity.

The region possesses a particularly wide range of potentially useful medicinal plants, more extensive indeed than available in many other parts of the country. This study indicates that, the fear of destruction of medicinal plants due to plant parts collected for the purpose of medicine is minimal as leaves were the leading plant parts sought in the area. Moreover, the harvest of whole plants is not practiced in the area, but the combined effect of factors faced a major threat to the survival of the mother plants, stock source.

This finding show us that there is lack of precision in the determination of doses in the study area since there were variations in the units of measurement and the quantity of plant parts used. Hence, the precision and standardization dosage as one drawback for the recognition of the traditional health care system.

Herbal preparations are crude and could potentially be toxic. Research is therefore needed to determine optimal doses and concentrations of the preparations and to identify the side effects of the remedies. Moreover, the efficacy of the preparations, techniques, and practices need to be investigated to identify promising plants for use in livestock development proposals.

The documentation and conservation of medicinal plants is highly recommended.

Training and awareness creation should be given to traditional healers and the local community respecting the management of medicinal trees and shrubs, to encourage and permit their cultivation on a large scale.

ACKNOWLEDGEMENTS

The author gratefully acknowledges Hawassa Agricultural Research Center (HARC), Hawassa, Ethiopia for financial and logistic support. I would like to express my deepest gratitude to the informants for sharing their time and knowledge genuinely. I give my appreciative to Mr. Melaku Wondafrash, Herbarium Specialist, for his priceless support in the identification of plant specimens at the National Herbarium, Addis Ababa University. And I like to appreciate Bensa Woreda Agricultural Office for its kind cooperative.

CITE THIS PAPER

Yibrah Tekle (2015) Medicinal Plants in the Ethno Veterinary Practices of Bensa Woreda, Southern Ethiopia. *Open Access Library Journal*,**02**,1-12.

REFERENCES

1. 1. DACA (2006) Standard Treatment Guidelines for Veterinary Practice.
2. Wondimu, T., Asfaw, Z. and Kelbessa, E. (2007) Ethnobotanical Study of Medicinal Plants around "Dheeraa" Town, Arsi Zone, Ethiop. Journal of Ethnopharmacology, 112, 152-161.
3. Zerhiun, W. and Mesfin, T. (1990) The Status of the Vegetation in the Lake Region of the Rift Valley of Ethiopia and Possibilities of Its Recovery. SINET: Ethiopian Journal of Science, 392, 97-120.
4. Nasir, T., Mohammed, A. and Nandikola, J. (2011) Ethnobotanical Survey of Medicinal Plants in the Southeast Ethiopia Used in Traditional Medicine. Spatula DD, 1, 153-158.
5. Ketema, T., Etana, D., Spiridoula, A., Adugna, T., Gebeyehu, G. and Jos, G.M. (2013) Ethno-Medicinal Study of Plants Used for Treatment of Human and Livestock Ailments by Traditional Healers in South Omo, Southern Ethiopia. Journal of Ethnobiology and Ethnomedicine, 9, 32.
6. Edwards, S., Tadesse, M., Demissew, S. and Hedberg, I., Eds. (2000) Flora of Ethiopia and Eritrea Part 1. Vol. 2. The National Herbarium Addis Ababa, Ethiopia, and Department of Systematic Botany, Uppsala, Sweden.
7. Giday, M. and Ameni, G. (2003) An Ethnobotanical Survey on Plants of Veterinary Importance in Two Woredas of Southern Tigray, Northern Ethiopia. SINET: Ethiop J Sci, 26, 123-126.
8. Kokwaro, J.O. (1976) Medicinal Plants of East Africa. East African Literature Bureau, Nairobi, 292-294.
9. McCorkle, C.M. (1995) Back to the Future Lessons from Ethnoveterinary Research, Development Extension for Studying and Applying Knowledge. Journal of the Agriculture, Food, and Human Values Society, 22, 52-80.
10. Sinha, P., Govil, J.N. and Singh, V.K. (2002) Diseases and Their Management, Recent Progress in Medicinal Plants. SCI Tech Publishing LLC, Houston, 1-105.
11. Kibebew, F. (2001) The Status of Availability of Data of Oral and Written Knowledge and Traditional Health Care in Ethiopia. In: Zewdu, M. and Demissie, A., Eds., Conservation and Sustainable Use of Medicinal Plants in Ethiopia, Institute of Biodiversity Conservation and Research, Addis Ababa, 107-119.
12. Teklehaymanot, T. and Giday, M. (2007) Ethnobotanical Study of Medicinal Plants Used by People in Zegie Peninsula, Northwestern Ethiopia. Journal of Ethnobiology and Ethnomedicine, 3, 12.
13. Alexiades, M.N. (1996) Advances in Economic Botany. Vol. 10. Selected Guidelines for Ethnobotanical Research: A Field Manual. The New York Botanical Garden, Bronx.
14. Gemechu, W., Girma, A., Temesgen, S. and Ensermu, K. (1997) Aspects of Farmers' Knowledge, Attitudes and Practices (KAP) of Animal Health Problems in the Central Highlands of Ethiopia. International Conference Ethnoveterinary Medecine, Pune.
15. Woldegerima, B., Abula, T. and Ragunathan, M. (2008) Ethnoveterinary Use of Medicinal Plants in Dabat District, Western Ethiopia. Pharmacognosy Magazine, 4, S93-S99.
16. Pound, B. and Gonfa, E. (2006) No. 3: Cattle in Southern Ethiopia: Participatory Studies in Wolaita and Konso Woredas. FARM-AFRICA Working Papers. http://www.farmafrica.org/downloads/resources/WP%203%20-%20Cattle%20in%20Southern%20Ethiopia.%20Participatory%20studies%20in%20Wolaita%20&%20Konso%20Woredas.pdf
17. Fullas, F. (2010) Ethiopian Medicinal Plants in Veterinary Healthcare: A Mini-Review. Ethiopian e-Journal for Research and Innovation Foresight, 2, 48-58.
18. Bekele, D., Asfaw, Z., Petros, B. and Tekie, H. (2012) Ethnobotanical Study of Plants Used for Protection against Insect Bite and for the Treatment of Livestock Health

Problems in Rural Areas of Akaki District, Eastern Shewa, Ethiopia. Topclass Journal of Herbal Medicine, 1, 40-52.

19. Bekele, A. and Musa, A. (2009) Ethnoveterinary Practice in Chiro District, Western 56 Hararge, Ethiopia. Pharmacologyonline, 1, 128-139.
20. Abebe, D. and Ayehu, A. (1993) Medicinal Plants and Enigmatic Health Practices of Northern Ethiopia. B.S.P.E, Addis Ababa, 419-431.
21. Tabuti, J.R.S., Dhillion, S.S. and Lye, K.A. (2003) Ethnoveterinary Medicines for Cattle (Bos indicus) in Bulamogi County, Uganda: Plant Species and Mode of Use. Journal of Ethnopharmacology, 88, 279-286.
22. Hedberg, I. and Edwards, S., Eds. (1989) Flora of Ethiopia and Eritrea. Pittosporaceae to Araliaceae. Vol. 3, The National Herbarium, Addis Ababa, Ethiopia, and Department of Systematic Botany, Uppsala, Sweden.
23. Hedberg, I. and Edwards, S., Eds. (1995) Flora of Ethiopia and Eritrea. Vol. 7: Poaceae (Graminae). The National Herbarium, Addis Ababa, Ethiopia, and Department of Systematic Botany, Uppsala, Sweden.
24. Hedberg, I., Kelbessa, E., Edwards, S., Demissew, S. and Persson, E., Eds. (2006) Flora of Ethiopia and Eritrea. Vol. 5. Plantaginaceae. The National Herbarium, Addis Ababa, Ethiopia, and Department of Systematic Botany, Uppsala, Sweden.
25. CSA (2010) Federal Democratic Republic of Ethiopia Population Census Commission Central Statistical Agency. The 2007 Population and Housing Census of Ethiopia Results for Southern Nation Nationalities, and People Region. Part I: Statistical Report on Population Size and Characteristics. August 2010, Addis Ababa.
26. Martin, G.J. (1995) Ethnobotany: A Conservation Manual. Chapman and Hall, London.
27. Jain, S.C., Jain, R. and Singh, R. (2009) Ethnobotanical Survey of Sariska and Siliserh Regions from Alwar District of Rajasthan, India. Ethnobotanical Leaflets, 13, 171-188.
28. Giday, M., Asfaw, Z. and Woldu, Z. (2009) Medicinal Plants of Meinit Ethnic Group of Ethiopia: An Ethnobotanical Study. Journal of Ethnopharmacology, 124, 513-521.
29. Mesfine, T. and Lemma, M. (2001) The Role of Traditional Veterinary Herbal Medicine and Its Constraints in the Animal Health Care System in Ethiopia. In: Zewdu, M. and Demissie, A., Eds., Conservation and Sustainable Use of Medicinal Plants in Ethiopia, Institute of Biodiversity Conservation and Research, Addis Ababa, 22-28.
30. Ensermu, K., Sebsebe, D., Zerihun, W. and Edwards, S. (1992) Some Threatened Endemic Plants of Ethiopia. In: Edwards, S. and Zemede, A., Eds., The Status of Some Plants in Parts of Tropical Africa, Botany 2000: NAPREC, Onograph Series No. 2, Addis Ababa University, Addis Ababa, 35-55.
31. Kebu, B., Ensermu, K. and Zemede, A. (2004) Indigenous Medicinal Utilization, Management and Threats in Fentale Area, Eastern Shewa, Ethiopia. Ethiopian Journal of Biological Scinces, 3, 1-7.
32. Pankhurst, R. (2001) The Status and Availability of Oral and Written Knowledge on Traditional Health Care. In: Medhin, Z. and Abebe, D., Eds., Conservation and Sustainable Use of Medicinal Plants in Ethiopia. Proceedings of the National Workshop on Biodiversity Conservation and Sustainable Use of Medicinal Plants in Ethiopia, IBCR, Addis Ababa, 92-106.
33. Cunningham, A.B. (1996) People, Park and Plant Use Recommendation for Multiple Uses Zones and Development Alternatives around Bwindi. Impenetrable National Park Uganda, People Plant Working Paper, UNESCO, Paris, Vol. 4, 18-23.
34. Gebrezgabiher, G., Kalayou, S. and Sahle, S. (2013) An Ethno-Veterinary Survey of Medicinal Plants in Woredas of Tigray Region, Northern Ethiopia. International Journal of Biodiversity and Conservation, 5, 89-97.
35. Yineger, H., Kelbessa, E., Bekele, T. and Lulekal, E. (2007) Ethnoveterinary Medicinal Plants at the Bale Mountains National Park, Ethiopia. Journal of Ethnopharmacology, 112, 55-70.

36. Assefa, B., Glatzel, G. and Buchmann, C. (2010) Ethnomedicinal Uses of Hagenia abyssinica (Bruce) J.F. Gmel. among Rural Communities of Ethiopia. Journal of Ethnobiology and Ethnomedicine, 6, 20.
37. Giday, M. (2001) An Ethnobotanical Study of Medicinal Plants Used by the Zay People in Ethiopia. Centrum för Biologisk Mångfald (CBM):s Skriftserie, 3, 81-89.
38. Mesfin, T. and Obsa, T. (1994) Ethiopian Traditional Veterinary Practices and Their Possible Contribution to Animal Production and Management. Revue Scientifique et Technique, 13, 417-424.
39. Abebe, G., Dawson, L.J., Detweiler, G., Gipson, T.A. and Sahlu, T. (2000) Hagenia abyssinica (Kosso) for Internal Parasite Control in Goats. In: Merkel, R.C., Abebe, G. and Goetsch, A.L., Eds., The Opportunities and Challenges of Enhancing Goat Production in East Africa, E (Kika) de la Garza Institute for Goat Research, Debub University, Awassa, 10-12 November 2000, 190-195.
40. Sori, T., Bekana, M., Adugna, G. and Kelbessa, E. (2004) Medicinal Plants in the Ethnoveterinary Practices of Borana Pastoralists, Southern Ethiopia. International Journal of Applied Research in Veterinary Medicine, 2.

Index